Lecture Notes in Computer Science 8969

Commenced Publication in 1973
Founding and Former Series Editors:
Gerhard Goos, Juris Hartmanis, and Jan van Leeuwen

More information about this series at http://www.springer.com/series/7407

Michel Daydé · Osni Marques
Kengo Nakajima (Eds.)

High Performance Computing for Computational Science – VECPAR 2014

11th International Conference
Eugene, OR, USA, June 30 – July 3, 2014
Revised Selected Papers

 Springer

Editors
Michel Daydé
IRIT, ENSEEIHT
Toulouse Cedex
France

Osni Marques
Lawrence Berkeley National Laboratory
Berkeley, CA
USA

Kengo Nakajima
Information Technology Center
The University of Tokyo
Tokyo
Japan

ISSN 0302-9743   ISSN 1611-3349 (electronic)
Lecture Notes in Computer Science
ISBN 978-3-319-17352-8   ISBN 978-3-319-17353-5 (eBook)
DOI 10.1007/978-3-319-17353-5

Library of Congress Control Number: 2015937402

LNCS Sublibrary: SL1 – Theoretical Computer Science and General Issues

Springer Cham Heidelberg New York Dordrecht London

Printed on acid-free paper

Springer International Publishing AG Switzerland is part of Springer Science+Business Media (www.springer.com)

Preface

VECPAR is an international conference series dedicated to the promotion and advancement of all aspects of high-performance computing for computational science, as an academic discipline and as a technique for real-world applications, extending the frontier of the state of the art and the state of practice. The audience and participants of VECPAR are researchers in academy, laboratories, and industry. The memory of the conference is preserved at http://vecpar.fe.up.pt.

The 11th edition of the conference, VECPAR 2014, was held in Eugene, OR, during June 30 – July 3, 2014. It was the sixth time the conference was held outside its "birth place," in Porto (Portugal), succeeding Valencia (Spain) in 2004, Rio de Janeiro (Brazil) in 2006, Toulouse (France) in 2008, Berkeley (USA) in 2010, and Kobe (Japan) in 2012.

The conference program consisted of 2 invited talks, 18 papers, and 11 posters. The invited talks were presented by John Shalf, "Exascale Programming Challenges: Adjusting to the New Normal for Computer Architecture," and Masaki Satoh, "A Super High-Resolution Global Atmospheric Simulation by the Nonhydrostatic Icosahedral Atmospheric Model Using the K Computer." In his talk, Dr. Shalf discussed challenges of programming future computing systems, with very high level of parallelism, and provided some highlights from the search for durable programming abstractions to more closely track emerging computer technology trends to guarantee the longevity of codes. In his talk, Dr. Satoh discussed a new type of the global atmospheric model called NICAM (non-hydrostatic icosahedral atmospheric model) that covers the Earth with a quasi-uniform mesh, and whose horizontal interval can be a subkilometer by using a high-end computer. He also gave an overview of recent results from super-high resolution simulations with NICAM using the K computer at the RIKEN AICS, in Kobe, Japan.

The major themes of the conference (thus the accepted papers and posters) were:

- Large-scale Simulations in CS&E
- Parallel and Distributed Computing
- Numerical Algorithms for CS&E
- Multiscale and Multiphysics Problems
- Data Intensive Computing
- Performance Analysis

The most significant contributions of VECPAR 2014 have been made available in the present book, edited after the conference, and after a second review of all orally presented papers. The first round of reviews was based on an eightpage extended abstract. Each paper was reviewed by three reviewers; in some cases, a fourth reviewer helped in the final decision. Out of 32 submissions, 18 were accepted for presentation. For the second round of reviews, authors were given a larger page budget, so they could better address reviewers' comments and suggestions. Finally, 17 were accepted for publication in this book.

In addition, three related events were organized in the first two days of the conference:

- The Ninth International Workshop on Automatic Performance Tuning (iWAPT 2014), whose contributions are also included in this book,
- Tutorial on Trilinos, a software library for solving large-scale mathematical problems arising in science and industry,
- Programming and Optimizing for the Intel® Xeon Phi™ Coprocessor, in which participants could learn about programming models and optimization for that architecture, complemented with hands-on work.

VECPAR 2014 took place at the University of Oregon and Hilton Conference Center, in Eugene, Oregon, USA. Paper submissions were managed with the EasyChair conference system; the conference website and registration process were managed by the University of Oregon.

The success of VECPAR and the long life of the series result from the work of many people. As in all previous occasions, a large number of collaborators were involved in the organization and promotion of the conference. Here, we would like to express our gratitude to Allen Malony and Sameer Shende, and the iWAPT organizers, in particular Franz Franchetti and Yusaku Yamamoto.

We thank all authors who have contributed to this book, for adhering to the deadlines and responding to the reviewers' comments and suggestions, and all members of the Scientific Committee, who greatly helped us with the paper selection process.

February 2015   Michel Daydé
                                                                 Osni Marques
                                                              Kengo Nakajima

Organization

Organizing Committee

Allen Malony	University of Oregon, USA
Sameer Shende	University of Oregon, USA

Steering Committee

Osni Marques (Chair)	Lawrence Berkeley National Laboratory, USA
Álvaro Coutinho	COPPE/UFRJ, Brazil
Michel Daydé	IRIT, France
Jack Dongarra	University of Tennessee, USA
Inês Dutra	University of Porto, Portugal
Kengo Nakajima	The University of Tokyo, Japan

Scientific Committee

Kengo Nakajima (Chair)	The University of Tokyo, Japan
Yifeng Cui (Vice-Chair)	San Diego Supercomputer Center, USA
Osni Marques (Vice-Chair)	Lawrence Berkeley National Laboratory, USA
Sameer Shende (Vice-Chair)	University of Oregon, USA
Reza Akbarinia	Inria, France
William Barth	TACC/University of Texas at Austin, USA
Taisuke Boku	University of Tsukuba, Japan
Jed Brown	Argonne National Laboratory, USA
Xiao-Chuan Cai	University of Colorado, USA
Xing Cai	Simula Research Laboratory, Norway
Christophe Calvin	CEA, France
Andrew Canning	Lawrence Berkeley National Laboratory, USA
Lucia Catabriga	Universidade Federal do Espírito Santo, Brazil
Li Chen	Tsinghua University, China
Edmond Chow	Georgia Institute of Technology, USA
Olivier Coulaud	Inria, France
Alvaro Coutinho	Federal University of Rio de Janeiro, Brazil
Jose Cuminato	Universidade de São Paulo, Brazil
José Cunha	Universidade Nova de Lisboa, Portugal
Claudio Curotto	Federal University of Paraná, Brazil
Michel Daydé	IRIT, France
Frédéric Desprez	Inria, France
Philippe Devloo	UNICAMP, Brazil
Tony Drummond	Lawrence Berkeley National Laboratory, USA

|---|---|
Ines Dutra	University of Porto, Portugal
Akihiro Fujii	Kogakuin University, Japan
Luc Giraud	Inria, France
Jorge Gonzalez-Dominguez	Universidade da Coruña, Spain
Ronan Guivarch	ENSEEIHT, France
Daniel Hagimont	ENSEEIHT, France
Abdelkader Hameurlain	IRIT, France
Hidehiko Hasegawa	University of Tsukuba, Japan
Mark Hoemmen	Sandia National Laboratories, USA
Toshiyuki Imamura	RIKEN AICS, Japan
Takeshi Iwashita	Kyoto University, Japan
Jean-Pierre Jessel	IRIT, France
Zhong Jin	Chinese Academy of Sciences, China
Takahiro Katagiri	The University of Tokyo, Japan
Harald Koestler	Universität Erlangen-Nürnberg, Germany
Jakub Kurzak	University of Tennessee, USA
Julien Langou	University of Colorado at Denver, USA
Stéphane Lanteri	Inria, France
Jean-Yves L'Excellent	Inria ENS Lyon, France
Sherry Li	Lawrence Berkeley National Laboratory, USA
Paul Lin	Sandia National Laboratories, USA
Thomas Ludwig	German Climate Computing Center, Germany
Piotr Luszczek	University of Tennessee, USA
Naoya Maruyama	RIKEN AICS, Japan
Hiroshi Nakashima	Kyoto University, Japan
Satoshi Ohshima	The University of Tokyo, Japan
Hiroshi Okuda	The University of Tokyo, Japan
Kenji Ono	RIKEN AICS, Japan
Christian Perez	Inria, France
Serge Petiton	Université de Lille, France
François-Henry Roeut	Lawrence Berkeley National Laboratory, USA
Tetsuya Sakurai	University of Tsukuba, Japan
Augusto Sousa	University of Porto, Portugal
Reiji Suda	The University of Tokyo, Japan
Frederic Suter	IN2P3 ENS Lyon, France
Daisuke Takahashi	University of Tsukuba, Japan
Osamu Tatebe	University of Tsukuba, Japan
Keita Teranishi	Sandia National Laboratories, USA
Miroslav Tuma	Academy of Sciences of the Czech Republic, Czech Republic
Paulo Vasconcelos	University of Porto, Portugal
Xavier Vasseur	CERFACS, France
Richard Vuduc	Georgia Institute of Technology, USA
Weichung Wang	National Taiwan University, Taiwan

Roland Wismuller	Universität Siegen, Germany
Rio Yokota	King Abdullah University of Science and Technology, Saudi Arabia

Invited Speakers

John Shalf	Lawrence Berkeley National Laboratory, USA
Masaki Satoh	The University of Tokyo, Japan

Sponsoring Organizations

The organizing committee is very grateful to the following organizations for their kind support to VECPAR 2014:

Fujitsu Limited
Intel
Rogue Wave Software
ParaTools

ParaTools

Posters

---

[1] Winner of the Best Poster Award.

Message from the Chairs of iWAPT 2014

The International Workshop on Automatic Performance Tuning (iWAPT) brings together researchers studying how to automatically adapt algorithms and software for high performance on modern machines. The workshop convened for the 9th consecutive year on July 1st, 2014 at the University of Oregon and Hilton Conference Center, Eugene, Oregon, USA.

The invited presentations *The End of Coding as We Know it* by Boyana Norris (University of Oregon) and *Auto-Tuning for Unreliable HPC* by Keita Teranishi (Sandia National Laboratories) addressed the challenges of auto-tuning in the world of ever-increasing code and hardware complexity and the resulting need for automation and fault tolerance.

The remaining presentations reinforced various aspects of this theme, spanning across a wide range of topics from service orchestration and checkpointing to the tuning of numerical linear algebra kernels and code generation. There were eight accepted technical papers, one reprint and one vision paper. The eight technical papers are included in these proceedings while the other papers were only distributed at VECPAR.

Many people and organizations helped to make this workshop a success. We are especially grateful to the VECPAR Organizing Committee, especially Osni Marques and Kengo Nakajima, for their logistical and intellectual support; the iWAPT Steering Committee, especially Reiji Suda and Takahiro Katagiri, for their guidance; and the Program Committee for volunteering their time to help assemble an outstanding program. Furthermore, the workshop would not be possible without the generous financial support of the Japan Science and Technology Agency, whose contributions have made Japan a leading international player in auto-tuning research. Lastly, we thank the invited speakers, authors, and meeting participants for their insights and thoughtful debate throughout the workshop.

February 2015

Osni Marques
Franz Franchetti
Yusaku Yamamoto

iWAPT 2014 Organizing Committees

Osni Marques (General Chair)	Lawrence Berkeley National Laboratory, USA
Toshiyuki Imamura (Finance Chair)	RIKEN AICS, Japan
Reiji Suda (Steering Committee Liaison)	The University of Tokyo, Japan
Hisayasu Kuroda (Web Chair)	Ehime University, Japan
Akihiro Fujii (Publicity Chair)	Kogakuin University, Japan

Steering Committee

Domingo J. Canovas	University of Murcia, Spain
Jonathan T. Carter	Lawrence Berkeley National Laboratory, USA
John Cavazos	University of Delaware, USA
Victor Eijkhout	Texas Advanced Computing Center, University of Texas, USA
Toshiyuki Imamura	The University of Electro-Communications, Japan
Takeshi Iwashita	Kyoto University, Japan
Takahiro Katagiri	The University of Tokyo, Japan
Ken Naono	Hitachi Ltd., Japan
Osni Marques	Lawrence Berkeley National Laboratory, USA
Markus Püschel	ETH Zurich, Switzerland
Reiji Suda	The University of Tokyo, Japan
Richard Vuduc	Georgia Institute of Technology, USA
R. Clint Whaley	The University of Texas at San Antonio, USA
Yusaku Yamamoto	Kobe University, Japan

Program Committee

Franz Franchetti (Program Chair)	Carnegie Mellon University, USA
Yusaku Yamamoto (Program Vice-Chair)	Kobe University, Japan
Siegfried Benkner	University of Vienna, Austria
I-Hsin Chung	IBM T.J. Watson Research Center, USA
Thomas Fahringer	University of Innsbruck, Austria
Takeshi Fukaya	RIKEN AICS, Japan
Torsten Hoefler	ETH Zurich, Switzerland
Jeremy Johnson	Drexel University, USA
Takahiro Katagiri	The University of Tokyo, Japan
Jakub Kurzak	University of Tennessee, USA
Boyana Norris	University of Oregon, USA
Satoshi Ohshima	The University of Tokyo, Japan
Louis-Noel Pouchet	Ohio State University, USA
Takao Sakurai	Hitachi Ltd., Japan
Daisuke Takahashi	University of Tsukuba, Japan
Hiroyuki Takizawa	Tohoku University, Japan
Teruo Tanaka	Kogakuin University, Japan
Masahiro Yasugi	Kyushu Institute of Technology, Japan

Contents

---

[2] Winner of the Best Paper Award.

Algorithms for GPU and Manycores

A Communication Optimization Scheme for Basis Computation of Krylov Subspace Methods on Multi-GPUs

Langshi Chen[1]([✉]), Serge G. Petiton[1,2], Leroy A. Drummond[3],
and Maxime Hugues[4]

[1] Maison de la Simulation, USR3441, Digiteo Labs Bât 565-PC 190,
91191 Gif-sur-Yvette, France
langshi.chen@etudiant.univ-lille.fr
[2] Laboratoire d'informatique Fondamentale de Lille, Université des Sciences et
Technologies de Lille, 59650 Villeneuve d'Ascq, France
Serge.Petiton@lifl.fr
[3] Lawrence Berkeley National Laboratory, One Cyclotron Road,
Berkeley, CA 94720, USA
ladrummond@lbl.gov
[4] INRIA Saclay, 1 rue Honor d'Estienne d'Orves, Bât Alan Turing,
91120 Palaiseau, France
maxime.hugues@lifl.fr

Abstract. Krylov Subspace Methods (KSMs) are widely used for solving large-scale linear systems and eigenproblems. However, the computation of Krylov subspace bases suffers from the overhead of performing global reduction operations when computing the inner vector products in the orthogonalization steps. In this paper, a hypergraph based communication optimization scheme is applied to Arnoldi and incomplete Arnoldi methods of forming Krylov subspace basis from sparse matrix, and features of these methods are compared in a analytical way. Finally, experiments on a CPU-GPU heterogeneous cluster show that our optimization improves the Arnoldi methods implementations for a generic matrix, and a benefit of up to 10x speedup for some special diagonal structured matrix. The performance advantage also varies for different subspace sizes and matrix formats, which requires a further integration of auto-tuning strategy.

Keywords: Krylov subspace · Auto-tuning · Arnoldi orthogonalization

1 Introduction

Krylov Subspace Methods (KSMs), such as GMRES and Arnoldi method, are kinds of iterative solvers frequently used in large-scale linear problems [1]. In KSMs, a basic and important part is to generate an orthogonal basis for the Krylov subspace. Arnoldi method is commonly adopted to form the basis [2,3], but it is proved to be time-consuming due to its blocking scalar product from its orthogonalization

© Springer International Publishing Switzerland 2015
M. Daydé et al. (Eds.): VECPAR 2014, LNCS 8969, pp. 3–16, 2015.
DOI: 10.1007/978-3-319-17353-5_1

process. In a parallel framework, the matrix-vector product in Arnoldi method also can cause a heavy communication cost. It is even worse in clusters equipped with accelerators like GPU, since the data exchange among GPUs is still expensive. Thus, efforts are made to reduce the communication in KSMs. Ghysels et al. [4] has proposed a pipelined variation of GMRES, hiding the global communication latencies by overlapping them with other communication or computations. Hoemmen [5] has implemented a Communication Avoiding version of the Power method for computing non-orthogonal bases, which replaces data exchange by redundant local computation. In this paper, we apply a hypergraph based communication optimization to parallel Arnoldi and incomplete Arnoldi orthogonalization methods. Together with the non-optimized Arnoldi and incomplete Arnoldi methods, the four algorithms are tested within a CPU-GPU framework. Our evaluation and comparison of performance concentrate only on the time spent in the computing of a Krylov subspace basis. While the number of restarts and the total time for obtaining a converged solution also depend on conditions such as the features of target matrices, which makes it difficult to have a general analysis. For example, methods like incomplete Arnoldi could generate fast a less orthogonal basis but later require more iterations and time to reach convergence.

2   Methods for Generating Krylov Subspace Basis

In order to obtain a vector basis for the Krylov subspace, the Gram-Schmidt orthogonalization based Arnoldi process is commonly used. In this paper, we focus on the Arnoldi process in the sparse matrix case. Arnoldi consists of a BLAS 2 bandwidth bounded Sparse matrix-vector multiplication (SpMV) operations, and a vector inner product operations which incurs a global data reduction across all the MPI processes. In order to reduce the communication overhead, we first introduce a variant of *Arnoldi* named *Incomplete Arnoldi*, which truncates the number of inner product operation to lower down the communication from global reduction. Then, we introduce our hypergraph based optimization and apply them to both *Arnoldi* and *Incomplete Arnoldi*. A time complexity analysis is given in Sects. 2.1, 2.2, 2.3 and 2.4 for the four algorithms we have studied here.

2.1   Arnoldi

Arnoldi method uses a Classic Gram-Schmidt (CGS) process to form a full-sized orthogonal subspace basis. The CGS is preferred to Modified Gram-Schmidt (MGS) because it is easier to implement in parallel and is better for the scalability of our implementations. We evaluate its computational complexity in Eq. 1.

$$T_{Arnoldi}(s, p, N, n) = \alpha(2sn/pL) \ + \alpha(3Ns^2 + 9Ns)/2pL$$
$$+ \beta(2s \log_2(p)) + G(Ns + 2s^2)(p - 1)/p \ (1)$$

The variable s is the size of Krylov subspace; p is the number of MPI processes; N is the row number of matrix and n is the number of nonzero entries. We divide the overall time into three parts: (1) $\alpha(2sn/pL)$ is the time of the matrix-vector multiplication in Gram-Schmidt process. The parameter α denotes the time per arithmetic operation, and L is the number of simultaneous parallel processing elements in each MPI process (e.g. maximal threads within a GPU). (2) $\alpha(3Ns^2 + 9Ns)/2pL$ is the time spent in the vector inner product of Gram-Schmidt process, which is quadratic to the subspace size s. (3) $\beta(2s\log_2(p)) + G(Ns + 2s^2)(p-1)/p$ is the communication overhead. It includes a latency part β and a bandwidth part G. The complexity analysis indicates that the communication overhead augments with the number of MPI processes p. When p is relatively large, the latency part will become the dominant increment to the communication cost.

2.2 IArnoldi(q)

Secondly, we review the *IArnoldi(q)* based on the Classic Gram-Schmidt process. *IArnoldi(q)* truncates the number of orthogonal vectors so that each new generated basis vector should only be orthogonal to its q previous basis vectors [6]. Its time complexity is evaluated in Eq. 2.

$$
\begin{aligned}
T_{IArnoldi(q)}(s,p,N,n,q) = {} & \alpha(2sn/pL) + \alpha(2q+3)Ns/pL \\
& + \beta[(s+q)\log_2(p)] + G[(Ns + 2qs)(p-1)/p] \quad (2)
\end{aligned}
$$

s, p, N, n is the same parameters in Eq. 1, and q is the number of vectors each new generated basis vector should be orthogonal. Thus, a small q value means a large extent of truncation and a less time both in computation and communication. The second term $\alpha(2q+3)Ns/pL$ reduces the time of inner products from $O(Ns^2/pL$ to $O(Nsq/pL)$. The time of latency part also drops from $\beta(2s\log_2(p))$ to $\beta(s\log_2(p))$ (we assume that $q \ll s$). However, it suffers from a less orthogonal basis, which incurs more iterations to reach convergence.

2.3 ArnoldiHG

ArnoldiHG is an optimized *Arnoldi* based on an hypergraph model proposed in [7]. As the Power iteration of SpMV in *Arnoldi* is communication bounded, we model it as an hypergraph. Each vertex in hypergraph refers to a single row in matrix A, and two vertex (row i and j) are connected by a edge if and only if the entry $A(i,j)$ is nonzero. The edge (i,j) means that in each power iteration of SpMV, the multiplication of row $A(i,;)$ and vector X requires the data $X(j)$ stored in row j. In a parallel implementation, the rows of A are partitioned and assigned to different processes (CPU or GPU computation element). Thus, an edge across two groups of vertex (rows) incurs a data exchange within processes, and this dependency is invariant during the power method iteration. Then the optimization takes two steps. It first finds an optimal partition of the rows in A to minimize the data exchange. Then the communication only occurs

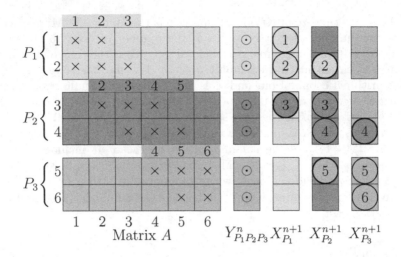

Fig. 1. Matrix A is divided into 3 partitions. When update X^{n+1} from Y^n, each partition only requires some X data depending on their nonzero entries. E.g. P_2 (Blue color) requires the data from $2,3,4,5$ row positions, because only column $2,3,4,5$ has nonzero entries. When forming $X_{P_2}^{n+1}$, data $3,4$ could be directly copied from P_2's own memory, but communication occurs when requires data $2,5$ which belongs to P_1, P_3. Thus for P_2, the minimal communication is receiving data $2,5$ and sending data $3,4$ (Color figure online)

among rows that have data dependency in each iteration of SpMV. Figure 1 gives an example of the optimization scheme. After the optimization, the communication overhead $\beta(2s\log_2(p)) + G(Ns + 2s^2)(p-1)/p$ in Eq. 1 is replaced by $\{\beta[sf(A,p) + s\log_2(p)] + G[sg(A,n) + 2s^2(p-1)/p]\}$, where the terms $f(A,p)$ and $g(A,n)$ depends on the structure and partition of matrix A. In a best case, we have $f(A,p) = O(1)$, $g(A,n) = O(n/p)$ which greatly reduces the communication overhead in SpMV part.

2.4   IArnoldiHG(q)

Similarly, *IArnoldiHG(q)* is the hypergraph optimized *IArnoldi(q)*. Due to the truncation and the hypergraph optimization, the second and third terms in Eq. 3 show a significant improvement at execution time.

$$T_{IArnoldi(q)}(s,p,N,n,q) = \alpha(2sn/pL) + \alpha(2q+3)Ns/pL$$
$$+ \beta[q\log_2(p) + sf(A,p)] + G[sg(A,n) + 2qs(p-1)/p] \quad (3)$$

2.5   Comparison of Four Algorithms

In Table 1, we compare the four algorithms in various aspects, like the execution time, orthogonality and scalability. Here the total execution time for computing a basis is divided into three parts as presented in Sects. 2.1, 2.2, 2.3 and 2.4. The orthogonality evaluates the quality of the basis, which shall affect the total

Table 1. Comparison of 4 parallel Arnoldi Algorithmic Implementations

Algo name	Arnoldi	ArnoldiHG	IArnoldi(q)	IArnoldiHG(q)
Computation	$O(2sn/pL + 3s^2N/2pL)$	$O(2sn/pL + 3s^2N/2pL)$	$O(2sn/pL + 2qsN/pL)$	$O(2sn/pL + 2qsN/pL)$
Latency	$O(2s\log_2(p))$	$O(sf(A,p) + s\log_2(p))$	$O(s\log_2(p))$	$O(sf(A,p) + q\log_2(p))$
Bandwidth	$O(Ns)$	$O(sg(A,n) + 2s^2)$	$O(Ns)$	$O(sg(A,n) + 2qs)$
Orthogonality	Full	Full	Depend q	Depend q
Strong scalability	Medium	High	Low	Medium
Weak scalability	Low	High	Low	High
Optimized communication	NO	Yes	NO	Yes

time for finding a converged solution in iterative methods. The Strong scalability reflects the parallelism of the computational workload, a more computational intensive algorithm shall have a better strong scalability. While the weak scalability measures the communication overhead, where a communication intensive algorithm has a worse performance. According to our analysis, *IArnoldiHG* has the best execution time due to the truncation and our hypergraph optimization. Both of the hypergraph optimized methods have better scalability because of their reduced communication overhead. Nevertheless, they are affected by the structure of the matrix which determines the factor $f(A,p)$ and $g(A,n)$. The truncated versions of *Arnoldi* method do not have a full orthogonality of its basis, which depends on the choice of value q. A relatively small q could reduce the execution time but reduce the orthogonality at the same time, which may be remedied by methods like reorthogonalization.

3 Experimentation

In the experiment, we test the four methods on two heterogeneous clusters. One is the cluster *Poincare* from *Maison de la Simulation*, which has 4 GPU nodes and each node consists of 2 Xeon CPU and 2 Nvidia K20m GPUs. Another cluster *HAPACS-TCA* resides in the University of Tsukuba. It has a total 64 GPU nodes, and each node contains 2 Xeon CPUs and 4 Nvidia K20x GPUs (See Table 2). According to this heterogeneous architecture, the four algorithms presented in Table 1 are also implemented under a CPU-GPU hybrid programming paradigm (See Fig. 2). The matrix A is evenly row partitioned and the workload is distributed over GPUs. Each GPU has its own CPU process to handle the communication operations via MPI. The GPU codes are written in CUDA 5.5 framework, and the MPI codes are compiled by openmpi 1.6.3. In order to

Table 2. Two GPU cluster's characteristics

Name	Nodes	CPU/node	GPU/node	GPU	Interconnect
Poincare	4	2	2	K20m	Infiniband
HA-PACS/TCA	64	2	4	K20x	Infiniband

test the influence of different matrix structure to our hypergraph model, we use sparse matrices with different structures. The Krylov Subspace size is also varied to test its effect on the truncated methods *IArnoldi* and *IArnoldiHG*.

3.1 The Test Matrices

In the experiment, we use four test sparse matrices. The Continuous Diagonal Matrix (C-Diag) in Fig. 3(a), has all its subdiagonals continuously aggregates around the main diagonal. It represents a major type of matrices generated by PDE applications, and it maps well to a row partition where each row part only communicates with its neighbours. The second matrix is the Equidistributed Diagonal Matrix (E-Diag) in Fig. 3(b), which is a contrast to the C-Diag matrix

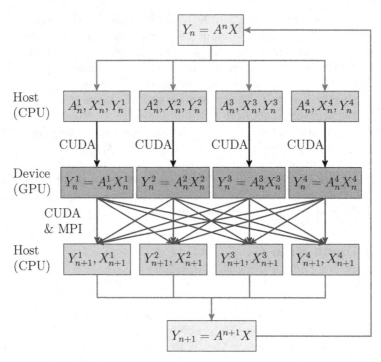

Fig. 2. A hybrid CPU-GPU programming scheme for the *SpMV* part of *Arnoldi*. At every iteration, a group of rows of the matrix A is attached to a CPU and a GPU. A CPU first delivers the computation task to a GPU. After the GPU has finished its work, each CPU gathers the Y vector from its own GPU, and it gathers Y vector or part of Y vector from other CPUs via MPI. Then a new X vector is formed in CPU for the next iteration

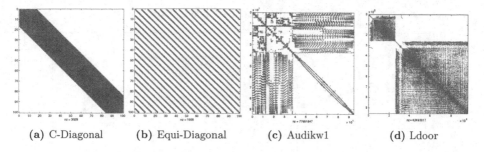

(a) C-Diagonal   (b) Equi-Diagonal   (c) Audikw1   (d) Ldoor

Fig. 3. (a) Continuous Diagonal Matrix, size of 960000, diagonals of 33; (b) Equidistributed Diagonal Matrix, size of 960000, diagonals of 33; (c) Audikw1, size of 943695, nonzeros of 77651847; (d) Ldoor, size of 952203 nonzeros of 42943817

where subdiagonals are distributed evenly across all columns. In equal row partition of E-Diag, each row part shall exchange data with distant rows, which may examine the worst case of hypergraph optimization model. Both of C-Diag and E-Diag matrices are generated in runtime, which could be extended to an arbitrary scale in the strong scaling and weak scaling tests. The third and fourth matrices are real matrices collected from the Florida University's Sparse matrix database as shown in Fig. 3(c) and (d). They have more complicated structure which is taken to test the influence of matrix structure on the hypergraph model in the strong scaling tests. We also use two different Sparse Matrix formats. One is the Compressed Sparse Row format (CSR format), the other is the ELLPACK format provided by *Nvidia*, and we also evaluate the potential impact of these formats in the overall performance of the implementations.

3.2   The Krylov Subspace Size

During the construction of the subspace basis, one of the most important parameters is the subspace size s (See Sect. 2). The value of s would significantly affect the performance of *IArnoldi* methods. Typically, a small value of s is preferred in the cases of *IArnoldi* and *IArnoldiHG* because the loss of orthogonality would be reduced. In the test, we also use some large values of s to have a complete range of tests and study the influence of the subspace size over the performance of time and scalability.

4   Results and Analysis

Figures 4 and 5 shows the scalability results of the four algorithms on C-Diag matrix presented in Sect. 3.1. The Krylov subspace size is set to 64, and we use a commonly adopted sparse matrix format CSR. In Fig. 4, the strong scalability of the four algorithms is not good, which is due to the low computation/communication ratio in the sparse matrix case. While the result shows a significant benefit from hypergraph optimization. When the algorithms are scaled to more than 100 GPUs, *ArnoldiHG* could achieve a nearly 10x speed up over

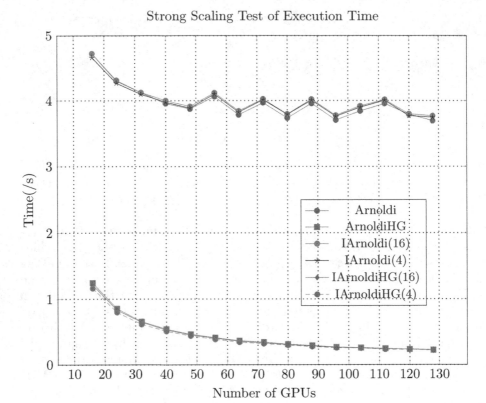

Fig. 4. Scalability Test on *HAPACS-TCA* with C-Diag matrix stored in CSR format double precision. Strong scaling with total matrix size 9600000 × 9600000, diagonals of 33 and Krylov subspace size 64

non-optimized *Arnoldi*. This benefit comes from the reduction of communication overhead in the C-Diag matrix, where each process (GPU) shall only exchange data with its neighbours in the hypergraph model. The weak scaling test in Fig. 5 endorses our model as well, where *ArnoldiHG* and *IArnoldiHG* show a perfect weak scalability over 100 GPUs. In contrast, the non-optimized methods suffer a lot from their communication overhead. In Fig. 6, we have a comparison for the two methods. When the number of GPUs is relatively small, *IArnoldiHG(4)* has the best performance. It comes from the benefit of *IArnoldiHG*'s truncation on the number of vector inner product operations. A smaller q value leads to a better performance gain (*IArnoldiHG(4)* outperforms *IArnoldiHG(16)*). However, this advantage disappears as the number of GPUs increases, and the performances of the three curves in Fig. 6 converge. It means that the dominance of computation is eliminated when p is large in the computation part of Table 1. As the communication is also optimized by hypergraph model, a performance convergence is expected. Thus, we prefer *IArnoldiHG* when p is small; otherwise *ArnoldiHG* shall be chosen as it has a better orthogonality.

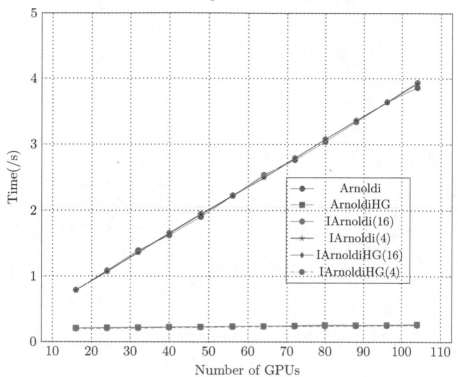

Fig. 5. Scalability Test on *HAPACS-TCA* with C-Diag matrix stored in CSR format double precision. Weak scaling with total matrix size 9600000 × 9600000, diagonals of 33 and Krylov subspace size 64

4.1 Different Structure of Input Matrix

The structure of sparse matrix may also affect the performance of hypergraph optimization. We will first test an E-Diag matrix both in strong scaling and weak scaling. E-Diag could be considered as a worst-case test for our optimization model, where each process should communicate with distant processes. In Fig. 7(a), we find a cross point at around 50 processes (GPUs). As the bandwidth part of *ArnoldiHG* $O(sn/p + 2s^2)$ is much lower than that of *Arnoldi*'s $O(Ns)$, the crossing point could be explained by the latency part of *ArnoldiHG* $O(sf(A,p) + slog_2(p))$ with $f(A,p) = min(p, diags)$. In our case, the number of subdiagonals is 33. When p increases to pass a particular value, the latency part of *ArnoldiHG* is surpassed by that of *Arnoldi* ($log_2(p)$), and a benefit of our optimization is expected. The weak scaling result also supports our analysis, where a crossing point is found in the same place around 50 GPUs. Besides the E-Diag matrix, we also test two real sparse matrices *Audikw1* and *Ldoor* presented in Fig. 3. As they have a fixed size, we only take the strong scaling test on them. In Fig. 8, we find that our hypergraph model optimization is effective on both of

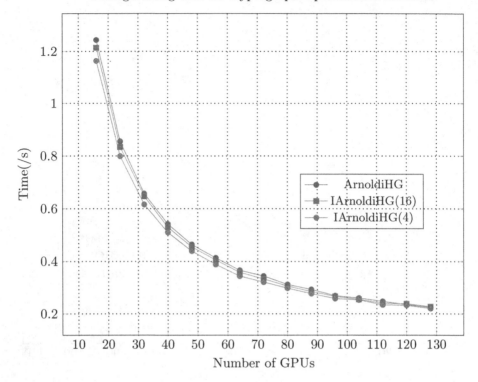

Fig. 6. Scalability Test of Hypergraph Optimization methods on *HAPACS-TCA* with C-Diag matrix stored in CSR format double precision. Strong scaling with total matrix size 9600000 × 9600000, diagonals of 33 and Krylov subspace size 64

the matrices with complicated structure. The matrix *Ldoor* gains more benefits from our optimization compared to the matrix *Audikw*. As the two matrices have a row size and number of nonzeros of the same order, the difference should come from their different structures. According to the Fig. 3, *Ldoor* has more of its nonzero entries aggregated around its main diagonals than that of *Audikw*, which shall reduce the communication among distant processes (GPUs in our test). Thus, we may have a conjecture that the benefit of our model depends on the structure of the matrix (e.g. the number of subdiagonals in E-Diag matrix) in some cases.

4.2   Varying the Size of the Krylov Subspace

Furthermore, we study the influence of the various Krylov subspace sizes on our hypergraph optimization. Since the Krylov subspace size is one of the most important parameters in many Krylov iterative methods, we choose three different values 8, 64, 256 in our test. A much larger value is not preferred in KSMs because it shall involve a very large workload. Instead, this value is generally set

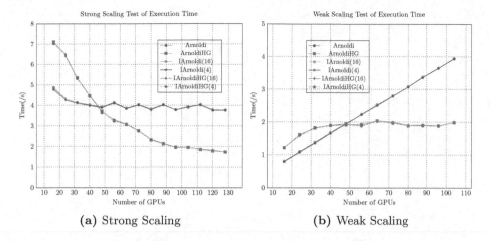

(a) Strong Scaling   **(b)** Weak Scaling

Fig. 7. Scalability Test on *HAPACS-TCA* with E-Diag matrix stored in CSR format double precision. (a) Strong scaling with total matrix size 9600000×9600000, diagonals of 33 and Krylov subspace size 64; (b) Weak scaling with each submatrix size 96000×96000, diagonals of 33 and Krylov subspace size 64

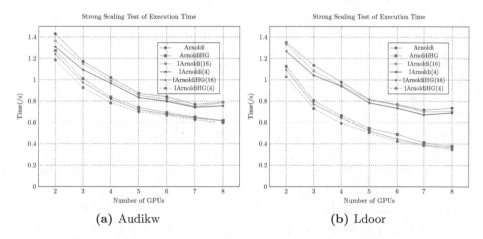

(a) Audikw   **(b)** Ldoor

Fig. 8. Strong Scalability Test on *Poincare* with Audikw and Ldoor stored in CSR format double precision. (a) Audikw matrix size of 943695, nonzeros of 77651847, and Krylov subspace size 64; (b) Ldoor matrix size of 952203 nonzeros of 42943817, and Krylov subspace size 64

to around 10 to 30. In Fig. 9, the x-axis denotes three testing Krylov subspace sizes, and the y axis is the time speedup of *ArnoldiHG*, *IArnoldi*, *IArnoldiHG* over the original *Arnoldi*. Here, we set the q value of the truncation to be $s/4$ with s equals the Krylov subspace size. We find that our *ArnoldiHG* performs better in small subspace size rather than large subspace size. It is due to an increase of computation/communication intensity when the value s augments ($O(3s^2N)$). Thus, the algorithm is computation-bounded rather than

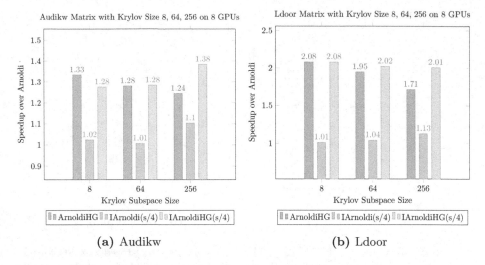

(a) Audikw   **(b)** Ldoor

Fig. 9. Krylov Subspace Size Test on *Poincare* with Audikw and Ldoor stored in CSR format double precision. (a) Audikw matrix size of 943695, nonzeros of 77651847; (b) Ldoor matrix size of 952203 nonzeros of 42943817

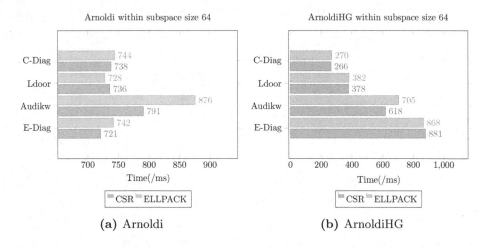

(a) Arnoldi   **(b)** ArnoldiHG

Fig. 10. Comparison of format CSR and ELLPACK on *Poincare* with 4 matrices with subspace size 64 running on 8 GPUs with double precision

communication-bounded, and the hypergraph optimization shall have a not significant performance impact. In terms of *IArnoldi*, the truncation of the computational workload has better performance impact when the subspace size s is larger, which leads to a better speedup over *Arnoldi*. Finally for *IArnoldiHG*, the two effects coexist, and the influence of subspace size varies according to many factors like the structure of the matrix. In Fig. 9, the test on *Audikw* shows a better speedup of *IArnoldiHG* within larger s values. While the *Ldoor* matrix shows a worse speedup of *IArnoldiHG* within larger s values.

4.3   Impact of Sparse Matrix Format

In [8], the influence of formats on the performance of SpMV has been evaluated. Similarly, we compare the two formats CSR and ELLPACK in our test. In our implementation, both of the CSR and ELLPACK format use the vectorization version which uses half a warp of 16 threads for each row of the matrix. Because the data storage of CSR is more contiguous than that of ELLPACK (padding zeros for some rows), the retrieval of data has a better efficiency. In Fig. 10, the implementation of format CSR runs slightly better than that of ELLPACK. In matrix *Audikw*, the gap is more evident because *Audikw* has a more irregular structure which causes much more padding of zeros for rows.

5   Conclusion

In this paper, we presented a hypergraph model to optimize the communication cost in computing the Krylov subspace basis from sparse matrices. The Classical Gram-Schmidt orthogonalization based Arnoldi method and its truncated version *Incomplete Arnoldi* have been chosen, and the optimization scheme have been applied to them. We study the time complexity and scalabilities of the four algorithmic implementation in a theoretical way and compare their features. According to the experiments on a CPU-GPU hybrid platform, we arrived on the following concluding remarks: (1) Our hypergraph based optimization is useful for communication-bounded kernels in Krylov subspace methods. (2) The nonzeros distribution pattern of sparse matrix has an influence on our optimization, where a matrix with more nonzeros aggregated around the main diagonal shall have more benefits from our scheme. (3) Factors like the subspace size and sparse matrix format also have affected the results of optimization, and an auto-tuning framework is expected to choose different parameters intelligently in the optimization. In order to further remove the communication bottleneck of the applications, we will consider the adoption of more aggressive strategies like Communication Avoiding method (e.g. TSQR, CAQR) [5] in future work.

Acknowledgments. The research presented in this work is partly supported by the ANR-JST Japanese-French project FP3C (Framework and Programming for Post-Petascale Computing). We thank Professor Tetsuya SAKURAI, Professor Taisuke BOKU and his team in University of Tsukuba for their support in providing the use of cluster HAPACS-TCA. We also thank the anonymous reviewers for their comments and recommendations, which help us continually improving the work.

References

1. Saad, Y.: Iterative Methods for Sparse Linear Systems, 2nd edn. Society for Industrial and Applied Mathematics, Philadelphia (2003)
2. Arnoldi, W.E.: The principle of minimized iterations in the solution of the matrix eigenvalue problem. Q. Appl. Math. **9**, 17–29 (1951)

3. Saad, Y.: Numerical Methods for Large Eigenvalue Problems. Society for Industrial and Applied Mathematics, Philadelphia (2011)
4. Ghysels, P., Ashby, T.J., Meerbergen, K., Vanroose, W.: Hiding global communication latency in the GMRES algorithm on massively parallel machines. SIAM J. Sci. Comput. **35**, C48–C71 (2013)
5. Hoemmen, M.: A communication-avoiding, hybrid-parallel, rank-revealing orthogonalization method. In: 2011 IEEE International Parallel Distributed Processing Symposium (IPDPS), pp. 966–977 (2011)
6. Saad, Y., Wu, K.: DQGMRES: a direct quasi-minimal residual algorithm based on incomplete orthogonalization. Numer. Linear Algebra Appl. **3**, 3–329 (1996)
7. Catalyurek, U.V., Aykanat, C.: Hypergraph-partitioning based decomposition for parallel sparse-matrix vector multiplication. IEEE Trans. Parallel Distrib. Comput. **10**, 673–693 (1999)
8. Hugues, M., Petiton, S.: Sparse matrix formats evaluation and optimization on a GPU. In: 2010 12th IEEE International Conference on High Performance Computing and Communications (HPCC), pp. 122–129 (2010)

Mixed-Precision Orthogonalization Scheme and Adaptive Step Size for Improving the Stability and Performance of CA-GMRES on GPUs

Ichitaro Yamazaki[✉], Stanimire Tomov, Tingxing Dong, and Jack Dongarra

Department of Electrical Engineering and Computer Science,
University of Tennessee, Knoxville, USA
iyamazak@icl.utk.edu

Abstract. The Generalized Minimum Residual (GMRES) method is a popular Krylov subspace projection method for solving a nonsymmetric linear system of equations. On modern computers, communication is becoming increasingly expensive compared to arithmetic operations, and a communication-avoiding variant (CA-GMRES) may improve the performance of GMRES. To further enhance the performance of CA-GMRES, in this paper, we propose two techniques, focusing on the two main computational kernels of CA-GMRES, tall-skinny QR (TSQR) and matrix powers kernel (MPK). First, to improve the numerical stability of TSQR based on the Cholesky QR (CholQR) factorization, we use higher-precision arithmetic at carefully-selected steps of the factorization. In particular, our mixed-precision CholQR takes the input matrix in the standard 64-bit double precision, but accumulates some of its intermediate results in a software-emulated double-double precision. Compared with the standard CholQR, this mixed-precision CholQR requires about 8.5× more computation but a much smaller increase in communication. Since the computation is becoming less expensive compared to the communication on a newer computer, the relative overhead of the mixed-precision CholQR is decreasing. Our case studies on a GPU demonstrate that using higher-precision arithmetic for this small but critical segment of the algorithm can improve not only the overall numerical stability of CA-GMRES but also, in some cases, its performance. We then study an adaptive scheme to dynamically adjust the step size of MPK based on the static inputs and the performance measurements gathered during the first restart loop of CA-GMRES. Since the optimal step size of MPK is often much smaller than that of the orthogonalization kernel, the overall performance of CA-GMRES can be improved using different step sizes for these two kernels. Our performance results on multiple GPUs show that our adaptive scheme can choose a near optimal step size for MPK, reducing the total solution time of CA-GMRES.

1 Introduction

The cost of executing software can be modeled by a function of its computational and communication costs (e.g., in terms of required cycle time or energy

© Springer International Publishing Switzerland 2015
M. Daydé et al. (Eds.): VECPAR 2014, LNCS 8969, pp. 17–30, 2015.
DOI: 10.1007/978-3-319-17353-5_2

consumption). For instance, the computational cost can be modeled based on the number of required floating point operations (flops), while the communication includes the synchronization and data transfer between the parallel processing units, as well as the data movement through the levels of the local memory hierarchy. On modern computers, communication is becoming increasingly expensive compared to computation. It is critical to take this hardware trend into consideration when designing high-performance software for new and emerging computers.

The Generalized Minimum Residual (GMRES) method [6] is a popular Krylov subspace projection method for solving a large-scale nonsymmetric linear system of equations. To address the current hardware trend, we studied a communication-avoiding variant of GMRES [5] on multicore CPUs with multiple GPUs [8]. Our experimental results demonstrated that CA-GMRES can obtain the speedups of up to two by avoiding some of the communication on such shared-memory computer architectures. Our experimental results also showed that both the performance and numerical stability of CA-GMRES depends on the two computational kernels, the orthogonalization (*Orth*) and matrix powers kernels (*MPK*). For example, compared with other orthogonalization schemes, the Cholesky QR (CholQR) factorization [7] obtained a superior performance based on the optimized BLAS-3 GPU kernels. Unfortunately, when the input matrix is ill-conditioned, CholQR can be numerically unstable, and CA-GMRES may not converge even with reorthogonalization. We also found that depending on the sparsity pattern of the coefficient matrix, *MPK* can be slower than the standard sparse-matrix vector multiply (*SpMV*) due to the computational and/or communication overheads traded for reducing the communication latency. This is especially true in CA-GMRES, where a relatively large step size is preferred by *Orth*.

To address the aforementioned limitations of CA-GMRES, in this paper, we first design and study a mixed-precision variant of CholQR that takes the input matrix in the standard 64-bit double precision but accumulates some of its intermediate results in software-emulated double-double precision [4]. Compared with the standard CholQR, our mixed-precision CholQR increases the computational cost by $8.5\times$ but the increase in its communication cost is less significant. Since the computation is becoming less expensive compared to the communication on new and emerging computers, we hope to improve the overall numerical stability of CA-GMRES using the higher-precision without a significant increase in the orthogonalization time. Case studies on different GPUs demonstrate that this mixed-precision CholQR can improve not only the overall stability of CA-GMRES but also, in some cases, its performance by allowing a larger step size, avoiding the reorthogonalization, and improving the solution convergence rate. We then study an adaptive scheme that uses different step sizes for *MPK* and *Ortho* and dynamically adjusts the step size of *MPK* at run time. We demonstrate that our adaptive scheme can find a near optimal step size based on the static input parameters and the performance measurements gathered during the first restart loop, and reduce the total solution time of CA-GMRES.

```
x̂ := 0 and v₁ := b/‖b‖₂.
repeat (restart-loop)
    Generate Krylov Subspace on GPUs (inner-loop):
    for j = 1, s + 1, 2s + 1, ..., m do
        MPK: Generate new vectors v_{k+1} := Av_k
            for k = j, j + 1, ..., min(j + s, m).
        BOrth: Orthogonalize V_{j+1:j+s+1} against V_{1:j}.
        TSQR: Orthogonalize the vectors within V_{j+1:j+s+1}.
    end for

    Solve Projected Subsystem on CPUs (restart):
    Compute the solution x̂ in the generated subspace,
        which minimizes its residual norm.
    Set v₁ := r/‖r‖₂, where r := b − Ax̂.
until solution convergence do
```

Fig. 1. CA-GMRES(s,m) pseudocode.

```
Step 1: Gram-matrix formation
for d = 1, 2, ..., n_g do
    B^{(d)} := V^{(d)T}_{1:s+1} V^{(d)}_{1:s+1}  on GPU
end for
B := Σ B^{(d)}  (global reduce)

Step 2: Cholesky factorization
R := chol(B) on CPU

Step 3: Orthogonalization
copy R to all the GPUs (broadcast)
for d = 1, 2, ..., n_g do
    V^{(d)}_{1:s+1} := V^{(d)}_{1:s+1} R^{-1}  on GPU
end for
```

Fig. 2. CholQR pseudocode.

The rest of the paper is organized as follows: In Sect. 2, we first review the CA-GMRES, *MPK*, and CholQR algorithms, and present their implementations on the multicore CPUs with multiple GPUs. Then in Sect. 3, we describe the mixed-precision CholQR and its implementation with the GPU. The performance of the mixed-precision CholQR and its effects on the performance of CA-GMRES are also presented in this section. Next, in Sect. 4, we describe our adaptive scheme for the *MPK*'s step size and present its effectiveness in selecting the near-optimal step size. We provide final remarks in Sect. 5.

2   Communication-Avoiding GMRES

The Generalized Minimum Residual (GMRES) method [6] is a popular Krylov subspace projection method for solving a nonsymmetric linear system of equations, $Ax = b$. The GMRES's j-th iteration generates the $(j + 1)$-th Krylov basis vector $\mathbf{v}_{j+1}$. This is done through a sparse matrix-vector multiply (*SpMV*) with the previously-generated basis vector $\mathbf{v}_j$, followed by the orthonormalization (*Orth*) of the resulting vector against all the previously-generated basis vectors $\mathbf{v}_1, \mathbf{v}_2, \ldots, \mathbf{v}_j$. As the iteration proceeds, this explicit orthogonalization of the basis vectors becomes increasingly expensive in terms of both computational and storage requirements.

To avoid the expensive costs of generating a large projection subspace, GMRES iteration is restarted after computing a fixed number $m + 1$ of basis vectors. Before restart, GMRES updates the approximate solution $\hat{\mathbf{x}}$ by solving a least-squares problem $\mathbf{g} := \arg\min_{\mathbf{t}} \|\mathbf{c} - H\mathbf{t}\|$, where $\mathbf{c} := V^T_{1:m+1}(\mathbf{b} - A\hat{\mathbf{x}})$, $H := V^T_{1:m+1} A V_{1:m}$, $\hat{\mathbf{x}} := \hat{\mathbf{x}} + V_{1:m}\mathbf{g}$, and $V_{1:m}$ is the matrix consists of the column vectors $\mathbf{v}_1, \mathbf{v}_2, \ldots, \mathbf{v}_m$. Compared with the coefficient matrix A, the projected matrix H, a by-product of the orthogonalization procedure, is smaller in dimension (i.e., $m \ll n$) and is in a Hessenberg form. Hence, the least-squares problem can be efficiently solved, requiring only about $3(m + 1)^2$ flops. On the other hand, for an n-by-n matrix A with $nnz(A)$ nonzeros, *SpMV* and *Orth* require a total of about $2m \cdot nnz(A)$ and $2m^3 n$ flops over the m iterations, respectively (i.e., $n, nnz(A) \gg m$). Hence, the solution time of GMRES is often dominated by the first step of generating the basis vectors. To accelerate the solution process using GPUs, we distribute the coefficient matrix A and the basis vectors $V_{1:m+1}$ in a 1D block row format among the GPUs. We then generate these basis vectors on the GPUs, while the least-square problem is solved on the CPUs.

Even with a single GPU, both *SpMV* and *Orth* require communication to move the data through the memory hierarchy of the GPU, while with multiple GPUs, additional communication is needed among the GPUs. CA-GMRES [5] aims to reduce this communication by redesigning the algorithm to replace *SpMV* and *Orth* with three new kernels – matrix powers kernel (*MPK*), block orthogonalization (*BOrth*), and tall-skinny QR (*TSQR*) – that generate and orthogonalize a set of s basis vectors at once. By avoiding the communication, even with one GPU, CA-GMRES can obtain a speedup of up to two [8]. Figure 1 shows the pseudocode of CA-GMRES (s, m). A more detailed description of our implementation can be found in [8].

In the rest of this section, we review the two main computational kernels of CA-GMRES, *TSQR* and *MPK*, improving whose performance is the focus of this paper.

2.1 Cholesky QR Factorization

To orthonormalize the tall-skinny matrix $V_{1:s+1}$ with $s + 1$ columns, we focus on *TSQR* based on the Cholesky QR (CholQR) factorization [7]. To describe our implementation of CholQR on multicore CPUs with multiple GPUs, we use $V_{1:s+1}^{(d)}$ to denote the local submatrix of $V_{1:s+1}$, distributed to the d-th GPU, and n_g is the number of available GPUs. To orthogonalize the $s + 1$ vectors $V_{1:s+1}$, CholQR first forms the Gram matrix, $B := V_{1:s+1}^T V_{1:s+1}$, through the local matrix-matrix product $B^{(d)} := V_{1:s+1}^{(d)T} V_{1:s+1}^{(d)}$ on the GPU, followed by the reduction $B := \sum_{d=1}^{n_g} B^{(d)}$ on the CPU. Next, the Cholesky factor R of B is computed on the CPU. Finally, the GPU orthogonalizes $V_{1:s+1}$ by a triangular solve $V_{1:s+1}^{(d)} := V_{1:s+1}^{(d)} R^{-1}$. Hence, all the required GPU-GPU communication is aggregated into a pair of messages between the CPU and GPUs, while all the GPU computation is based on BLAS-3. As a result, both intra and inter GPU communication can be optimized. Figure 2 shows these three steps of CholQR. Unfortunately, the condition number $\kappa(B)$ of the Gram matrix B is the square of the condition number $\kappa(V_{1:s+1})$ of the input matrix $V_{1:s+1}$ (i.e., $\kappa(B) = \kappa(V_{1:s+1})^2$). This often causes numerical instability, especially in CA-GMRES, where even using the Newton basis [1], the vector $\mathbf{v}_j$ can converge to the principal eigenvector of A, and $\kappa(V_{1:s+1})$ can be large.

2.2 Matrix Powers Kernel

For *SpMV* on multiple GPUs, the communication of the distributed vector elements through the PCI Express bus could become a bottleneck. To reduce this bottleneck, given a starting vector $\mathbf{v}_j$, *MPK* first communicates all the

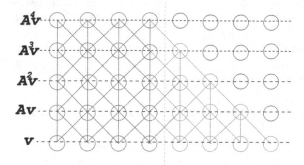

Fig. 3. Illustration of *MPK* for a tridiagonal matrix A with the starting vector **v** and the step size $s = 4$. The blue circles represent the local elements of the vectors to be computed on this GPU and the red circles are the required non-local vector elements. The GPU first communicates the red elements of **v** on the s-level overlap, then independently performs *SpMV* (Color figure online).

required vector elements of $\mathbf{v}_j$ on the s-level overlap at once so that each GPU can independently compute the local components of the s matrix-vector products $A\mathbf{v}_j, A^2\mathbf{v}_j, \ldots, A^s\mathbf{v}_j$ without further communication [2]. Figure 3 illustrates our implementation of *MPK* for a tridiagonal matrix A. As a result, *MPK* reduces the communication latency by a factor of s, but introduces the overheads to store, communicate, and perform computation on the s-level overlap. Though *MPK* often improve the performance of *SpMV* using a small step size s, its optimal step size may be much smaller than that of *BOrth* or *TSQR* due to the overheads associated with *MPK*. See [8] for the detailed discussion of our implementation of *MPK* and its performance.

3   Mixed-Precision CholQR

To improve the numerical stability of CholQR in the working 64-bit double precision, we use a software-emulated quadruple precision at the first two steps of CholQR, while the last step is in the working precision. This is motivated by the fact that the condition number of the Gram matrix B is the square of the condition number of the input matrix $V_{1:s+1}$. Hence, the numerical stability should be improved by using the higher-precision arithmetic for forming and factorizing the Gram matrix (see [9] for the detailed numerical analysis and studies of the mixed-precision CholQR). Here, in Sect. 3.1, we first describe our implementation of the mixed-precision CholQR on the multicore CPUs with the GPU. Then, in Sect. 3.2, we present its performance. Finally, in Sect. 3.3, we study the effects of using the higher-precision on the performance of CA-GMRES.

3.1   Implementation

When the target hardware does not support a desired higher precision, software emulation is needed. For instance, double-double (dd) precision emulates the

double-double operation	# of double precision instructions			
	Add/Substitute	Multiply	FMA	Total
Multiply (double-double input)	5	3	1	9
Multiply (double input)	3	1	1	5
Addition (IEEE-style)	20	0	0	20
Addition (Cray-style)	11	0	0	11

Fig. 4. Number of double-precision instructions in double-double operations.

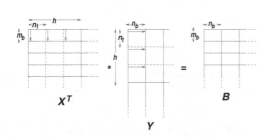

```
double regC[m_b][n_b], regA[m_b], regB
for ℓ = 1, ... h/n_t do
    for j = 1 ... n_b
        regA[i] = x_{ℓ*n_t,j}
    end for
    for j = 1, ..., n_b do
        regB = y_{ℓ*n_t,j}
        for i = 1 ... m_b
            regC[i][j] += regA[i] * regB
        end for
    end for
end for
```

Fig. 5. *InnerProds* implementation (arrow shows data access by a GPU thread).

Fig. 6. *InnerProds* pseudocode.

quadruple precision by representing each numerical value by an unevaluated sum of two double precision numbers, and is capable of representing the 106 bits precision, while the double-precision number is of 53 bits precision. There are two standard implementations [4] of adding two numerical values in double-double precision, $a + b = \hat{c} + e$, where e is the round-off error; one satisfies the IEEE-style error bound ($e = \delta(a + b)$ with $|\delta| \leq 2\epsilon_{dd}$ and $\epsilon_{dd} = 2^{-105}$), and the other satisfies the weaker Cray-style error bound ($e = \delta_1 a + \delta_2 b$ with $|\delta_1|, |\delta_2| \leq \epsilon_{dd}$). Figure 4 lists the computational costs of the double-double operations required by our mixed-precision CholQR (dd-CholQR). The standard CholQR in double precision (d-CholQR) performs about a half of its total flops at Step 1 and the other half at Step 3. On the other hand, compared with the input matrix $V_{1:s+1}$, the Gram matrix B is much smaller in its dimension (i.e., $s \ll n$), and CholQR spends only a small portion of its flops and orthogonalization time, computing its Cholesky factor at Step 2. Hence, using the Cray-style double-double precision for Steps 1 and 2, our dd-CholQR performs about 8.5× more computation than d-CholQR. On the other hand, the increase in communication is less significant; our intra-GPU communication is about the same, only writing the s-by-s output matrix in double-double precision while reading and writing the n-by-s input matrix $V_{1:s+1}$ in double precision ($s \ll n$). In addition, we communicate twice more data between the GPUs ($16s^2$Bytes with $s \approx 10$), but with the same latency.

Though CholQR performs only a half of the total flops at Step 1, its orthogonalization time can be dominated by Step 1. This is because though

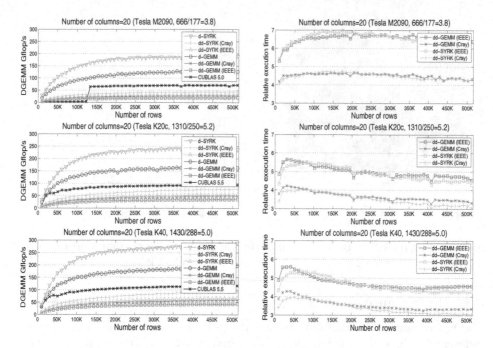

Fig. 7. Performance of standard and mixed-precision *InnerProds* in double precision.

the other half of the total flops is performed at Step 3, solving the triangular system with many right-hand-sides at Step 3 exhibits a high parallelism and can be implemented efficiently on a GPU. On the other hand, at Step 1, computing each element of the Gram matrix requires a reduction operation on two n-length vectors. These inner-products (*InnerProds*) are communication-intensive and exhibit only limited parallelism. Hence, Step 1 often becomes the bottleneck, where standard implementations fail to obtain high-performance on the GPU. In our *batched* implementation of a matrix-matrix multiply (GEMM) to compute *InnerProds*, $B := X^T Y$, each thread block computes a partial product, $B^{(i,j,k)} := X^{(k,i)T} Y^{(k,j)}$, where $X^{(k,i)}$ and $Y^{(k,j)}$ are the h-by-m_b and h-by-n_b blocks of X and Y, respectively.[1] Within the thread block, each of its n_t threads computes its partial result in the local registers (see Fig. 5 for an illustration, and Fig. 6 for the pseudocode, where $\underline{x}_{\ell,j}$ is the (ℓ, j)-th element of $X^{(k,i)}$). Then, each thread block performs the binary reduction of the partial results among its threads, summing n_r columns at a time using the shared memory to store $n_t \times (m_b \times n_r)$ numerical values. The final result is computed by another binary reduction among the thread blocks. Our implementation is designed to reduce the number of synchronizations among the threads while relying on the CUDA runtime and the parameter tuning to exploit the data locality. For the symmetric (SYRK) multiply, $B := V^T V$, the thread blocks compute only a triangular part

---

[1] In the current implementation, the numbers of rows and columns in X and Y are a multiple of h, and multiples of m_b and n_b, respectively, where n_b is a multiple of n_r.

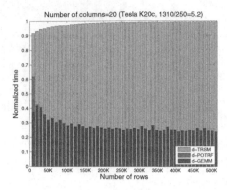 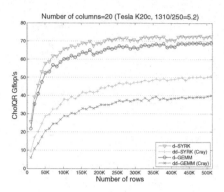

Fig. 8. d-CholQR time breakdown.   **Fig. 9.** d/dd-CholQR performance.

of B and reads $V^{(k,j)}$ once for computing a diagonal block. Our performance studies in the next subsection are based on this batched kernel.

3.2   Performance

Figure 7 compares the standard and mixed-precision *InnerProds* performance on different GPUs, where the mixed-precision *InnerProds* reads the input matrix in the standard 64-bits double precision, but accumulates its intermediate results into the output matrix in the double-double precision. Each GPU has a different relative cost of communication to computation, and on top of each plot, we show the ratio of the double-precision peak performance (Gflop/s) over the shared memory bandwidth (GB/s) (i.e., flop/B to obtain the peak). This ratio tends to increase on a newer architecture, indicating a greater relative communication cost. We tuned our kernel for each matrix dimension on each GPU in each precision (see the five tunable parameters h, m_b, n_b, n_r, and n_t in Sect. 3.1), and the figure shows the optimal performance. Based on the memory bandwidth and the fixed number of columns in the figure, the respective peak performances of the standard d-GEMM are 442, 625, and 720Gflop/s on the M2090, K20c, and K40 GPUs. Our d-GEMM obtained 29, 26, 28 % of these peak performances and speedups of about 1.8, 1.7, and 1.7 over CUBLAS 5.5 on these three GPUs. In addition, though it performs 16× more floating-point instructions, the gap between the standard d-GEMM and the mixed-precision dd-GEMM tends to decrease on a newer architecture, and dd-GEMM is only less than four times slower on K20c. We also see that by taking advantage of the symmetry, both d-SYRK and dd-SYRK improve the performance of d-GEMM and dd-GEMM, respectively.

Figure 8 shows the breakdown of the standard d-CholQR orthogonalization time on two eight-core Intel Sandy Bridge CPUs with one NVIDIA K20c GPU. Because of our efficient implementation of *InnerProds*, only about 30 % of the orthogonalization time is now spent in *d-InnerProds*. As a result, while the mixed precision *dd-InnerProds* was about four times slower than *d-InnerProds*, Fig. 9 shows that the mixed-precision dd-CholQR is only about 1.7 or 1.4 times slower

Name	Source	$n/1000$	nnz/n
cant	FEM Cantilever	62.4	64.2
shipsec1	FEM Ship secion	140.8	87.3
dielFilterV2real	FEM in Electromagnetic	1157.5	41.9
G3_circuit	Circuit simulation	1585.4	4.8

Fig. 10. Test matrices used for test cases with CA-GMRES.

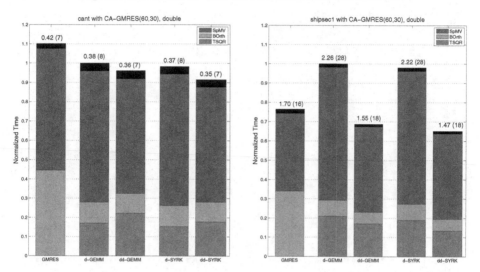

Fig. 11. Performance comparison of CA-GMRES using dd-CholQR (with dd-GEMM or dd-SYRK) against CA-GMRES using d-CholQR with (d-GEMM or d-SYRK) and GMRES using CGS: On top of each bar shows total time in seconds and restart count. To obtain the solution convergence, the reorthogonalization was used with d-CholQR, while it was not needed with dd-CholQR.

than the standard d-CholQR when GEMM or SYRK is used for *InnerProds*, respectively. In other words, if the reorthogonalization is avoided using the higher-precision, then dd-CholQR may obtain a performance competitive to that of d-CholQR with reorthogonalization. For the mixed-precision dd-CholQR, the Cholesky factorization in the double-double precision is computed using MPACK[2] on the CPU, while for d-CholQR, we use the threaded version of MKL for the Cholesky factorization in the double precision.

3.3 Case Studies with CA-GMRES

Figure 11 shows the solution time of CA-GMRES using the standard d-CholQR and the mixed-precision dd-CholQR on two eight-core Sandy Bridge CPUs with a single K20c. To maintain the numerical stability and obtain the solution convergence, the full-reorthogonalization was needed with d-CholQR, while it was

---

[2] http://mplapack.sourceforge.net.

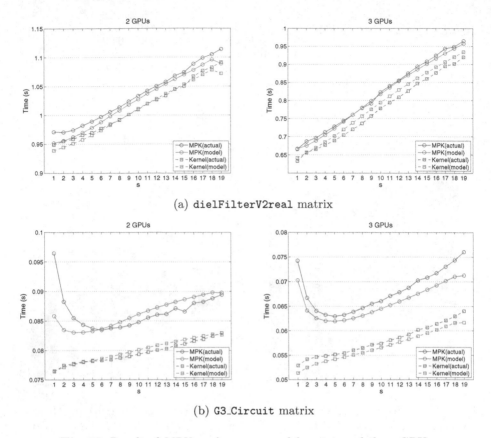

(a) dielFilterV2real matrix

(b) G3_Circuit matrix

Fig. 12. Result of *MPK* performance model on two and three GPUs.

not needed with dd-CholQR. For *BOrth*, we used the classical Gram-Schmidt (CGS) process [3] with reorthogonalization, which obtains high performance with the GPU [8]. The solution time is normalized using the corresponding solution time of GMRES that uses CGS with reorthogonalization for orthogonalizing its basis vectors. Figure 10 shows the properties of our test matrices that were downloaded from the University of Florida Sparse Matrix collection.[3] We see that using dd-CholQR, even with the computationally expensive software emulation, the solution time was reduced not only because the reorthogonalization was avoided but also because CA-GMRES converged in fewer iterations.

4   Adaptive Step Size for Matrix Powers Kernel

Most of CA-GMRES implementations including ours [8] use the same step size s for *MPK*, *BOrth*, and *TSQR*, while the optimal s for *MPK* is typically smaller than that of *BOrth* or *TSQR* due to the computational and/or communication

---

[3] http://www.cise.ufl.edu/research/sparse/matrices/.

overheads associated with MPK. To address this performance difference, we first adapted our implementation such that MPK uses a smaller step size $\hat{s}$ than the step size s used for $BOrth$ and $TSQR$. Hence, to generate the s basis vectors, we invoke MPK $s/\hat{s}$ times using the step size $\hat{s}$ before calling $BOrth$ and $TSQR$. In addition, instead of having a different $\hat{s}$ for MPK as a user-specified input parameter, we design an adaptive scheme to dynamically adjust the step size $\hat{s}$ of MPK based on the static inputs (i.e., the sparsity pattern of the coefficient matrix A and the maximum step size s) and the performance measurements gathered during the first restart-loop of CA-GMRES. In particular, for our experiments, we use the following performance model:

$$MPK \text{ time} = \text{Inter-communication time} + \text{Kernel time},$$

where we let

$$\text{Inter-communication time} = \text{Latency} + \frac{\text{Communication volume}}{\text{Bandwidth}}, \text{ and}$$

$$\text{Kernel time} = \frac{\text{Flop count}}{\text{flop/s}} + \# \text{ of random data accesses} \times \text{Data access time},$$

and "Kernel time" consists of the computation and intra-GPU communication time. In our experiments, "Communication volume" and "Flop count" are computed based on the sparsity pattern of the coefficient matrix A, while "# of random data accesses" is approximated by the aggregated number of non-local vector elements accessed by MPK. On the other hand, we computed "Latency," "Bandwidth," "flop/s," and "Data access time" based on the measured time of the reduction for the dot-products, point-to-point communication for $SpMV$, flop count and time required by $SpMV$, and data copy on the GPU, respectively. All the performance measurements are collected during the first restart loop of CA-GMRES. In practice, we often use GMRES iteration for the first restart loop (i.e., $s = 1$). This is because to maintain the numerical stability, MPK generates the Newton basis [1] whose shifts can be computed during the first restart. Since these shifts are not available for the first restart loop, to maintain the numerical stability, GMRES iteration is used. Hence, with the proposed adaptive scheme, we gather both the numerical and performance statistics of the given problem during the first restart loop. Then, based on these statistics, the input parameters are adjusted to enhance both the performance and stability of CA-GMRES for the remaining loops.

Figure 12 shows the effectiveness of the performance model to capture the performance of MPK for two sparse matrices on Intel Sandy Bridge CPUs with three NVIDIA Tesla M2090 GPUs. The properties of our test matrices from the University of Florida Sparse Matrix collection are shown in Fig. 10. Since we use the performance model to select a good step size, the model only needs to capture the performance trend and not the exact performance. In addition, in many cases, the performance of MPK does not change significantly around the optimal step size. The figure demonstrates that for both matrices, the model was successful in capturing the performance trends and selecting a near-optimal step

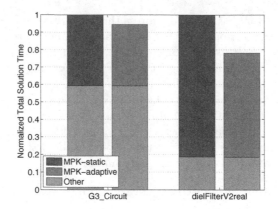

Fig. 13. Effects of adaptive *MPK* step size on performance of CA-GMRES.

size. In particular, for the `dielFilterV2real` matrix, due to the overhead associated with *MPK*, the standard *SpMV* was faster than *MPK*. Our performance model could capture this and select the step size of one for *MPK*.

Figure 13 shows the effects of the adaptive step size on the performance of CA-GMRES, where the static scheme uses the fixed step size for *MPK*, *BOrth*, and *TSQR* that obtains the near-optimal performance of CA-GMRES. Though the improvement was not significant, this is based on the near-optimal performance of *MPK*. We expect the benefit of the adaptive scheme to increase on the computer where the communication cost is higher (e.g., a GPU cluster). Finally, in all the test cases, it only required marginal overheads to gather the performance measurements.

5   Conclusion

We proposed a mixed-precision orthogonalization scheme to improve the numerical stability of CA-GMRES. When the target hardware does not support a desired higher precision, software emulation is needed. We showed that though the software emulation could significantly increase the computational cost, the increase in the communication cost is less significant. As a result, the overhead of using the software emulation is decreasing on a newer GPU architecture where the cost of the computation is decreasing compared to the cost of the communication. Our case studies on multicore CPUs with a GPU demonstrated that though it requires about 8.5× more computation, using a higher-precision for this small but critical segment of CA-GMRES can improve not only its overall stability but also, in some cases, its performance.

In this paper, we only studied the effects of a higher-precision on a single GPU. On multiple GPUs of a compute node, the performance of CA-GMRES depends more on the performance of the GPU kernels (i.e., intra-GPU communication) than the inter-GPU communication [8]. Hence, similar benefits of using a higher-precision are expected on the multiple GPUs. We will study its effects on a system with a greater communication latency (e.g., distributed GPUs or CPUs) where the performance improvement by using the higher-precision arithmetic may be greater. We are also studying the use of mixed-precision in eigensolvers where the orthogonality can be more crucial, and are writing an extended paper focusing on the numerical properties of our mixed-precision scheme [9]. Finally, it is of our interest to apply or extend recent mixed precision efforts (e.g., reproducible BLAS[4] and precision tuning[5]) for our studies.

In this paper, we also studied an adaptive scheme to adjust the step size of *MPK* on multiple GPUs. Our performance results demonstrated that our adaptive scheme can find a near optimal step size based on the static input parameters and the performance measurements gathered during the first restart loop, and reduce the total solution time of CA-GMRES. Our *MPK* is currently optimized only for the inter-GPU communication which is relatively inexpensive on a node. We are looking to optimize *MPK* on a GPU, which should increase the benefit of the adaptive step size. We also plan to study the effectivness of the adaptive schme on a larger system with greater communication cost (e.g., a distributed system), where a greater benefit of the adaptive scheme is expected (in term of time or memory).

Acknowledgments. This material is based upon work supported by the U.S. Department of Energy (DOE), Office of Science, Office of Advanced Scientific Computing Research (ASCR), and was founded in part by National Science Foundation under Grant No. ACI-1339822, DOE Grant #DE-SC0010042: "Extreme-scale Algorithms & Solver Resilience (EASIR)," Russian Scientific Fund, Agreement N14-11-00190, and NVIDIA. This research used resources of the Keeneland Computing Facility at the Georgia Institute of Technology, which is supported by the National Science Foundation under Contract OCI-0910735. We thank Maho Nakata, Daichi Mukunoki, and the members of the DOE EASIR project for helpful discussions.

References

1. Bai, Z., Hu, D., Reichel, L.: A Newton basis GMRES implementation. IMA J. Numer. Anal. **14**, 563–581 (1994)
2. Demmel, J., Hoemmen, M., Mohiyuddin, M., Yelick, K.: Avoiding communication in computing Krylov subspaces. Technical report UCB/EECS-2007-123, University of California Berkeley EECS Department, October 2007
3. Golub, G., van Loan, C.: Matrix Computations, 4th edn. The Johns Hopkins University Press, Baltimore (2012)
4. Hida, Y., Li, X., Bailey, D.: Quad-double arithmetic: algorithms, implementation, and application. Technical report LBNL-46996 (2000)
5. Hoemmen, M.: Communication-avoiding Krylov subspace methods. Ph.D. thesis, University of California, Berkeley (2010)
6. Saad, Y., Schultz, M.: GMRES: A generalized minimal residual algorithm for solving nonsymmetric linear systems. SIAM J. Sci. Stat. Comput. **7**, 856–869 (1986)

---

[4] http://www.eecs.berkeley.edu/~hdnguyen/rblas.
[5] http://crd.lbl.gov/groups-depts/ftg/projects/current-projects/corvette/precimonious.

7. Stathopoulos, A., Wu, K.: A block orthogonalization procedure with constant synchronization requirements. SIAM J. Sci. Comput. **23**, 2165–2182 (2002)
8. Yamazaki, I., Anzt, H., Tomov, S., Hoemmen, M., Dongarra, J.: Improving the performance of CA-GMRES on multicores with multiple GPUs. In: The Proceedings of the IEEE International Parallel and Distributed Processing Symposium (IPDPS), pp. 382–391 (2014)
9. Yamazaki, I., Tomov, S., Dongarra, J.: Mixed-precision Cholesky QR factorization and its case studies on multicore CPUs with multiple GPUs. Submitted to SIAM J. Sci. Comput. (2014)

Heterogenous Acceleration for Linear Algebra in Multi-coprocessor Environments

Azzam Haidar[1], Piotr Luszczek[1]([✉]), Stanimire Tomov[1], and Jack Dongarra[1,2,3]

[1] University of Tennessee Knoxville, Knoxville, USA
luszczek@eecs.utk.edu
[2] Oak Ridge National Laboratory, Oak Ridge, USA
[3] University of Manchester, Manchester M13 9PL, UK

Abstract. We present an efficient and scalable programming model for the development of linear algebra in heterogeneous multi-coprocessor environments. The model incorporates some of the current best design and implementation practices for the heterogeneous acceleration of dense linear algebra (DLA). Examples are given as the basis for solving linear systems' algorithms – the LU, QR, and Cholesky factorizations. To generate the extreme level of parallelism needed for the efficient use of coprocessors, algorithms of interest are redesigned and then split into well-chosen computational tasks. The tasks execution is scheduled over the computational components of a hybrid system of multi-core CPUs and coprocessors using a light-weight runtime system. The use of light-weight runtime systems keeps scheduling overhead low, while enabling the expression of parallelism through otherwise sequential code. This simplifies the development efforts and allows the exploration of the unique strengths of the various hardware components.

1 Programming Models for the Off-load Mode

The Intel Xeon Phi coprocessor is a hardware accelerator that made its debut in the late 2012 as a platform for high-throughput technical computing, sometimes known under an alternative name of Many Integrated Cores (MIC). The common mode of operation for the device is called off-load but the stand-alone and reverse off-load are also possibilities. When in off-load mode, the device receives work from the host processor and reports back as soon as the computational task completes. Any such assignment of work proceeds and completes without the host device (commonly an Intel CPU such as Sandy Bridge or Ivy Bridge) being involved. The CPU may monitor the activity of communication and/or computation through an event-based interface and can also pursue its own computational activities between events. This is very similar to the operation of hardware accelerators based on

This research was partially supported by the National Science Foundation under Grants OCI-1032815, ACI-1339822, and Subcontract RA241-G1 on NSF Prime Grant OCI-0910735, DOE under Grants DE-SC0004983 and DE-SC0010042, and Intel.

M. Daydé et al. (Eds.): VECPAR 2014, LNCS 8969, pp. 31–42, 2015.
DOI: 10.1007/978-3-319-17353-5_3

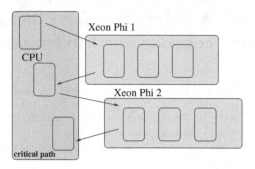

Fig. 1. DLA algorithm as a collection of BLAS-based tasks represented as rectangles and their dependences represented as arrows. The algorithm's critical path is, in general, scheduled on the CPUs, and large data-parallel tasks on the Xeon Phi coprocessors.

compute-capable GPUs and FPGAs that are specialized for certain types of workloads beyond what could be achieved on standard multicore CPUs. In fact, Xeon Phi is often considered to be an alternative to the hardware accelerators from AMD and NVDIA despite the fact that there exist many technical differences between the three (Fig 1).

The off-load mode for the Xeon Phi devices has direct support from the compiler in that it is possible to issue requests to the device and ascertain the completion of tasks directly from the user's C/C++ code. The support for this mode of operation is offered by the Intel compiler through Phi-specific pragma directives: offload, offload_attribute, offload_transfer, and offload_wait [4]. This is very closely related to the off-load directives now included in the OpenMP 4 standard. In fact, the two are syntactically and semantically equivalent, barring the difference in the "omp" prefix for the OpenMP syntax. A similar standard for GPUs is called OpenAcc. A summary of various programming methods on Xeon Phi is provided in Table 1.

For many scientific applications, the offload model offered by the Intel compiler and OpenMP 4 is sufficient. Until recently, this was not the case for a port of the MAGMA library to the Xeon Phi because of the very rich functionality that MAGMA inherits from both its CUDA and OpenCL ports. We had to use the LLAPI (Low-Level API) based on Symmetric Communication InterFace

Table 1. Programming models for the Intel Xeon Phi coprocessors and their current status and properties.

Programming model/API	Status	Portability	Overhead	Language support
SCIF	Mature	No	None	No
COI	Mature	Yes	Minimal	Yes
OpenMP 4.0	Early	Yes	Varies	Yes
OpenCL	Experimental	Yes	Minimal	No

(SCIF) that offers, as the name suggests, a very low-level interface to the host and device hardware. The use of this API is discouraged for most workloads as it tends to be error-prone and offers very little abstraction on top of the hardware interfaces. The motivation to use SCIF is to take advantage of the capability of asynchronous events that allows the user for low-latency messaging between the host and the device as well as to notify about completion of kernels on Xeon Phi. This enabled the possibility of hiding the cost of data transfer between the host and the device which requires the transfer of submatrices to overlap with the computation. The direct access to the DMA (Direct Memory Access) engine allowed us to maximize the bandwidth of data transfers over the PCI Express bus. The only requirement was that the memory regions for transfers to be page-aligned and pinned to guarantee their fixed location in the physical memory.

With the continuous improvements in the APIs that conceptually reside above SCIF, the overheads and functionality afforded by SCIF is no longer exclusive, and we are able to achieve very much comparable performance and asynchronous interface using higher-level APIs, while gaining portability as an important added bonus.

2   Efficient and Scalable Programming Model Across Multiple Devices

In this section, we describe a programming model that raises the level of abstraction above the hardware specifics while still allowing us to capture the strengths of the various hardware components in a heterogeneous system and develop highly efficient algorithms. We present the accompanying software stack and the techniques developed for the effective use of both single and multi Xeon Phi coprocessors. Our proposed techniques consider both the higher ratio of execution and the hierarchical memory model of the new emerging coprocessors.

2.1   Task Distribution Based on Hardware Capability

Programming models that raise the level of abstraction are of great importance for reducing software development efforts. A traditional approach has been to organize algorithms in terms of BLAS calls, where hardware specific optimizations would be hidden in BLAS implementations such as Intel's MKL or AMD's ACML. This is illustrated in Algorithm 1 where the factorization is split on two phases: the panel factorization phase which is a sequence of Level 1 BLAS call and thus memory bound operations followed by the trailing matrix update which is Level 3 BLAS call and thus compute intensive operations. This is still used but has shown some drawbacks on new architectures. In particular, parallelization is achieved using a fork-join approach since only the trailing matrix update, can be performed efficiently in parallel (fork) but a synchronization is needed before performing the next call (join). The number of synchronizations thus can become a prohibitive bottlenecks for performance on highly parallel devices such as the

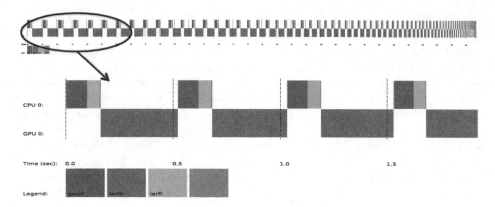

Fig. 2. Execution trace of DGEQRF (QR factorization) based on Algorithm 1 and using 1 Xeon Phi cards (Color figure online).

MICs. This type of programming has been popularized under the Bulk Synchronous Processing name [13,14]. The synchronization do not allow to overlap the panel computation with the trailing matrix update and thus can results in slow performance especially that the panel factorization is memory bound and thus performs close to sequential. We depict in Fig. 2 the execution trace of the QR decomposition as implemented based on the technique described in Algorithm 1. It is clear that the panel factorization phase (*"geqrf+larft"* red and orange color of Fig. 2) cannot be overlapped with the trailing matrix update (*"larfb"* green color of Fig. 2).

Algorithm 1. Two-phase traditional blocked factorization of $A = [P_1, P_2, \ldots]$.

> **for** $P_i \in \{P_1, P_2, \ldots, P_n\}$ **do**
> > PanelFactorize $_{\text{on CPU}}(P_i)$
> > TrailingMatrixUpdate$_{\text{on MIC}}(A^{(i)})$

Instead, the algorithms (like matrix factorizations) are broken into computational tasks (e.g., panel factorizations followed by trailing submatrix updates) and pipelined for execution on the available hardware components (see below). Moreover, particular tasks are scheduled for execution on the hardware components that are best suited for them. Thus, this task distribution based on *hardware capability* allows the user for the efficient use of each hardware component. In the case of DLA factorizations, the less parallel panel tasks are scheduled for execution on multicore CPUs, and the parallel updates mainly on the MICs. We illustrate this in Algorithm 2 and depict the execution trace of the QR decomposition on Fig. 3. Now the panel factorization (which is memory bound phase *"geqrf+larft"* red and orange color of Fig. 3) can be overlapped with the remaining of the trailing matrix

---

Algorithm 2. Two-phase factorization of $A = [P_1, P_2, \ldots]$ with lookahead of depth 1. Matrix A and the result are assumed to be on the MIC memory.

PanelStartReceiving$_{\text{on CPU}}(P_1)$;
for $P_i = P_1, P_2, \ldots$ **do**
    PanelFactorize$_{\text{on CPU}}(P_i)$;
    PanelSend$_{\text{to MIC}}(P_i)$;
    TrailingMatrixUpdate$_{\text{on MIC}}(P_{i+1})$;
    PanelStartReceiving$_{\text{on CPU}}(P_{i+1})$;
    TrailingMatrixUpdate$_{\text{on MIC}}(P_{i+2}, \ldots)$;

---

update $(P_{i+2}, \ldots$ "*larfb*" green color of Fig. 3). The cost of the panel phase can be hidden with the update on the MIC and thus we expect performance to reach close to the Level 3 BLAS peak of the MIC.

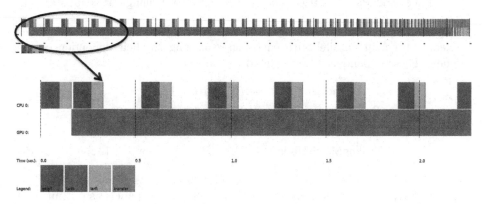

Fig. 3. Execution trace of DGEQRF (QR factorization) based on Algorithm 2 and using 1 Xeon Phi cards (Color figure online).

2.2 Task Based Runtime Model

The scheduling of tasks for execution can be static or dynamic. In either case, the small and not easy to parallelize tasks from the critical path (e.g., panel factorizations) are executed on CPUs, and the large and highly parallel task (like the matrix updates) mostly on the MICs.

The use of multiple coprocessors complicates the development using static scheduling. Instead, the use of a light-weight runtime system is preferred as it can keep scheduling overhead low, while enabling the expression of parallelism through sequential-like code. The runtime system relieves the developer from keeping track of the computational activities that, in the case of heterogeneous systems, are further exacerbated by the separation between the address spaces of the main memory of the CPU and the MICs. Our runtime model is build on the

QUARK [15] superscalar execution environment that has been originally used with great success for linear algebra software on just multicore platforms [8]. The conceptual work though could be replicated within other models such as StarPU [1], OmpSS [11], Cilk [2], and Jade [12], to just mention a few. A detailed comparison of some of these systems was attempted before [7] with very little performance difference on CPU-only systems and a difference on the order of few percentage points was observed [3,9,10] when memory-bound (rather than compute-bound) kernels were scheduled by the system.

We will now use Algorithm 2 to illustrate the task based execution model. There are two types of communication tasks $Task_{comm}$: PanelSend, PanelStartReceiving, and two computational tasks $Task_{exe}$: PanelFactorize, TrailingMatrixUpdate. In a general sense these tasks are asynchronous in that they do not have prescribed moment in time to execute, rather they execute when both all their input dependences are satisfied and the appropriate hardware entity is available. Formally, the task space $\mathcal{T}$ for the factorization is defined as:

$$\mathcal{T} = \{\text{PanelSend, PanelStartReceiving, PanelFactorize, TrailingMatrixUpdate}\} \quad (1)$$

the tasks are sequential and free of side-effects which gives the runtime scheduler to execute them in parallel with each other as long as their dependences are satisfied. These dependences are defined as follows:

$$P_i \rightarrow \quad\quad \text{PanelSend} \quad\quad \rightarrow P_i \quad \text{for } i = 1 \dots n \quad (2)$$
$$P_i \rightarrow \text{PanelStartReceiving} \rightarrow P_i \quad \text{for } i = 1 \dots n \quad (3)$$
$$P_i \rightarrow \quad\quad \text{PanelFactorize} \quad\quad \rightarrow P_i \quad \text{for } i = 1 \dots n \quad (4)$$
$$P_i \rightarrow \text{TrailingMatrixUpdate} \rightarrow P_{i+1}, P_{i+2} \dots \quad \text{for } i = 1 \dots n \quad (5)$$

where all the tasks take panel P_i as input. The panel tasks produce panel P_i as output and the update tasks produces all the subsequent panels as output.

2.3   Improving Offload Mode Communication

It is well known that the off-load transfer mode copies only continuous chunks of data from and to the coprocessors. However most of the scientific application algorithms require to exchange data with $2D$ or $3D$ storage and thus this may create an issue when using the off-load transfer mode. In particular, the one-sided factorizations (Cholesky, LU, and QR) require to send the panel to the CPU and then receive it later after being factorized by the CPU. A simple implementation loop over one direction and call the off-load section to send/receive a contiguous vector. Such implementation behaves poorly and as a result the communication will become expensive and slow down the algorithm. Indeed, another alternative is to copy the $2D$ panel to a contiguous temporary space on the MIC, and then to send it and vice versa. Hence, there are two points that need to be taken into consideration. Firstly, the copy needs to be implemented as a multi-threaded operation in order to hide its cost. For that, we implemented a parallel copy that uses all of the 240 hardware threads of the MIC to perform the copy. This might be against the common wisdom that multi-threading is of little help for bandwidth-limited operations such as a memory copy. However, it is not the case on the MIC, where the clock frequency of the compute cores is twice as low as that of the main memory on the compute card – the exact opposite of what is the case for the Intel x86 multicore processors. In addition to the low frequency, the current MIC hardware is to a large degree an in-order architecture with dual-pipeline execution and single-issue fetch/decode units [6] which poses constraints on the amount of bandwidth that can be utilized by a single core. These can be overcome in multiple ways, including the use of streaming loads and having multiple threads issuing requests for data simultaneously. Furthermore, when the MIC copies data to or from the temporary space, it should be the only kernel running, otherwise, it would interfere with other kernels executing at the same time and this may slow down both of the kernels. To that end, we represented the copy kernel as a task with high priority and the scheduler was responsible for executing it as soon as possible and to handle the dependences such that no other kernel will be running at the same time. Experiments showed that when using these optimizations the performance of the off-load communication mode is comparable to both the SCIF and the COI mode with a variance of less than 5 %.

2.4   Data Distribution to Minimize Communication

Data distribution formats for multi-device computations can drastically affect the performance. In particular, swapping rows (the dlaswp routine) in LU, or the dlarfb trailing matrix update routine in QR, may require unnecessary data movements in certain data formats. Therefore, to minimize the amount of communication between devices, our implementation uses a $1D$ block cyclic distribution. Indeed, using the well known $2D$ block cyclic distribution among multi-devices will enforce an extra amount of communication between them in order to perform the dlaswp in LU, while using $1D$ block cyclic distribution will not need any of these communication. Another example is the dlarfb routine used in the QR factorization to perform $(I - VT^TV^T)\tilde{A}$. Here V holds the Householder reflectors generated during the panel factorization at step k, and $\tilde{A}$ is the trailing matrix at step k. A $2D$ block cyclic distribution will require a sum between the devices in order to compute $V^T\tilde{A}$, while a $1D$ block cyclic distribution will again not need any communications. Note that the overall workload is well spread among the multi-coprocessors when using $1D$ block cyclic distribution.

2.5   Trading Extra Computation for Higher Execution Rate

The optimization discussed here is MIC-specific but is often valid for any hardware architecture with multilayered memory hierarchy. The dlarfb routine used by the QR decomposition consists of two dgemms and one dtrmm. Since coprocessors are better at handling compute-bound tasks, for computational efficiency, we replace the dtrmm by dgemm, yielding 5–10 % performance improvement. For the Cholesky factorization, the trailing matrix update requires a dsyrk. Due to uneven storage, the multi-device dsyrk cannot be assembled purely from regular dsyrk calls on each device. Instead, each block column must be processed individually. The diagonal blocks require special attention. One can use a dsyrk to update each diagonal block, and a dgemm to update the remainder of each block column below the diagonal block. The small dsyrk operations have little parallelism and therefore their execution is inefficient on MICs. This can be improved to some degree by using pragma to run several dsyrk's simultaneously. Nevertheless, because we have copied the data to the device, we can consider the space above the diagonal to be a scratch workspace. Thus, we update the entire block column, including the diagonal block, writing extra data into the upper triangle of the diagonal block, which is subsequently ignored. We do extra computation for the diagonal block, but gain efficiency overall by launching fewer BLAS kernels on the device and using the more efficient dgemm kernels, instead of small dsyrk kernels, resulting in overall 5–10 % improvement in performance.

2.6 Scaling with Respect to Multiple Coprocessor Cards

To achieve scalability of the computation when new MIC cards are added to the system, our programming model offers a number of advantages. The underlying messaging layer at the hardware level is the PICexpress interface. In its current form, it is not meant for scale-up operation and even the high-end installations can accommodate up to about 10 cards in a single shared memory server node. Working within this limitation, the task-based dataflow representation offers plentiful opportunities for parallelism and dynamic scheduling to adapt to a number of hardware components in an asynchronous fashion. The scaling across the cards is further supported by the 1D block cyclic data distribution, that is essential for a balanced distribution of work. Finally, our dynamic scheduling system handles simultaneously CPU cores and coprocessor cards' workload and is capable of managing a one-to-one match between the two, which more than adequate given the aforementioned limitations of the PCIexpress bus.

3   Experimental Results

We present performance results on an Intel dual-socket multicore system with three Intel Xeon Phi cards. Each CPU processor is eight-core Intel Xeon E5-2670 (Sandy Bridge), running at 2.6 GHz, and has a 24 MB shared Level 3 cache. Each core has a private 256 KB Level 2 and 64 KB Level 1 caches. The system is equipped with 52 GB of memory. Its theoretical peak in double precision is 332 Gflop/s. The Intel Xeon Phi cards have 15 GB memory each, running at 1.09 GHz, and yielding a double precision theoretical peak of 1, 046 Gflops.

On the CPU side we use the MKL (Math Kernel Library) [5], version 11.00.03. The Intel Xeon Phi is running the MPSS 2.1.5889-16 software stack and the icc 13.1.1 20130313 compiler. These come with the composer_xe_2013.3.163 suite.

Figures 4, 5, and 6 show the performance results for the Cholesky, LU, and QR factorizations respectively. The figures show a scalability study for up to 3

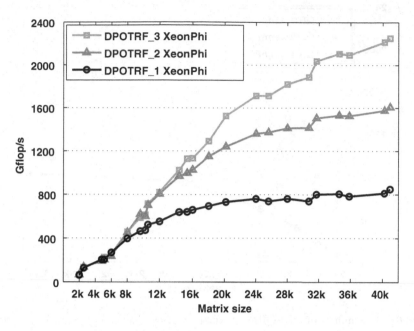

Fig. 4. Performance of DPOTRF (Cholesky factorization) on up to 3 Xeon Phi cards.

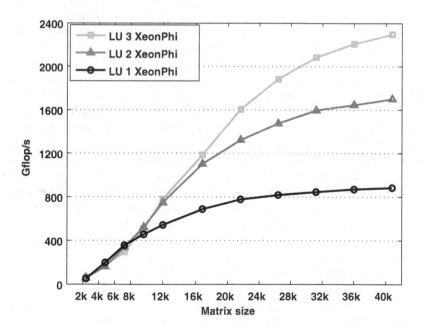

Fig. 5. Performance of DGETRF (LU factorization) on up to 3 Xeon Phi cards.

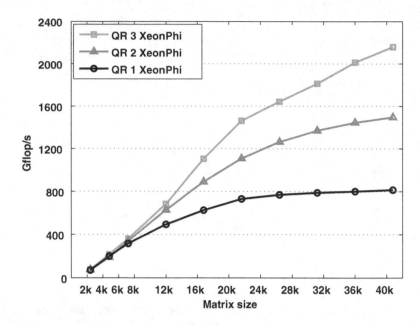

Fig. 6. Performance of DGEQRF (QR factorization) on up to 3 Xeon Phi cards.

Xeon Phi devices. The first observation is the large matrix sizes (beyond 5000) required to take advantage of the benefits that the devices offer and, consequently, outperform the peak performance of the CPU. Similarly, adding the second Xeon Phi is beneficial for matrix sizes larger than 10,000 for Cholesky, 12,000 for LU, and QR factorizations. Finally, the addition of the third Xeon Phi benefits all three factorizations only beyond matrices of size 16,000. This behavior is to be expected from a compute-oriented device that is connected to the CPU through a high-latency, low-bandwidth bus such as the PCI Express. Each matrix panel is factorized on the CPU and for that must make its way from the Xeon Phi device to the CPU and back, thus suffering the communication penalty twice. While the dynamic scheduling allows us to hide this overhead at the beginning of the factorization when the trailing matrix updates carry enough of a computational load, the final steps are squarely dominated by the panel computation and very little can be done about it since the Xeon Phi is a throughput oriented device and in our attempts delivered low performance for latency-bound workloads such as the panel factorization.

As far as scaling and parallel efficiency are concerned, Figs. 4, 5, and 6 show that once the matrix sizes grow beyond the aforementioned threshold, the scaling from one to two and from two to three Xeon Phi devices remains steady and progresses as expected.

Another important aspect of the performance behavior that we observed on our Xeon Phi cards can be seen in Fig. 4. The figure shows extra data points to underscore the variability of the performance with respect to the problem size. In particular, when the matrix sizes are not divisible by a particular value, a blocking factor for the underlying BLAS library, the resulting performance might not follow a smooth path and experience variations. However, we are finding this behavior to continuously become less of a burden with every new release of the software stack for the Xeon Phi card.

Finally, our results for the single Xeon Phi card match the performance numbers obtained by Intel's own MKL library in off-load mode.

4 Conclusions and Future Work

We designed algorithms and a programing model for developing high-performance dense linear algebra in co-processors environments. Further, despite the complexity of the hardware, acceleration was achieved at a surprisingly low software development effort using a high-level methodology of developing hybrid algorithms. In particular, we obtained high fraction of the peak performance for the entire heterogeneous system. The promise shown so far motivates and opens opportunities for future research and extensions, e.g., tackling more complex algorithms and hybrid hardware. When a complex algorithm needs to be executed on a complex heterogeneous system, scheduling decisions have a dramatic impact on performance. Therefore, new scheduling strategies must be designed to fully benefit from the potential of future large-scale machines. In particular, taking our algorithms and the programming paradigm beyond a single node is also an important topic of the future research.

Acknowledgements. This research was supported in part by the National Science Foundation under Grants OCI-1032815, ACI-1339822, and Subcontract RA241-G1 on NSF Prime Grant OCI- 0910735, DOE under Grants DE-SC0004983 and DE-SC0010042, and Intel Corporation.

References

1. Augonnet, C., Thibault, S., Namyst, R., Wacrenier, P.-A.: StarPU: a unified platform for task scheduling on heterogeneous multicore architectures. Concur. Comput. Pract. Exp. **23**(2), 187–198 (2011)
2. Blumofe, R.D., Joerg, C.F., Kuszmaul, B.C., Leiserson, C.E., Randall, K.H., Zhou, Y.: Cilk: an efficient multithreaded runtime system. SIGPLAN Not. **30**, 207–216 (1995)
3. Haidar, A., Ltaief, H., Luszczek, P., Dongarra, J.: A comprehensive study of task coalescing for selecting parallelism granularity in a two-stage bidiagonal reduction. In: Proceedings of the IEEE International Parallel and Distributed Processing Symposium, Shanghai, China, 21–25 May 2012, pp. 25–35. IEEE Computer Society (2012)
4. Intel® Xeon Phi™ coprocessor system software developers guide. http://software.intel.com/en-us/articles/
5. Math Kernel Library. http://software.intel.com/intel-mkl/
6. Jeffers, J., Reinders, J.: Intel® Xeon Phi™ Coprocessor High-Performance Programming. Morgan Kaufmann Publishers, San Francisco (2013)

7. Kurzak, J., Ltaief, H., Dongarra, J.J., Badia, R.M.: Scheduling dense linear algebra operations on multicore processors. Concur. Comput. Pract. Exp. **21**(1), 15–44 (2009)
8. Kurzak, J., Luszczek, P., YarKhan, A., Faverge, M., Langou, J., Bouwmeester, H., Dongarra, J.: Multithreading in the PLASMA Library. In Handbook of Multi and Many-Core Processing: Architecture, Algorithms, Programming, and Applications. Computer and Information Science Series. Chapman and Hall/CRC, 26 April 2013
9. Ltaief, H., Luszczek, P., Dongarra, J.: Enhancing parallelism of tile bidiagonal transformation on multicore architectures using tree reduction. In: Wyrzykowski, R., Dongarra, J., Karczewski, K., Waśniewski, J. (eds.) PPAM 2011, Part I. LNCS, vol. 7203, pp. 661–670. Springer, Heidelberg (2012)
10. Luszczek, P., Ltaief, H., Dongarra, J.: Two-stage tridiagonal reduction for dense symmetric matrices using tile algorithms on multicore architectures. In: Proceedings of IPDPS 2011: IEEE International Parallel and Distributed Processing Symposium, Anchorage, Alaska, USA, 16–20 May 2011, pp. 944–955. IEEE Computer Society (2011)
11. Pérez, J.M., Badia, R.M., Labarta, J.: A dependency-aware task-based programming environment for multi-core architectures. In: Proceedings of the 2008 IEEE International Conference on Cluster Computing, Tsukuba, Japan, 29 September–1 October 2008, pp. 142–151. IEEE (2008)
12. Rinard, M.C., Scales, D.J., Lam, M.S.: Jade: a high-level, machine-independent language for parallel programming. Computer **26**(6), 28–38 (1993). doi:10.1109/2. 214440
13. Valiant, L.G.: Bulk-synchronous parallel computers. In: Reeve, M. (ed.) Parallel Processing and Artificial Intelligence, pp. 15–22. Wiley, New York (1989)
14. Valiant, L. G.: A bridging model for parallel computation. Commun. ACM **33**(8) (1990). doi:10.1145/79173.79181
15. YarKhan, A.: Dynamic task execution on shared and distributed memory architectures. Ph.D. thesis, University of Tennessee, December 2012

A Study of SpMV Implementation Using MPI and OpenMP on Intel Many-Core Architecture

Fan Ye[1,2]($\boxtimes$), Christophe Calvin[1], and Serge G. Petiton[2,3]

[1] CEA/DEN/DANS/DM2S, CEA Saclay, 91191 Gif-sur-Yvette Cedex, France
[2] Maison de la Simulation, USR3441,
Digiteo Labs Bât 565, 91191 Gif-sur-Yvette, France
fan.ye@cea.fr
[3] Laboratoire d'Informatique Fondamentale de Lille,
Université des Sciences et Technologies de Lille, 59650 Villeneuve d'Ascq, France

Abstract. The Sparse Matrix-Vector Multiplication (SpMV) is fundamental to a broad spectrum of scientific and engineering applications,
such as many iterative numerical methods. The widely used Compressed
Sparse Row (CSR) sparse matrix storage format was chosen to carry on
this study for sustainability and reusability reasons.

We parallelized for Intel Many Integrated Core (MIC) architecture a
vectorized SpMV kernel using MPI and OpenMP, both pure and hybrid
versions of them. In comparison to pure models and vendor-supplied
BLAS libraries across different mainstream architectures (CPU, GPU),
the hybrid model exhibits a substantial improvement.

To further assess the behavior of hybrid model, we attribute the inadequacy of performances to vectorization rate, irregularity of non-zeros,
and load balancing issue. A mathematical relationship between the first
two factors and the performance is then proposed based on the experimental data.

1 Introduction

The SpMV is vital to scientific and engineering applications. It is the essential
operation of many iterative linear and eigen solvers such as Conjugate Gradient
(CG) and Generalized Minimum Residual (GMRES). In this paper, we take Intel
Xeon Phi coprocessor as the underlying system for revealing some idiosyncrasies
in an efficient SpMV implementation. A simplified way to view this many-core
architecture is a chip-level SMP which offers remarkably high bandwidth. The
prototype C0 codenamed Knights Corner (KNC) has 61 cores, each featuring
a 512-bit wide vector unit and being capable of running up to 4 HW threads.
These factors enable such single chip to yield over 1 TFlops double precision
peak performance.

Due to sparse matrices' nature of irregularity, the memory subsystem often
appears as the main bottleneck of SpMV's efficiency in terms of FLOPS
(FLoating-point Operations Per Second). Furthermore, in a shared memory
context with a large count of cores such as MIC, the scalability behavior is

© Springer International Publishing Switzerland 2015
M. Daydé et al. (Eds.): VECPAR 2014, LNCS 8969, pp. 43–56, 2015.
DOI: 10.1007/978-3-319-17353-5_4

not obvious which may depend on issues like data locality and access pattern. A common approach to address these problems is to propose a new sparse matrix storage format [10]. However, some certain techniques used in new formats may become less pertinent as the targeting architecture evolves. They may need to be adapted accordingly. Another potential downside of a new format is that it is hard to implement it in a large numerical package such as PETSc [1] or Trilinos [7], and thus be not easy to integrate or interface in large scientific applications. Both PETSc and Trilinos adopt the CSR (Compressed Row Storage) as the underlying sparse format. We want to study the SpMV kernel within the context of linear or eigen solvers, making the availability of these numerical packages prominent to us. As a result, we chose to use CSR format.

The preceding studies [4,5] hold pessimistic views of hybrid fashion compared to a unified MPI approach. The related literatures usually underline the importance of network performance in explaining the gap between different models. Therefore the high on-chip bandwidth of MIC drives us to investigate the potential benefit of using hybrid programming. We refer the hybrid execution here to a scenario where the coprocessor resources (cores and caches) are divided into several separate domains and each domain is governed by one MPI process and shared by a number of OpenMP threads.

To set an architectural baseline, we also perform the tests over the same matrix suite on dual Intel Sandy-Bridge octa-core processors, as well as the NVIDIA Tesla K20 GPU, using the vendor-supplied BLAS libraries.

The outline of this paper is structured as follows: the Sect. 2 details the architectural features of MIC, the Sect. 3 discusses different dimensions of parallelisms of a vectorized SpMV kernel, the Sect. 4 is devoted to the experimental environment and results, the Sect. 5 concentrates on the performance analysis and modeling, and the Sect. 6 concludes.

2   Architectural Overview of MIC

The Intel Xeon Phi coprocessor is x86-based many-core architecture. It has 61 cores connected via a 512-bit bidirectional ring interconnect. There are 8 memory controllers supporting up to 16 GDDR5 channels. The core's memory interface are 32-bit wide with two channels which sustains a total bandwidth 8.4 GB/s per core. The STREAM Triad benchmark achieves around 160 GB/s on this architecture with ECC turned on.

There are two levels of cache memory. The level one cache has 32 KB instruction cache and 32 KB data cache. Associativity was increased to 8-way, with a 64 byte cache line. The bank width is 8 bytes. Data return is out-of-order. The L1 cache has a load-to-use latency of 1 cycle which allows an integer value loaded from the cache to be used in the next clock by an integer instruction. The L2 cache has a unified 512 KB capacity. Each core can access to all other L2 cache via the ring interconnection which makes a collective L2 cache size up to 32 MB. The L2 organization comprises 64 bytes per way with 8-way associativity, 1024 sets, 2 banks, 32 GB (35 bits) of cacheable address range and a raw latency of 11 clocks [8].

The vector processing units (VPU) of each core contains 32×512 bit SIMD registers which accommodates eight 64-bit values or sixteen 32-bit values. The VPU supports Fused Multiply-Add (FMA) operations which, for benchmarking purposes, counted as two floating operations. All these factors enable one single chip to yield over 1 TFlops double precision peak performance.

3 Sparse Matrix Vector Product Implementations for CSR Format

The CSR [11] comprises of 3 arrays, *row_ptrs*, *col_inds*, and *vals*, representing respectively the position of the first nonzero element of each row stored inside of *vals*, the column indices of every single nonzero element stored in *vals*, and the nonzero entries of the matrix in row-major order. Taking the standard CSR format as a starting point, we derived a vectorized kernel for SpMV.

Algorithm 1. Vectorized multiplication of the kth row (zero-based) of a matrix stored in CSR format (in *row_ptrs*, *col_inds*, and *vals*) with the vector x.

$reg\_y \leftarrow 0$
$start \leftarrow row\_ptrs[k]$
$end \leftarrow row\_ptrs[k+1]$
for $i = start$ to end **do**
    $writemask \leftarrow (end - i) > 8 \; ? \; \texttt{0xff} : (\texttt{0xff} \gg (8 - end + i))$
    $reg\_ind \leftarrow load(writemask, \&col\_inds[i])$
    $reg\_val \leftarrow load(writemask, \&vals[i])$
    $reg\_x \leftarrow gather(writemask, reg\_ind, x)$
    $reg\_y \leftarrow fmadd(reg\_x, reg\_val, reg\_y, writemask)$
    $i = i + 8$
end for
$y[k] = reduce\_add(reg\_y)$

3.1 Vectorized Kernel

For CSR, a natural way to parallelize the SpMV is to assign the subsets of rows to different execution units. The elementary operation is then shrinked into the product of a compressed sparse vector with a dense vector. By using the SIMD instruction we insert at the lowest dimension a parallelism resulting from the vectorization. In this direction we propose the row-wise vectorized kernel for SpMV, which is similar to recent work on SpMV for MIC [10]. The Algorithm 1 delineates the SIMDized kernel that handles the row-wise multiplication. The *writemask* functions as a shifting window ensuring only the lower portion of vector being operated when there're less than 8 nonzero elements left in a row. The "8" in Algorithm 1 implies 8 double precision floating numbers which fill the 512-bits SIMD units in MIC. It is worth noting that, to ensure the correctness of results, the CSR used here must not be a simplified form for symmetric matrices.

3.2   Hierarchical Exploitation of Hardware Resources

The second dimension of parallelism is built upon the number of cores. Along with the hierarchical memory subsystem, these resources can be exploited by processes and threads spawned and managed by the multiprocessing techniques. In most cases, it is easier to implement with the pure model than the hybrid one. In this paper we discuss both pure and hybrid implementations. We expect to promote the efficiency of data access and alleviate the scaling pressure occurred in the pure model by mixing different approaches.

We define the SpMV process $y \leftarrow Ax + y$ as two phases:

1. The computing phase, where all elements of y should be calculated.
2. The communication phase, where y is copied to x.

The communication phase occurred usually in a iterative solver where the SpMV process needs to be repeated until the convergence of the solver. Because the memory space is unified for all threads, the communication phase of pure OpenMP can't be started before the termination of computing phase. However, with the participation of MPI, these two phases could be partially overlapped. In our study, we collect the computing phase timings corresponding to the slowest MPI process of each execution. These timing data were used to deduce the performance of SpMV. To obtain statistically meaningful results, we iterated 100 times for each measure of SpMV timing.

In terms of implementations, some conventional optimizations are applied to both pure and hybrid cases, such as software prefetching, and streaming stores. The rows of matrix are distributed so that each process receives the same number of nonzero elements. The minimal unit of partioning is one row. We also altered the number of processes and threads and attempted exhaustively all possible combinations of processes and threads to seek the best configuration of maximizing the performance for each test matrix. In particular, we don't take advantage of matrix symmetry to achieve better performance. All matrices are considered equally as non-symmetric ones. For the sake of better cache usage and to avoid oversubscription of threads, it is important to configure properly the processes' and threads' affinity. What's more, the highest numbered core of MIC (61th) should be left unused so it may process the interference from OS threads.

4   Experimental Results

4.1   Matrix Suite

In practice, we have 3 principles in selecting the test matrices [6]. Firstly, we prefer the matrices that have been used in previous literatures. Secondly, the matrices should have a larger volume in memory than 30 MB, which is the aggregate L2 cache size of Xeon Phi, in order to neutralize the promotion in temporal locality induced by repeated runs of SpMV kernel. Lastly, our future study of eigensolvers requires the matrices to be square. We also include a dense 8000 × 8000 matrix (*dense8000*) expressed in CSR format. We outline the basic characteristics of 18 selected matrices in Table 1.

Table 1. List of main characteristics of test matrices. nnz is the number of nonzero elements. nrow is the square matrix dimension.

Name	Dim (K)	nnz (M)	nnz/nrow	Name	Dim (K)	nnz (M)	nnz/nrow
mixtank_new	29.957	1.995	66.597	sme3Db	29.067	2.081	71.595
mip1	66.463	10.353	155.768	ldoor	952.203	46.522	48.858
rajat31	4690.002	20.316	4.332	Si41Ge41H72	185.639	15.011	80.863
nd6k	18.000	6.897	383.184	pdb1HYS	36.417	4.345	119.306
cage15	5154.859	99.199	19.244	bone010	986.703	71.666	72.632
crankseg_2	63.838	14.149	221.637	dense8000	8	64.000	8000
ns3Da	20.414	1.680	82.277	pwtk	217.918	11.634	53.389
in-2004	1382.908	16.917	12.233	torso1	116.158	8.517	73.318
circuit5M	5558.326	59.524	10.709				

4.2 Experimental Environment

Different SpMV kernels were conducted and compared on various architectures.

- Intel MIC, pre-production of KNC prototype C0, 61 cores running at 1.2 GHz, 16 GB GDDR5 memory with ECC enabled, installed with MPSS v3.1.
- Dual-socket Intel Xeon E5-2670, 16 core running at 2.6 GHz, 64 GB DDR memory with ECC enabled.
- NVIDIA K20 GPU, 2496 cores running at 0.7 GHz, 5 GB GDDR5 memory with ECC enabled.

On MIC, SpMV kernels of pure OpenMP, hybrid MPI/OpenMP, MKL (Intel Math Kernel Library, v11.1) were tested. On CPU, only the MKL SpMV routine was tested. On GPU, cuSPARSE (NVIDIA CUDA Sparse Matrix library, v2.6) kernel was tested. All tested vendor-supplied BLAS libraries use the CSR sparse format.

4.3 OpenMP and MKL Performances

The SpMV is one of the challenging instances that is known to be memory bandwidth bound. Its streaming memory access pattern makes the cores hard to run at full speed. Adding the number of threads helps to hide the latency due to data miss. But the increase of virtual cores might leads to memory contention and network congestion, thus exhibits a poor scaling performance. At core level, the vectorization is necessary for improving performance. However, it also burdens more on memory subsystem for the vector instructions consume much more data than scalar ones. The load of data is less efficient for x than for col_inds or $vals$. A unified address of x for all cores may drive the problem even more severe. We implemented on MIC a multithreaded SpMV kernel using OpenMP. The MKL version was also tested as it is based on OpenMP threading environment therefore comparable to our implementation. We varied the number of threads

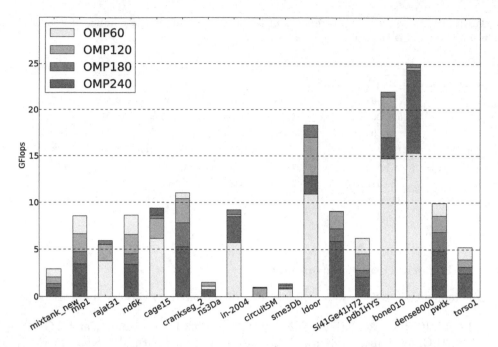

Fig. 1. Performances of OpenMP version of SpMV kernel. For each test matrix the performances are plotted in different colors depending on the number of threads used (Color figure online).

(from 1 to 4 threads per core) while measuring the performances. For each matrix we plot in Figs. 1 and 2 the bars of performance among which from top to bottom the performances corresponding to different thread configuration (1, 2, 3 or 4 threads per core) are shown in a descending order. All bars started from 0 GFlops. The lower part of the bars may be covered by other bars with smaller magnitude except the smallest one.

From these two figures we observe a similar behavior of both implementations on different matrices. None of them performs better in average than the other one, except that the MKL tends to have better performance when using more threads per core. We argue that's because of its better thread scheduling and some low-level optimizations. Both implementations are nowhere near the theoretical or achievable peak performance of MIC architecture.

4.4   Hybrid MPI/OpenMP Performances

To better deal with the issue of thread scaling and alleviate the memory contention, we propose to implement the hybrid MPI/OpenMP SpMV kernel. We expect to promote the efficiency of multithreading, scaling and cache utilization. In this case, we evenly divide the cores' domain according to common resources (cores, caches) and place one MPI process for each subdomain. In each subdomain, we spawn the same number of threads. The experiments were conducted

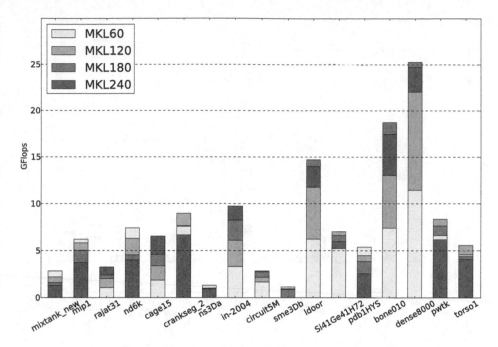

Fig. 2. Performances of MKL version of SpMV kernel. For each test matrix the performances are plotted in different colors depending on the number of threads used (Color figure online).

using all possible combinations of processes and threads with careful pinnings. Every subdomain governed by one MPI process is guaranteed to have the same and integer number of cores. And the highest numbered core of MIC is always free from application threads. All threads participate in the parallelization of vectorized SpMV kernel. Only the master thread manages the communication. The hybrid algorithm is described in Algorithm 2. We plot the gain of hybrid model against pure OpenMP in the Fig. 3. Over the entire matrix suite, the hybrid model exhibits a substantial performance improvement except in one case (*cage15*).

Algorithm 2. Hybrid MPI/OpenMP algorithm. Each MPI process accommodates the same number of OpenMP threads.

Distribute row blocks (rowptrs, colinds, vals) of A so that each MPI process receives approximately same number of nonzero elements

Replicate x on all MPI processes, allocate y (same size of x) on all MPI processes

Apply locally the vectorized SpMV kernel using OpenMP multithreading with "guided" scheduling

Gather the results from other MPI processes and update the local portion of y

---

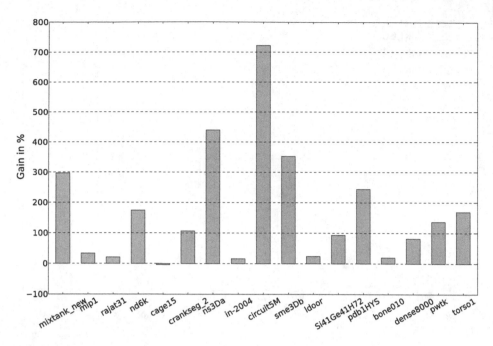

Fig. 3. Gain in percentage of hybrid MPI/OpenMP SpMV kernel against the pure OpenMP one.

4.5   Performances of SpMV Kernel on Various Architectures

Finally, the performances of different SpMV kernels will be presented here. The Fig. 4 delineates the performances of hybrid model versus vendor-supplied BLAS libraries across a variety of architectures. In most cases, the hybrid model outruns the other ones. Since we used the CSR format for all architectures, the results do not represent the inherent capacity of some architecture such as GPU. But it shows a path to better exploit the MIC architecture. We notice in some cases that CPU still achieved better performances. We will try to understand this phenomenon in Sect. 5.

5   Performance Analysis and Modeling

The experimental results reveal a considerable advantage of hybrid model over the pure ones. However, not being able to determine in advance the optimal combination of MPI processes and OpenMP threads invalidates this approach simply because the best results are irreproducible. As a consequence, it is imperative to devise a method to sketch the behavior of the machine. We will discuss qualitatively the reasons of performance improvement and the primary performance restraining factors, from where we develop a mathematical relationship that quantifies the effects of different factors. The effectiveness of the deduced model will be verified at the end of this section.

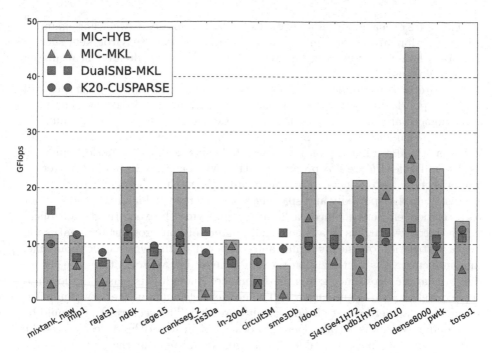

Fig. 4. Performances of different SpMV kernels on various architectures.

5.1 Performance Analysis

First thing to understand is the performance improvement due to the mixture of MPI and OpenMP. We argue that's mainly because of the promotion of data locality and thread scalability. The promoted data locality improves the data reusability in terms of better cache utilization. It also mitigates the memory contention. More specifically, the vector x is replicated, thus avoiding contention when large number of threads read elements of x. This would also be possible in pure OpenMP via thread-private variables. However, that means the replication of x has to be made on all threads. The memory usage would be varied if number of threads changes. In the case of hybrid model, the x is only replicated on each process and shared by threads belonging to that process which creates us a higher degree of flexibility. In addition, the rows of matrix A is distributed to different memory regions. Therefore, these data are spatially local to the process domain. By carefully binding the processes to the physical cores, the data are stored uniformly in the memory space. Therefore it is more likely to generate a higher aggregate bandwidth in the on-chip ring network.

The scaling factor should also be considered. In a large many-core system, the multithreading overheads such as loop scheduling overheads may not be linear when the number of threads grows. However, in hybrid model, each process keeps a relatively small number of threads making it easier to scale.

There are other potential advantages for hybrid model as well. For example, it is straightforward to implement it in a numerical software environment such as Trilinos, where the underlying MPI/OpenMP modules are already encapsulated and ready to use.

In spite of some performance improvement, the hybrid SpMV kernel still performs poorly in some cases compared to other implementations. The poor performances are likely to occurred in matrices having a low average number of nonzero elements[1], as seen in Table 1 for *rajat31*, *cage15*, *in-2004*, *circuit5M*. However, these are not the only matrices behaving badly. The performances of *ns3Da* and *sme3Db* are not promising either whereas they have decent average numbers of nonzero elements. Further research shows that their nonzero elements are distributed dispersedly and sparsely along the rows, which may lead to poor vectorization efficiency. Since each thread performs the vectorized multiplication between two arrays at a time, small number of nonzeros per row makes the vector instruction overheads significant compared to the whole execution time, thus inducing a low vectorization rate. In general, low vectorization rate shakes the foundation of producing high performance. Though this reasoning does not applied well in *ns3Da* and *sme3Db*. The nonzero elements of these two matrices are not only numerous but also uniformly spreaded. The vectorization should be well conducted unless certain operation described in Algorithm 1 decelerates the computation. The most probable explanation would be that the *gather* instruction appeared in line 8 of Algorithm 1 cancels out the high vectorization rate because of its long latency. The irregularity of nonzero elements makes the load of x inefficient, thus causing the unexpected cache misses and eventually bad performance.

Besides these two factors, there should be one more concern linked to the message passing programming, which is the load balancing issue. In the hybrid model, the processes are independent and the last terminated process determines the global performance. In our case, this issue is connected to the row partitioning policy. Using a dynamic instead of static row distributing strategy may improve load balancing. We will include this study in our future works.

5.2 Performance Modeling

Definition 1. *For a given matrix, let the nnz be the number of nonzero elements. If t is the execution time of SpMV computing phase defined in Sect. 3.2, then the performance P of a hybrid SpMV kernel is defined as*

$$P = \frac{2 \; nnz}{t_{max}}$$

where t_{max} is the execution time of the last terminated MPI process.

---

[1] The average number of nonzero elements is defined as the quotient of total number of nonzero elements over the row dimension.

If nnz_{glob} is the total number of nonzero elements in a given matrix, then the global performance P_{glob} is given by

$$P_{glob} = \frac{2\ nnz_{glob}}{t_{max}}$$

To properly model the performance, we want to seperate the load balancing factor from the vectorization rate and nonzero elements' irregularity. Since the t_{max} is the time of the slowest MPI process, it is more accurate to obtain its local performance P_{local} by applying the same formula:

$$P_{local} = \frac{2\ nnz_{local}}{t_{max}} = \frac{nnz_{local}}{nnz_{glob}} P_{glob}$$

where nnz_{local} is the number of nonzero elements of the row block assigned to the slowest process. We measured the execution time of the slowest process with its rank recorded. The ranking information helps to identify the row blocks assigned to the processes. Since thread is the minimal execution unit which performs vector instructions, the per thread performance is more meaningful for characterizing the indicators discussed in the last subsection.

Definition 2. *If P is the aggregate performance of n_{thd} number of threads, then the per-thread performance is estimated as*

$$P_{thd} = \frac{P}{n_{thd}}$$

Assume the n_{thd} is the number of threads spawned within the slowest process, then the local per-thread performance is

$$P_{thd} = \frac{P_{local}}{n_{thd}}$$

In this context, two indicators are proposed to quantify the SIMD efficiency as well as the impact of nonzero elements' dispersion. The first one is the average number of nonzero elements. There are at least three features helping to set up the functional relationship between this indicator and the per-thread performance. All these features are discussed without the interference of the second indicator.

1. If the number of nonzero elements equals to 0, the performance should also be 0. However, as the average number of nonzero elements starts from 0, the impact of vector instruction overheads might diminish rapidly.
2. Bigger the average number of nonzero elements is, less amplification of performance is gained.
3. The performance should have an upper bound as the number of nonzero elements is extremely large.

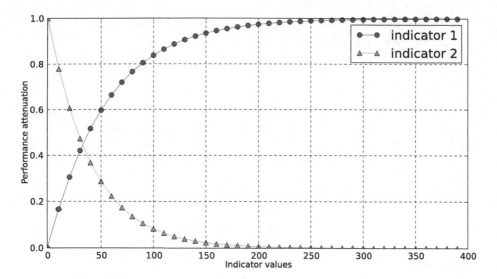

Fig. 5. The relationships between two indicators and the performance. The first indicator is the average number of nonzero elements per row. The second indicator is the average number of occurrences when the distance between any pair of contiguous nonzero elements within a row is greater than 2.

Such relationship could be delineated by the blue curve in Fig. 5. With more nonzero elements in a row we have generally better performance, if given a fixed level of nonzero elements' dispersion. The second indicator should then be able to describe the "level of dispersion". A direct solution is to estimate the cache misses. However, the cache behavior in modern architecture depends on, including but not limited to, cache capacity, cache associativity, cache line width, cache levels, and replacement policy. It is highly unpredictable using a low-cost model. Considering its complexity, this method is less practical. We are searching for a convenient and simple approach to establish the second indicator. It turns out that a simple trait of matrix, which based on the distances between each pair of contiguous nonzero elements in a row, is an effective indicator. It refers to the average number of occurrences when such distance is greater than 2. Similar to the first indicator, it averages over all studied rows. The red convex decreasing curve with triangle markers in Fig. 5 depicts the attenuation caused by the second indicator to the performance. According to the graphs of two indicators, we give the functional form of regression model in Eq. 1, where $\hat{P}_{thd}$ is the estimated per-thread performance, $\overline{nnz}$ is the first indicator and the $\overline{d}$ is the second indicator.

$$\hat{P}_{thd}(\overline{nnz}, \overline{d}) = \alpha \left[1 - \exp\left(-\frac{\overline{nnz}}{\epsilon_1} \right) \right] \exp\left(-\frac{\overline{d}}{\epsilon_2} \right) \tag{1}$$

The experimental data were collected from the slowest processes of different matrices listed in Table 1. These data were processed to obtain P_{local}, n_{thd}, $\overline{nnz}_{local}$, $\overline{d}_{local}$ for the use of regression analysis.

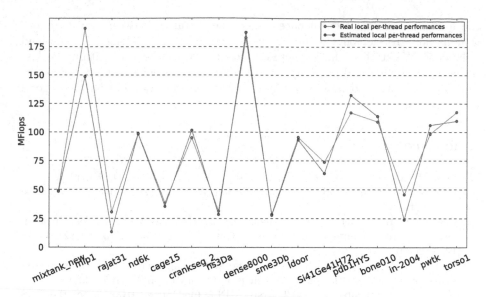

Fig. 6. The real and the estimated local per-thread performances over a set of test matrices.

The Fig. 6 draws the real and the estimated local per-thread performances over a set of test matrices. The local performance were collected from the slowest process of each execution. It is based on this set of data that we obtained the following coefficients. The estimated values are:

$$\widehat{\alpha} = 187.5, \ \widehat{\epsilon_1} = 55, \ \widehat{\epsilon_2} = 40$$

where the $\widehat{\alpha}$ corresponds to the per-thread performance (in MFlops) of *dense8000*. Considering its large average number of nonzero elements per row and continuous nonzero elements, it should execute almost optimally.

6   Conclusions

The SpMV is the key kernel that constitutes the main process in many iterative numerical methods. In this paper, we investigate two programming models, the pure OpenMP and the hybrid MPI/OpenMP. Starting from a vectorized CSR SpMV kernel, we proposed different ways of parallelizing it. A set of evaluations on various mainstream architectures (Intel Dual-Socket Sandy Bridge, NVIDIA K20 GPU) was conducted by using not only our own implementations but also the vendor supplied BLAS libraries. The results suggest that the hybrid MPI/OpenMP model is very promising on Intel MIC architecture. It can help to reduce the scaling overheads and promote data locality compared to the pure models, thus improving substantially the performance.

In order to better understand the performance of hybrid model, we identified 3 performance indicators, namely the average number of nonzero elements, the average number of occurrences when the distance between any two contiguous nonzero elements within a row is greater than 2, along with the load balancing. We studied the impacts of the first two indicators within the last terminated process and came up with a regression model based on the experimental data. We also estimated the regression coefficients and obtained good fitting results.

References

1. Balay, S., Brown, J., Buschelman, K., Gropp, W.D., Kaushik, D., Knepley, M.G., McInnes, L.C., Smith, B.F., Zhang, H.: PETSc Web page (2013). http://www.mcs.anl.gov/petsc
2. Berrendorf, R., Nieken, G.: Performance characteristics for OpenMP constructs on different parallel computer architectures. Concurrency Pract. Exp. **12**(12), 1261–1273 (2000)
3. Bull, J.M.: Measuring synchronisation and scheduling overheads in OpenMP. In: Proceedings of First European Workshop on OpenMP, pp. 99–105 (1999)
4. Cappello, F., Etiemble, D.: MPI versus MPI+OpenMP on IBM SP for the NAS benchmarks. In: Proceedings of the 2000 ACM/IEEE Conference on Supercomputing, Supercomputing 2000. IEEE Computer Society, Washington, DC (2000). http://dl.acm.org/citation.cfm?id=370049.370071
5. Chow, E., Hysom, D.: Assessing performance of hybrid MPI/OpenMP programs on SMP clusters. Technical report, Lawrence Livermore National Laboratory (2001)
6. Davis, T.A., Hu, Y.: The university of florida sparse matrix collection. ACM Trans. Math. Softw **38**(1), 1:1–1:25 (2011). http://doi.acm.org/10.1145/2049662.2049663
7. Heroux, M., Bartlett, R., Hoekstra, V.H.R., Hu, J., Kolda, T., Lehoucq, R., Long, K., Pawlowski, R., Phipps, E., Salinger, A., Thornquist, H., Tuminaro, R., Willenbring, J., Williams, A.: An overview of trilinos. Technical report, SAND2003-2927, Sandia National Laboratories (2003)
8. Intel: Intel Xeon Phi Coprocessor System Software Developers Guide. Technical report (2012)
9. Kourtis, K., Goumas, G., Koziris, N.: Exploiting compression opportunities to improve SpMxV performance on shared memory systems. ACM Trans. Architec. Code Optim. **7**(3), 16:1–16:31 (2010)
10. Liu, X., Smelyanskiy, M., Chow, E., Dubey, P.: Efficient sparse matrix-vector multiplication on x86-based many-core processors. In: Proceedings of the 27th International ACM Conference on International Conference on Supercomputing, ICS 2013, pp. 273–282. ACM, New York (2013). http://doi.acm.org/10.1145/2464996.2465013
11. Saad, Y.: Iterative Methods for Sparse Linear Systems, 2nd edn. Society for Industrial and Applied Mathematics, Philadelphia (2003)
12. Williams, S., Oliker, L., Vuduc, R., Shalf, J., Yelick, K., Demmel, J.: Optimization of sparse matrix-vector multiplication on emerging multicore platforms. In: Proceedings of the 2007 ACM/IEEE Conference on Supercomputing, SC 2007, pp. 38:1–38:12. ACM, New York (2007). http://doi.acm.org/10.1145/1362622.1362674

SIMD Implementation of a Multiplicative Schwarz Smoother for a Multigrid Poisson Solver on an Intel Xeon Phi Coprocessor

Masatoshi Kawai[1,2]($\boxtimes$), Takeshi Iwashita[3,4], and Hiroshi Nakashima[3]

[1] Graduate School of Informatics, Kyoto University, Kyoto, Japan
kawai@sys.i.kyoto-u.ac.jp
[2] JSPS Research Fellow, Tokyo, Japan
[3] ACCMS Kyoto University, Kyoto, Japan
[4] JST CREST, Sendai, Japan

Abstract. In this paper, we discuss an efficient implementation of the three-dimensional multigrid Poisson solver on a many-core coprocessor, Intel Xeon Phi. We have used the modified block red-black (mBRB) Gauss-Seidel (GS) smoother to achieve sufficient degree of parallelism and high cache hit ratio. We have vectorized (SIMDized) the GS steps in the smoother by introducing a partially SIMDizing technique based on loop splitting. Our numerical tests demonstrate that our implementation performs 35.5 % better than the conventional mBRB-GS smoother implementation on Xeon Phi.

1 Introduction

Discrete Poisson equation problems often appear in various computational science simulations. When the problem is associated with spatially varying diffusion coefficients, it is commonly solved using the finite difference method. The finite difference discretization results in a linear system of equations, that can require a large amount of computational effort, especially for a large-scale simulation problem. Consequently, there is a demand for a fast linear solver for the discrete Poisson equation problem.

A (geometric) multigrid [2] solver is one of the most popular linear iterative solvers for a linear system derived from the finite difference discretization of the Poisson equation. It has a convergence property suitable for large-scale problems. The multigrid solver's convergence rate is independent from the problem size when it is applied to the linear system derived from the homogeneous discrete Poisson equation. Consequently, we have developed a fast geometric multigrid Poisson solver. In this paper, we have investigated the performance of our solver on an Intel Xeon Phi coprocessor [7], which is a recently developed processor.

The Intel Xeon Phi coprocessor is based on Intel MIC architecture, and includes many relatively lower performance cores in its package. The current version of the processor, which we have used in our research, consists of 60 cores. Its peak performance reaches 1TFlops (DP). Moreover, the Xeon Phi coprocessor

© Springer International Publishing Switzerland 2015
M. Daydé et al. (Eds.): VECPAR 2014, LNCS 8969, pp. 57–65, 2015.
DOI: 10.1007/978-3-319-17353-5_5

has a good performance per watt ratio [3], and its programming model is easier than that for GPU. Various useful applications and tools developed for general purpose multi-core processors (such as MPI and OpenMP) can be used, because each processing core in Xeon Phi is based on X86 architecture. Xeon Phi is currently used as the accelerator for the host CPU [4], but standalone CPU models will be developed in the future. Because of these features, it is predicted that Xeon Phi will play an important role in future computational science.

To efficiently implement a multigrid Poisson solver on the Xeon Phi, we should consider the following key issues.

1. Large degree of thread parallelism: Xeon Phi has larger numbers of cores than a general multi-core processor, and it can simultaneously execute 240 threads using Intel Hyper-Threading technology.
2. Data locality: The processor uses a general cache based memory architecture. Therefore, data locality is important for attaining a high cache hit ratio.
3. Convergence rate: Similarly to implementations on other processors, the convergence rate of the solver has a significant effect on its performance.
4. Vectorization ($SingleInstructionMultipleData$ (SIMD) instructions): Xeon Phi has a relatively wide SIMD engine. Therefore, SIMD instructions should be effectively used in the analysis to let the processor achieve its full potential.

In this paper, we mainly discuss the parallel smoother in the multigrid solver, paying special attention to these key issues. The other components of the multigrid solver can be straightforwardly parallelized and vectorized using a domain decomposition approach.

In VECPAR 2012, we reported a parallel smoother called the modified block red-black Gauss-Seidel (mBRB-GS) smoother [6]. It is a multiplicative Schwarz smoother. The Schwarz smoother is parallelized by applying red-black ordering to cuboid blocks of the problem domain [5], and multiple Gauss-Seidel (GS) iterations are performed in each red or black block. Because the second or later GS iterations in the block are performed on-cache, high data locality is achieved in the smoothing step. Moreover, analytical investigation and numerical tests showed that the smoother attains a sufficient degree of thread parallelism and fast convergence. Accordingly, the mBRB-GS smoother has desirable characteristics in three out of the four key issues mentioned above, and it can be regarded as a candidate for a parallel smoother for the Xeon Phi coprocessor. However, the innermost loop of the smoother consists of sequential GS steps, and it cannot be straightforwardly vectorized.

Our solution to this problem is a partial SIMDization (vectorization) of the GS loop which we split into six simpler loops. Five of the loops are made do-all and thus SIMDizable. The loop-splitting itself is a classic technique for vector processors [1]. However, our revisit has various new aspects such as its application to the SIMD mechanism and the cache-awareness that is essential for scalar many-core processors. We conducted numerical tests on the Xeon Phi coprocessor to compare the effectiveness of the developed solver with the solver based on the conventional gridpoint-wise red-black GS smoother that is naturally vectorized.

2 Parallelized Multigrid Solver for the Three-Dimensional Poisson Equation

2.1 Poisson Equation Problem and Multigrid Solver

We used a 7-point finite difference scheme to solve the three-dimensional Poisson equation. This leads to the following linear system of equations.

$$A\phi = \rho, \tag{1}$$

where ρ is the discretized given source, ϕ is the unknown vector, and A is the coefficient matrix. The row of the linear system of (1) corresponding to the grid-point (i, j, k) is written as

$$a_{i,j,k} * \phi_{i,j,k-1} + b_{i,j,k} * \phi_{i,j-1,k} + c_{i,j,k} * \phi_{i-1,j,k} + \phi_{i,j,k}$$
$$+ e_{i,j,k} * \phi_{i+1,j,k} + f_{i,j,k} * \phi_{i,j+1,k} + g_{i,j,k} * \phi_{i,j,k+1} = \rho_{i,j,k} \tag{2}$$

where (i, j, k) represents the grid coordinates. We use the geometric multigrid method to solve the linear system.

The multigrid method consists of the smoother, residual calculator, restriction and prolongation operators and coarsest grid solver. Among these components, we have focused our analysis on the smoother. When we consider the parallelization of the multigrid solver, the residual calculation, the restriction, and the prolongation can be straightforwardly parallelized using domain decomposition. However, it is difficult to parallelize some smoothers. For example, the GS smoother cannot be naturally parallelized. Moreover, a smoother has a significant impact on the performance of the multigrid solver. It greatly affects the convergence of the solver, and its total computational effort is larger than the other components. Consequently, this paper mainly discusses the smoother and its vectorization for the many-core processor.

2.2 Modified Block Red-Black Gauss-Seidel Smoother

In our analysis, we have used the mBRB-GS smoother, which is a multiplicative Schwarz smoother. In this smoother, the entire grid is decomposed into subdomains based on block red-black (BRB) ordering. The entire grid is divided into multiple blocks, and then the red-black ordering is applied to the blocks. Each red or black block is treated as a subdomain in the Schwarz smoother. Multiple sequential GS steps are performed in each subdomain (red/black block), which is smaller than the cache size. Consequently, the second and subsequent GS steps in each subdomain are executed on-cache, which results in high cache hit ratio (good data locality). The degree of parallelism of the smoother is given by the number of blocks of each color. In general, the size of the entire grid is larger than the cache size, and the degree of parallelism is expected to be sufficiently large. Moreover, it was reported in [6] that the multigrid solver using mBRB-GS converges more quickly than when using hybrid Jacobi and GS, or red-black GS smoothers. Consequently, the mBRB-GS smoother is considered to be a promising parallel smoother candidate for the multigrid solver on the Intel Xeon Phi coprocessor.

3 Efficient Implementation of the GS Smoother on Xeon Phi

To develop an efficient multigrid Poisson solver on the Xeon Phi coprocessor, we should consider the vectorization (SIMDization) of the smoother in addition to the parallelization. This is because of the relatively wide SIMD engine of the coprocessor compared with general multi-core processors. However, the mBRB-GS smoother uses GS iterations, which cannot be naturally vectorized. We now introduce an implementation method that makes a compiler generate a partially SIMDized binary code.

Algorithm 1 shows the ordinary Fortran implementation of the GS method. When it is used in the mBRB-GS smoother, NX, NY and NZ correspond to the block sizes along the x, y and z axes, respectively.

In the program code, the innermost loop has a loop carried dependence caused by the term $c(i, j, k) * phi(i - 1, j, k)$ (highlighted in red), which usually prevents the compiler from SIMDizing the loop.

Algorithm 1. Ordinary implementation of the GS method

```
1  do k = 1, NZ
     do j = 1, NY
       do i = 1, NX !This loop is not SIMDized
         phi(i,j,k)= rho(i,j,k)                                        &
                   + a(i,j,k)*phi(i,j,k-1) + b(i,j,k)*phi(i,j-1,k) &
6                  + c(i,j,k)*phi(i-1,j,k) + e(i,j,k)*phi(i+1,j,k) &
                   + f(i,j,k)*phi(i,j+1,k) + g(i,j,k)*phi(i,j,k+1) )
       enddo
     enddo
   enddo
```

Although this dependence is essential to the GS method, it does not necessarily inhibit the SIMDization of the whole loop. In fact, as shown in Algorithm 2, if we apply loop-splitting so that we have six separated loops for each of the additive terms referencing phi, five loops out of the six are free from the loop carried dependence and thus easily and well SIMDized. Note that the loop-splitting does not always improve the loop performance because of additional operations, which in this case are load/store operations of the scratchpad array tmp. However, the negative impact is expected to be minor and the following positive effects are also possible.

First, the cost of the additional accesses to tmp is minimized by tuning the size NX so that tmp is always resident in the first level cache. For the mBRB-GS smoother that has an inherent blocking feature, this tuning is almost automatic and does not require further cache-aware blocking. Second, each of the five dependence-free loops is so simple that all compilers can easily grasp the structure of the loop body, so it is strongly expected that the loop body is efficiently SIMDized, even with the restricted SIMD architecture of Xeon Phi. That is, the relatively small number of streams (four for the first and three for others) makes it feasible for the compilers to exploit SIMD load/store and alignment instructions, while the common right-hand side structure of $x + y * z$ perfectly fits to the fused multiply-add instructions that are the other source of the high peak performance of Xeon Phi.

Algorithm 2. Partially SIMDized implementation of the GS method

```
do k = 1, NZ
  do j = 1, NY
    !$DEC SIMD
    do i = 1, NX
5       tmp(i) = rho(i,j,k) + a(i,j,k)*phi(i,j,k-1)
    enddo
    !$DEC SIMD
    do i = 1, NX
      tmp(i) = tmp(i) + b(i,j,k)*phi(i,j-1,k)
10      enddo

    !There are SIMDized phi(i+1,j,k), phi(i,j+1,k) and phi(i,j,k+1) loop.

    do i = 1, NX !This loop is not SIMDized
15      phi(i,j,k) = tmp(i) + c(i,j,k)*phi(i-1,j,k)
    enddo
  enddo
enddo
```

4 Numerical Tests

4.1 Numerical Test Conditions

We examined the performance of the multigrid Poisson solver with mBRB-GS using an implementation method for partial SIMDization on an Intel Xeon Phi 7120 coprocessor. The fundamental specifications are listed in Table 1. On the coprocessor, up to 240 threads can run on 60 cores. The program code was written in Fortran compiled by Intel Composer 14.0.0 with the options -O3 -openmp -mmic -no-opt-prefetch. It was run on Xeon Phi using its native execution mode. The test problem had 512^3 grid points. The multigrid solver has converged when the relative residual norm was less than 10^{-7}. In our numerical tests, we evaluated the performance of the multigrid Poisson solver with red-black GS (RB-GS), mBRB-GS based on the implementation methods in Algorithms 1 and 2. The block size of the mBRB-GS smoother $NX \times NY \times NZ$ was $512 \times 2 \times 2$, and there were two GS iterations for each block in a smoothing step.

4.2 Performance Evaluation of Proposed Implementation Method

Figure 1 shows the relative speedup of calculation time of the entire multigrid solver compared with the sequential mBRB-GS (Algorithm 1). These results confirm that

Table 1. Specifications of Xeon Phi

Processor	Model	7120(KnightsCorner)
	Number of cores	61
	Clock frequency	1.24 GHz
	L1D-cache size	32 KByte/core
	L2-cache size	512 KByte/core
Memory	Technology	GDDR5
	Size	16 GByte

mBRB-GS outperforms RB-GS on numerical tests conducted on the Xeon Phi coprocessor. In our analysis, RB-GS is implemented using stride memory access. This is one of the most popular implementation methods for RB-GS, because it does not require the array to be reordered for the unknowns. This is convenient for other multigrid components such as the restriction. However, stride memory access is not advantageous in terms of the cache hit ratio when compared with the contiguous method. Consequently, RB-GS is inferior to mBRB-GS in terms of performance, although it can be naturally vectorized. This result is similar to the numerical results on a general multi-core processor [6].

Next, we compared the two implementations of mBRB-GS. Although the partially SIMDized GS implementation (Algorithm 2) outperforms the ordinary method (Algorithm 1) on the numerical test using 240 threads, it is inferior to the ordinary implementation when using 120 or less threads. Using the Intel Vtune Amplifier, we found that Algorithm 2 suffers from two types of processor stalls. One is the VPU_STALL_REG event detected when a read-after-write hazard stalls the SIMD instruction pipeline. The other is the PIPELINE_AGI_STALL that corresponds to the stall of a load/store instruction. It is caused by the latency of the corresponding instruction to provide (a source of) the address to be accessed. We consider that the number of stalls increases because of the reduced number of instructions involved in a loop in Algorithm 2. The hyper-threading technology reduces the impact of these stalls on the performance, by interleaving instructions from multiple threads to hide the latency between the pair of instructions. Thus, Algorithm 2 performs 19.1 % better than Algorithm 1 in the case of 240 threads.

The other important observation obtained from our analysis of Algorithm 2 using the Vtune Amplifier is that the ratio of SIMD instructions to all executed instructions is only 25.9 %. This is much smaller than we expected for the five simple SIMDizable loops. We examined the object code for the loops and found that a significantly large portion of instructions in the loop body is occupied by calculations of the addresses of each element of the three or four arrays referred to in each loop. That is, Intel Composer generates fairly redundant codes for address calculations of three-dimensional array elements. This potential inefficiency might be hidden under the powerful out-of-order superscalar mechanism of ordinary Xeons with multiple integer units which simultaneously works together

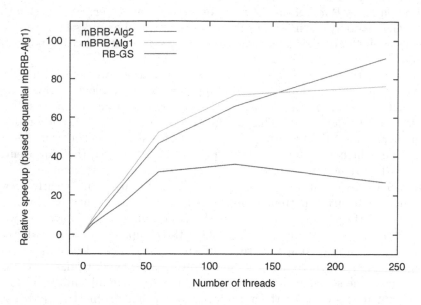

Fig. 1. Comparison of RB and two implementations of the mBRB-GS smoother

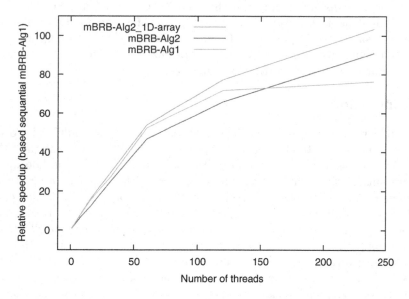

Fig. 2. Comparison of three implementations of the mBRB-GS smoother for the multi-grid Poisson solver

with SIMD floating-point units. However, this inefficiency is revealed when it is executed on Xeon Phi, because of its in-order two-way superscalar mechanism. This means that only one integer instruction stream can be processed when it has instructions tightly dependent on each other and/or the SIMD floating-point unit is in use.

To reduce the overhead of the address calculation, we made the three-dimensional arrays one-dimensional by applying the well-known array flattening technique. Then, most operations are made loop-invariant and are explicitly moved outside the loop body. This handmade optimization significantly reduced the number of non-SIMD instruction executions for address calculations, resulting in a much higher SIMD instruction ratio of 53.8 %. Then, the improvement of the SIMD instruction ratio directly effected the higher performance of three mBRB-GS implementations, as shown in Fig. 2. As the figure clearly shows, Algorithm 2 with array flattening improves the performance of the 240-thread case and outperforms Algorithm 1 by 35.5 %. It also improves performance when using 120 threads or less, and is the best of the three implementations in all cases.

Finally, we briefly compared the performances of Xeon Phi and an ordinary multi-core HPC server node of dual Xeon E5-2670 SandyBridge processors. We measured the server node performance using Algorithm 1 and 2 and found that the differences between them are insignificant. This is most likely because of the narrower 256-bit wide SIMD mechanism and the powerful out-of-order superscalar mechanism. On the other hand, an important observation is that Algorithm 2 with array flattening on the single Xeon Phi coprocessor using 240 threads outperforms Algorithm 1 on the dual-Xeon server using 16 threads by 34.1 %. This demonstrates its high potential, even for hardly-SIMDizable kernels. It also shows the importance of architecture-aware code tuning, which we expect to be incorporated into automated optimizations of future compilers for many-core processors with wider SIMD mechanisms.

5   Conclusion

In this paper, we discussed an efficient three dimensional multigrid Poisson solver, working on an Intel Xeon Phi coprocessor. To effectively use the SIMD instructions of Xeon Phi, we introduced the partially SIMDized method for the GS iterations in the multigrid solver, using the mBRB-GS smoother. In our implementation, the innermost loop is split into six loops, each of which corresponds to one additive term in a 7-point finite difference scheme. Because five of these loops are free from loop carried dependence, they can be SIMDized. The loop-splitting itself is a classic technique for vector processors. However, our revisit has various new aspects such as its application to the SIMD mechanism and the cache-awareness that is essential for scalar many-core processors. Moreover, using detailed performance profiling and analysis, we found that the reduction of address calculations in the loop body by using array flattening significantly improved performance. Overall, the partially SIMDized implementation attained a 35.5 % better performance than the conventional implementation of the mBRB-GS smoother in the 240-thread execution on Xeon Phi.

References

1. Allen, R., Kennedy, K.: Automatic translation of fortran programs to vector form. ACM Trans. Program. Lang. Syst. **9**(4), 491–542 (1987)
2. Briggs, W., Henson, V., McCormick, S.: A Multigrid Tutorial, 2nd edn. SIAM, Philadelphia, PA (2000)
3. Dokulil, J., et al.: High-level support for hybrid parallel execution of c++ applications targeting intel® xeon phi coprocessors. Procedia Comput. Sci. **18**, 2508–2511 (2013)
4. Heinecke, A., et al.: Design and implementation of the linpack benchmark for single and multi-node systems based on intel® xeon phi coprocessor. In: Parallel and Distributed Processing, pp. 126–137. IEEE (2013)
5. Iwashita, T., Shimasaki, M.: Block red-black ordering: a new ordering strategy for parallelization of ICCG method. Int. J. Parallel Prog. **31**, 55–75 (2003)
6. Kawai, M., Iwashita, T., Nakashima, H., Marques, O.: Parallel smoother based on block red-black ordering for multigrid poisson solver. In: Daydé, M., Marques, O., Nakajima, K. (eds.) VECPAR. LNCS, vol. 7851, pp. 292–299. Springer, Heidelberg (2013)
7. Reinders, J.: An overview of programming for intel® xeon processors and intel® xeon phi coprocessors (2012)

Performance Optimization of the 3D FDM Simulation of Seismic Wave Propagation on the Intel Xeon Phi Coprocessor Using the ppOpen-APPL/FDM Library

Futoshi Mori[1,2(✉)], Masaharu Matsumoto[3], and Takashi Furumura[1,2]

[1] Interfaculty Initiative in Information Studies, The University of Tokyo, 1-1-1 Yayoi,
Bunkyo-Ku, Tokyo 1130032, Japan
[2] Earthquake Research Institute, The University of Tokyo, 1-1-1 Yayoi,
Bunkyo-Ku, Tokyo 1130032, Japan
{f-mori,furumura}@eri.u-tokyo.ac.jp
[3] Information Technology Center, The University of Tokyo, 5-1-5 Kashiwanoha,
Kashiwa, Chiba 2778589, Japan
matsumoto@cc.u-tokyo.ac.jp

Abstract. We evaluate the performance of a parallel 3D finite-difference method (FDM) simulation of seismic wave propagation using the Intel Xeon Phi coprocessor. Since a continued decrease in the byte/flop ratio of future machines is forecast, program optimization with a decrease byte/flop ratio was applied by fusing the original major kernel and omitting the storing and loading of intermediate variables. We confirm that 1) MPI/OpenMP hybrid parallel computing with hyper-threading is more efficient than pure MPI parallel computing and 2) the performance of the FDM simulation with a splitting of triple DO loops is 1.3 times faster than the modified code with triple DO loops, while no performance acceleration is achieved with a fused double DO-loop calculation. We consider that loop distribution optimization is effective for prefetching and the thread parallelization of each loop by its use and reuse on cache data.

Keywords: Intel Xeon Phi coprocessor · Seismic wave propagation · Performance optimization · Pure MPI · MPI/OpenMP hybrid parallel computing · FDM

1 Introduction

Recent efforts in high-performance computing have focused on achieving increases in speed by using massively parallel computing with multicore and many-core processors. Recently, the Intel Xeon Phi coprocessor, a new type of many-core architecture coprocessor has received significant attention with regard to multicore parallel computing. This is because since June 2013, the Tianhe-2 supercomputer at the National University of Defense Technology, China, which uses the Intel Xeon Phi coprocessor, received the first prize in the Top 500 ranking of the world's fastest supercomputers [1]. With such rapid changes occurring in computing strategy, users must design code suitable for the new architecture every time a new parallel-architecture machine is developed. To avoid complications, we have developed a standard parallel finite-difference method (FDM)

© Springer International Publishing Switzerland 2015
M. Daydé et al. (Eds.): VECPAR 2014, LNCS 8969, pp. 66–76, 2015.
DOI: 10.1007/978-3-319-17353-5_6

library (ppOpen-APPL/FDM) as part of the ppOpen-HPC project [2], which will enable users to smoothly and easily transport existing FDM applications to newly developed architectures of current and future machines.

Multicore and many-core processors support different types of parallel-computing programming environments. Generally, the message passing interface (MPI) has been considered optimal for process-level coarse parallelism [3], while OpenMP is considered optimal for fine-grain loop-based parallelism [4]. For current massively parallel computing that combines MPI and OpenMP parallelization, the construction of a hybrid program may be more efficient when using multicore and many-core processors. MPI/OpenMP hybrid parallelization can reduce the communication overhead of MPI but can lead to a large OpenMP overhead due to thread process creation and increased memory bandwidth contention. For multicore architecture, such as the T2 K Open Supercomputer and the Fujitsu FX10 systems installed at the Information Technology Center of the University of Tokyo, many researchers have demonstrated that using MPI/OpenMP hybrid parallel computing is more effective than using MPI alone (known as pure MPI parallel computing) [5–8]. The advantage of MPI/OpenMP hybrid parallelism becomes more important in massively parallel computing when multicore and many-core processors are being used.

Another important issue for high-performance parallel computing of FDM simulation is the restriction of the memory bandwidth, relative to the CPU speed (i.e., bytes/flop or B/F). For example, the recent trend of supercomputer architecture has the B/F ratio dropping drastically from 4 to 0.5 for the Earth Simulator and the K-computer [9]. The B/F ratio of the next-generation computers is expected to fall even further. To overcome this challenge, it is necessary to optimize the performance of the FDM simulation, for example by more efficiently utilizing of cache memory.

In this study, we tuned the performance of the ppOpen-APPL/FDM library to make it suitable for the Intel Xeon Phi coprocessors. In the following section, we briefly explain the FDM simulation of seismic wave propagation using the ppOpen-APPL/FDM library. Next we present an overview of the Intel Xeon Phi coprocessor and demonstrate the performance of the parallel FDM simulation of seismic wave propagation in 3D, and go on to compare its parallel performance with those based on pure MPI and MPI/OpenMP hybrid parallel computing. Then, we described the performance tuning used to improve efficiency for lower B/F ratios and evaluate the results. Finally, we present our conclusions and remarks regarding future studies.

2 Overview of the Seismic Wave Propagation Simulation Using the ppOpen-APPL/FDM Library

Here, we briefly explain the procedure of the FDM simulation of seismic wave propagation using the ppOpen-APPL/FDM library. This simulation explicitly solves an equation of motion in 3D as follows:

$$\dot{u}_p^{n+\frac{1}{2}} = \dot{u}_p^{n-\frac{1}{2}} + \frac{1}{\rho}\left(\frac{\partial \sigma_{xp}^n}{\partial x} + \frac{\partial \sigma_{yp}^n}{\partial y} + \frac{\partial \sigma_{zp}^n}{\partial z} + f_p^n\right)\Delta t, \quad (p = x, y, z), \tag{1}$$

where σ_{pq}, f_p, ρ, and Δt are stress, body force, density, and time step, respectively. $\dot{u}_p^{n+1/2}$ and $\dot{u}_p^{n-1/2}$ are the particle velocities at time $t = (n \pm 1/2)\Delta t$.

Hereafter, we refer to this calculation procedure as "update-velocity". The stress component for the next time step at $t = (n + 1)\Delta t$ is updated using the spatial derivative of the velocity component derived from Eq. (1) and multiplying the elastic constants (denoted as "update-stress") as follows:

$$\sigma_{pq}^{n+1} = \sigma_{pq}^n + \left[\lambda\left(\frac{\partial \dot{u}_x^{n+\frac{1}{2}}}{\partial x} + \frac{\partial \dot{u}_y^{n+\frac{1}{2}}}{\partial y} + \frac{\partial \dot{u}_z^{n+\frac{1}{2}}}{\partial z}\right)\delta_{pq} + \mu\left(\frac{\partial \dot{u}_p^{n+\frac{1}{2}}}{\partial q} + \frac{\partial \dot{u}_q^{n+\frac{1}{2}}}{\partial p}\right)\right]\Delta t, \quad (p, q = x, y, z), \tag{2}$$

where λ and μ are Lamé's constants and δ_{pq} denotes the Kronecker delta.

The spatial derivatives in Eqs. 1 and 2 are calculated using a staggered-grid fourth-order-central FDM. Calculations of Eqs. 1 and 2 are repeated explicitly for the time desired for seismic wave propagation. The structures of the programs for the "update-velocity" and "update-stress" kernels with triple DO loops (i, j, k) are schematically illustrated in Fig. 1. Note that these four calculation kernels are the time-intensive part of this FDM simulation, respectively taking 20.9 % (difference-velocity), 20.4 % (difference-stress), 17.3 % (update-stress) and 12.7 % (update-velocity) of total calculation cost on the Intel Xeon Phi coprocessor (Fig. 2).

(a) Update-velocity kernel

```
do K=Nz00,Nz01
 do J=Ny00,Ny01
  do I=Nx00,Nx01
   ...
   Vx(I,J,K)=Vx(I,J,K)+...
   Vy(I,J,K)=Vy(I,J,K)+...
   Vz(I,J,K)=Vz(I,J,K)+...
  end do
 end do
end do
```

(b) Update-stress kernel

```
do K=Nz00,Nz01
 do J=Ny00,Ny01
  do I=Nx00,Nx01
   ...
   Sxx(I,J,K)=Sxx(I,J,K)+...
   Syy(I,J,K)=Syy(I,J,K)+...
   Szz(I,J,K)=Szz(I,J,K)+...
   Sxy(I,J,K)=Sxy(I,J,K)+...
   Sxz(I,J,K)=Sxz(I,J,K)+...
   Syz(I,J,K)=Syz(I,J,K)+...
  end do
 end do
end do
```

Fig. 1. Structure of the (a) "update-velocity" and (b) "update-stress" kernels of the seismic wave propagation simulation using the ppOpen-APPL/FDM.

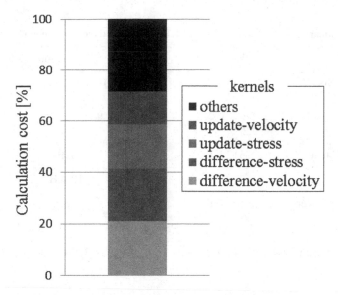

Fig. 2. Total of calculation cost of each kernel on the Intel Xeon Phi coprocessor for an FDM calculation for 128^3 grid points using 16 MPI parallel processes (Color figure online).

3   Overview of the Intel Xeon Phi Coprocessor

The Intel Xeon Phi 5110P coprocessor has 60 physical cores with a clock speed of 1.053 GHz, and each core offers four hyper-threading computations at the hardware level. Thus, it yields 240 logical cores for thread-parallel computing with the use of an 8-GB shared memory. The processor connected to a host computer via a PCI Express bus, has dual 8-core Intel Xeon E5-2670 processors with a 2.60-GHz clock speed and a 128-GB shared main memory. Table 1 summarizes the specifications of the host PC and the Intel Xeon Phi 5110P coprocessor.

We used the Intel Fortran Compiler, version 14.0.2, and the Intel MPI Library, version 4.1.0, with the compiler options "mpiifort –mmic –O3 –align array64byte" when conducting parallel computing using MPI alone (pure MPI parallel computing). In addition to those, we specified the OpenMP option ("–openmp") when conducting MPI/OpenMP hybrid parallel computing.

We evaluated the performance of the parallel FDM simulation on the native computing mode of the Intel Xeon Phi coprocessor. We compared the performance of the pure MPI and MPI/OpenMP hybrid parallelization using different combinations of MPI process (P) numbers and OpenMP thread processes (T) numbers. The calculation time for each kernel (update-velocity and update-stress) was evaluated from 200 time-step calculations. The size of the 3D FDM simulation is $240 \times 240 \times 240$ grid points which requires almost the maximum memory (8 GB) of the Intel Xeon Phi 5110P coprocessor available for single-precision arithmetic calculation, although the calculation time is

Table 1. Summary of specifications of the host PC cluster and the Intel Xeon Phi coprocessor

	PC Cluster	Intel Xeon Phi 5110P
CPU Clock Speed	2.60 GHz (Xeon E5-2670)	1.053 GHz
Number of Cores	8	60
Threads /Core	2	4
Size of L1 Cache	512 KB	32 KB
Size of L2 Cache	2 MB	512 KB
Size of Shared Memory	128 GB	8 GB
Peak Performance	332.8 GFLOPS	1.01 TFLOPS
Peak Memory Bandwidth	51.2 GB/sec	320 GB/sec (ECC off)

Table 2. Three-dimensional domain decomposition for the MPI parallel computing processes

Number of MPI processes	8	16	30	60	120	240
Partition number of X direction (grid points / process)	2 (120)	4 (60)	5 (48)	6 (40)	10 (24)	12 (20)
Partition number of Y direction (grid points / process)	2 (120)	2 (120)	3 (80)	5 (48)	6 (40)	10 (24)
Partition number of Z direction (grid points / process)	2 (120)	2 (120)	2 (120)	2 (120)	2 (120)	2 (120)

dependent on the number of grid points. Table 2 shows the partition number for each direction and the number of grid points per process corresponding to the different numbers of MPI processes used in this performance evaluation.

4   Performance in Pure MPI and MPI/OpenMP Hybrid Parallel Computing

Figure 3 shows the results of a strong-scaling test of pure MPI parallel computing for the two kernels. The calculation time decreases linearly from 28.2 to 3.04 s and from 35.3 to 4.3 s for the update-velocity and update-stress kernels, respectively, when going from 8 to 60 MPI processes. This is achieved 9.2 times and 8.2 times faster in the update-velocity and update-stress kernels, respectively, when using 60 MPI processes. However, the parallel performance is saturated when using more than 60 MPI processes. The 60-core speed-up ratio indicates a super-linear speed advancement. The number of grid points per process using 8 cores is $120 \times 120 \times 120$, and that using 60 cores is $40 \times 48 \times 120$. In these cases, the available local L1 cache size increases with the increase in cores because the

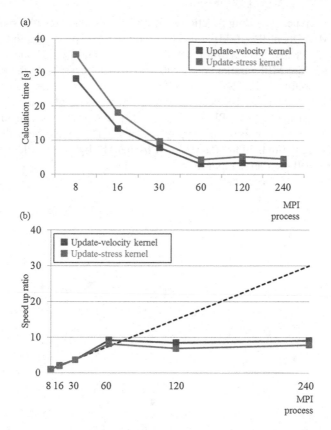

Fig. 3. Strong-scaling test of pure MPI parallel computing showing (a) the reduction in calculation time by the increase in the number of MPI processes and (b) the speed-up ratio. Blue and red lines denote the performance of the update-velocity and the update-stress kernels, respectively. The black dashed line indicates ideal speed-up as a function of the number of cores (Color figure online).

number of grid points with respect to the local cache decreases. Therefore, the calculation using 60 cores becomes faster than that using fewer than 60 cores. Moreover, the performance when using more than 60 cores is saturated because of the restriction of the memory bandwidth to CPU.

We then conducted a strong-scaling test of MPI/OpenMP hybrid parallel computing for the two kernels (Fig. 4), for comparison with the results obtained with pure MPI parallel computing (Fig. 3). For displaying the results of this test [Fig. 4 (a)], we selected the combinations of MPI process numbers and OpenMP thread numbers that had the shortest calculation times. We then inserted the OpenMP directives in the ppOpen-APPL/ FDM prior to executing the outermost DO loops of the kernels and then measured the calculation times. The results of the strong-scaling test of the MPI/OpenMP hybrid parallel computing case show that the calculation time scales down monotonically for up to 240 logical cores. We note that the calculation time for MPI/OpenMP hybrid parallel computing is shorter than that for pure MPI parallel computing using 240 logical cores due to the use of hyper-threading functions. Using 240 logical cores, the speed-up ratio [Fig. 4(b)] for MPI/OpenMP hybrid parallel computing (8 MPI processes and 30 OpenMP threads, denoted "P8T30") is 1.57 for the update-velocity and 1.64 for the update-stress over those using pure MPI parallel computing (60 MPI processes). In the case of pure MPI parallel computing using 240 cores (P240T1), 4 MPI processes are assigned to a physical core. In other words, the local cache of the core is shared with the 4 processes. On the other hand, in the case of MPI/OpenMP hybrid parallel computing using 240 cores (P8T30), 1 MPI process is assigned to 7.5 physical cores. Therefore, the cache efficiency of the P8T30 is greater than of the P240T1. MPI/OpenMP hybrid parallel computing is faster than pure MPI parallel computing because of its utilization of large-scale thread-parallel computing with hyper-threading functions.

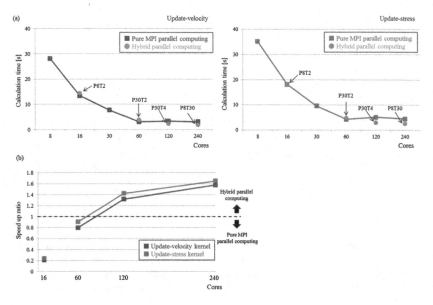

Fig. 4. Comparison of the strong scaling of MPI/OpenMP hybrid parallel computing (green points) and pure MPI parallel computing (blue and red lines) for the (a) update-velocity and update-stress kernels and (b) the speed up ratio comparison between 60 MPI parallel processes and each hybrid parallel computing. In (a), the horizontal axis is the number of cores and the vertical axis is the calculation time. In (b), the black dashed line indicates the reference line, which parallel computing is faster. In MPI/OpenMP hybrid parallel computing, we selected the shortest calculation time for each combination of MPI process (P) numbers and OpenMP thread (T) numbers (Color figue online).

5 Performance Evaluation of the Program for Decreasing the Byte/Flop Ratio

The FDM simulation of seismic wave propagation using ppOpen-APPL/FDM requires high memory bandwidth. Thus the memory bandwidth (bytes) is larger than the computational cost of the processors (flop). For example, the required B/F ratios for calculating the update-velocity and the update-stress kernels are 2.7 and 1.7, respectively, much larger than those required by current processors (0.01–0.1). In the original code, the values of each spatial derivative are stored in a corresponding array. Thus, to obtain good computational performance on today's powerful processors, we need to either make full use of the cache memory of the processors or modify the program structure to reduce the required B/F ratio, or both.

Figure 5 shows a procedure we propose to effectively decrease the required B/F ratio of the FDM simulation of seismic wave propagation. In the original simulation code, we first calculated each spatial derivative of the velocity and stress components and stored them in memory. These derivatives were then used for the kernels of update-stress and update-velocity, requiring large B/F ratios in order to load and store the large number of variables. To overcome this challenge, we modified the FDM code to merge the derivative and update calculations, thereby avoiding the need to store and load variables during the calculation. As a result, the required B/F ratios for both the update-velocity and the update-stress kernels dropped dramatically, to 0.4.

```
(a) Original code

do K=1,NZ
  do J=1,NY            Calculate Derivatives
    do I=1,NX                  kernel
    ...
    DxSxx(I,J,K)=Sxx(I,J,K)-Sxx(I-1,J,K)
                -Sxx(I+1,J,K)-Sxx(I-2,J,K)
    end do
  end do
end do


do K=Nz00,Nz01
  do J=Ny00,Ny01             Update-velocity
    do I=Nx00,Nx01                kernel
    ...
    Vx(I,J,K)=Vx(I,J,K)+(DxSxx(I,J,K)+...
    Vy(I,J,K)=Vy(I,J,K)+(DxSxy(I,J,K)+...
    Vz(I,J,K)=Vz(I,J,K)+(DxSxz(I,J,K)+...
    end do
  end do
end do
```

```
(b) Modified code (Calculate Derivatives kernel + Update-velocity kernel)

do K=NZ00, NZ01
  do J=NY00, NY01
    do I=NX00, NX01

    ! Calculate Derivatives kernel
    DxSxx0=(Sxx(I+1,J,K)-Sxx(I  ,J,K))*C40/dx-(Sxx(I+2,J,K)-Sxx(I-1,J,K))*C41/dx
    DxSxy0=(Syy(I,J,K)  -Sxy(I-1,J,K))*C40/dx-(Sxy(I+1,J,K)-Sxy(I-2,J,K))*C41/dx
    DxSxz0=(Sxz(I,J,K)  -Sxz(I-1,J,K))*C40/dx-(Sxz(I+1,J,K)-Sxz(I-2,J,K))*C41/dx
    ...

    ! Update-velocity kernel
    Vx(I,J,K)=Vx(I,J,K)+(DxSxx0+DySxy0+DzSxz0)*ROX*DT
    Vy(I,J,K)=Vy(I,J,K)+(DxSxy0+DySyy0+DzSyz0)*ROY*DT
    Vz(I,J,K)=Vz(I,J,K)+(DxSxz0+DySyz0+DzSzz0)*ROZ*DT
    end do
  end do
end do
```

Fig. 5. Procedure for decreasing the B/F ratio: (a) original code and (b) modified code. The required B/F of the update-velocity kernel (2.7) and the update-stress kernel (1.7) kernels can be reduced dramatically to 0.4 in each kernel. In (b), C40, C41, ROX, ROY, ROZ, dx and DT are constants.

Figure 6 displays the comparison of the performance of the original and modified code in MPI/OpenMP hybrid parallel computing. In Fig. 6, we show the shortest calculation times for each combination of MPI process (P) numbers and OpenMP thread (T) numbers. For parallel computing up to 60 physical cores, the modified code is slower than the original code. However, with the much larger scale parallel simulation using

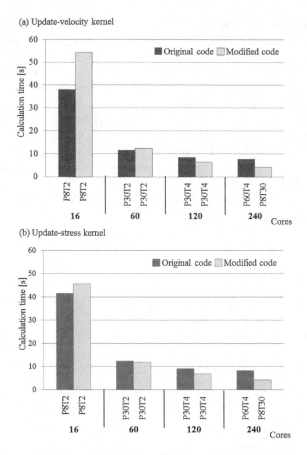

Fig. 6. Comparison of MPI/OpenMP hybrid parallel computing based on the original code and on the modified B/F-reduction code for the (a) update-velocity and (b) update-stress kernels. The horizontal axis is the number of cores and the vertical axis is the calculation time. On the horizontal axis, we indicate the shortest calculation time for each combination of MPI process (P) numbers and OpenMP thread (T) numbers in the original and the modified code (Color figure online).

240 logical cores, the modified code doubles the speed of the original code because of the moderation of the restriction on the memory bandwidth achieved by decreasing the B/F ratio.

6   Performance Optimization of Kernel Loops

The 3D FDM simulation kernels consist of triple DO loops with respect to the i, j, and k directions. It is well recognized that the fusion of many DO loops yields efficient thread-parallel computing performance with increasing loop length while decreasing the overhead of thread-parallel computing [10]. We examined the effectiveness of loop fusion for MPI/OpenMP hybrid parallel computing by using 240-core hyper-thread parallel computing, with the results as given in Fig. 7. There, we indicate the shortest

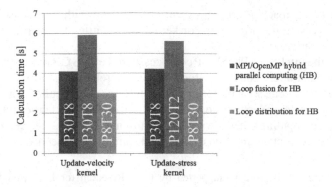

Fig. 7. Effect of loop fusion and loop distribution with MPI/OpenMP hybrid parallel computing (HB) using 240-core hyper-thread computation in the modified code: update-velocity (left) and update-stress (right) kernels. The blue bar indicates triple DO loops, the red bar indicates double DO-loop fusion, and the green bar indicates loop distribution of the triple DO loops. The horizontal axis is the kernel, and the vertical axis is the calculation time. We indicate the shortest calculation time for each combination of MPI process (P) numbers and OpenMP thread (T) numbers in the original and the modified code (Color figure online).

calculation time for each combination of MPI process (P) numbers and OpenMP thread (T) numbers. With the double DO-loop fusion, the calculation for P30T8 is 1.45 times slower for the update-velocity kernel, and that for P120T2 is 1.33 times slower for the update-stress kernel, compared with the triple DO loop (P30T8).

Loop distribution is the inverse process of loop fusion and divides a loop into multiple loops to improve the spatial locality. We investigated the loop distribution of the triple DO loops for MPI/OpenMP hybrid parallel computing by using 240-core hyper-thread parallel computing; the resulting calculation (for P8T30) is 1.38 times faster for the update-velocity kernel, and 1.13 times faster for the update-stress kernel, than that of the triple DO loops (P30T8), as is shown in Fig. 7. Loop distribution optimization is applied to prefetch and to the thread parallelization of each loop by its use and reuse on cache data.

7   Conclusion and Future Work

We have evaluated the parallel performance of the ppOpen-APPL/FDM for the 3D FDM simulation of seismic wave propagation implemented on the Intel Xeon Phi coprocessor, and have confirmed that MPI/OpenMP hybrid parallel computing is more effective than pure MPI parallel computing for many (i.e., 120–240) logical cores with hyper-threading processing. Furthermore, we have presented modified code that decreases the B/F ratio, thereby improving performance in MPI/OpenMP hybrid parallel computing. The performance of the FDM simulation with a splitting of triple DO loops was found to be faster than the modified code. We conclude that loop distribution optimization is effective for prefetching and the thread parallelization of each loop by its use and reuse on cache data. Since the B/F ratio is expected to continue decreasing in future machines, the program optimization with a decreased B\F ratio was applied by fusing the original major kernel and omitting the storing and loading of intermediate variables.

At present, the Intel Xeon Phi 5110P coprocessor has only 8 GB of local memory and is connected to other Intel Xeon Phi coprocessors via the relatively slow PCI Express bus. This is the current bottleneck for performing realistic large-scale applications of the FDM simulation using the Intel Xeon Phi coprocessor. To overcome this bottleneck, a comprehensive parallel computing scheme with the concurrent use of the Intel Xeon Phi coprocessor on the host computer, referred to as offload computing, is needed.

Acknowledgments. This work is supported by Core Research for Evolution Science and Technology (CREST), the Japan Science and Technology Agency (JST), Japan. We are grateful to Professor Tsuruoka at the Earthquake Research Institute, The University of Tokyo, for providing the Intel Xeon Phi coprocessor computing environment. We also thank our anonymous reviewers for a number of constructive comments for the revision of the manuscript.

References

1. Top 500, http://www.top500.org
2. ppOpen-HPC project, http://ppopenhpc.cc.u-tokyo.ac.jp/wordpress/
3. MPI Web Site: http://www.mcs.anl.gov/research/projects/mpi/
4. OpenMP Web Site: http://www.openmp.org/
5. Noronha, R., Panda, DK.: Improving Scalability of OpenMP Applications on Multi-core Systems Using Large Page Support. In: Parallel and Distributed Processing Symposium 2007 IPDPS 2007, pp. 1–8 (2007)
6. Tsuji, M., Sato, M.: Performance evaluation of OpenMP and MPI hybrid programs on a large scale multi-core multi-socket cluster: T2 K Open Supercomputer. Parallel Process. Workshops 2009, pp. 206–213 (2009)
7. Nakajima, K.: OpenMP/MPI hybrid parallel multigrid method on fujitsu FX10 supercomputer system. In: IEEE International Conference on Cluster Computing Workshops, pp. 199–206 (2012)
8. Furumura, T.: Large-scale parallel simulation of seismic wave propagation and strong ground motions for the past and future earthquakes in Japan. J. Earth Simulator **3**, 29–38 (2005)
9. Satoh, M., Tomita, H., Yashiro, H., Miura, H., Kodama, C., Seiki, T., Noda, A., Yamada, Y., Goto, D., Sawada, M., Miyoshi, T., Niwa, Y., Hara, M., Ohno, T., Iga, S., Arakawa, T., Inoue, T., Kubokawa, H.: The non-hydrostatic icosahedral atmospheric model: description and development. Prog. Earth Planet. Sci. **1**, 1–18 (2014)
10. Katagiri, T., Ito, S., Ohshima, S.: Early Experiences for Adaptation of Auto-tuning by ppOpen-AT to an Explicit Method. In: Proceedings of MCSoC2013, pp. 153–158 (2013)

Large-Scale Applications

Machine-Learning-Based Load Balancing for Community Ice Code Component in CESM

Prasanna Balaprakash[1,2]([✉]), Yuri Alexeev[2], Sheri A. Mickelson[1],
Sven Leyffer[1], Robert Jacob[1], and Anthony Craig[3]

[1] Mathematics and Computer Science Division, Argonne National Laboratory,
Argonne, IL, USA
pbalapra@mcs.anl.gov
[2] Leadership Computing Facility, Argonne National Laboratory, Argonne, IL, USA
[3] UCAR, Seattle, WA, USA

Abstract. Load balancing scientific codes on massively parallel architectures is becoming an increasingly challenging task. In this paper, we focus on the Community Earth System Model, a widely used climate modeling code. It comprises six components each of which exhibits different scalability patterns. Previously, an analytical performance model has been used to find optimal load-balancing parameter configurations for each component. Nevertheless, for the Community Ice Code component, the analytical performance model is too restrictive to capture its scalability patterns. We therefore developed machine-learning-based load-balancing algorithm. It involves fitting a surrogate model to a small number of load-balancing configurations and their corresponding runtimes. This model is then used to find high-quality parameter configurations. Compared with the current practice of expert-knowledge-based enumeration over feasible configurations, the machine-learning-based load-balancing algorithm requires six times fewer evaluations to find the optimal configuration.

1 Introduction

The Community Earth System Model (CESM) is one of the most widely used climate models in the world. Results from this model are a major part of the Intergovernmental Panel on Climate Change assessment reports [1]. CESM1.1.1 consists of six model components—atmosphere, ocean, sea-ice (CICE), land, river, and land-ice models—that communicate through a coupler. Each of the CESM model components has different scalability patterns and performance

The submitted manuscript has been created by the UChicago Argonne, LLC, Operator of Argonne National Laboratory (Argonne) under Contracts No. DE-AC02-06CH11357 and DE-FG02-05ER25694 with the U.S. Department of Energy. The U.S. Government retains for itself, and others acting on its behalf, a paid-up, nonexclusive, irrevocable worldwide license in said article to reproduce, prepare derivative works, distribute copies to the public, and perform publicly and display publicly, by or on behalf of the Government. The NCAR is sponsored by the National Science Foundation.

M. Daydé et al. (Eds.): VECPAR 2014, LNCS 8969, pp. 79–91, 2015.
DOI: 10.1007/978-3-319-17353-5_7

characteristics. In this paper, we focus on static load-balancing of computation, which is usually simple to implement with negligible overhead, making it suitable for "fine-grained" parallelism consisting of many small tasks. Previously, the load-balancing problem has been formulated as a mixed-integer nonlinear optimization problem and solved by using the optimization solver MINOTAUR [2]. This is a heuristic method that consists of gathering benchmarking data, calibrating a performance model using the data, and making decisions about optimal allocation by using the model. The performance model predicts the execution time of the program running in parallel as a function of problem size and the number of processors employed. Nonetheless, several challenges in intramodel load balancing for the CICE computations occur only where sea ice is located and the sun is shining. This restriction presents a load-balance problem because processors are allocated across the entire Earth grid and several locations on the grid that do not have any sea ice [3]. The poor fit of the CICE results in inefficient processor allocations to all components—incorrect allocation of the CICE affects all other allocations because the total number of processors available to components is a fixed number. This is the primary motivation for us to develop sophisticated approaches for load balancing the CICE component of the CESM.

Recently, machine-learning methods [4] have received considerable attention for tuning performance of large scientific codes and kernels on high-performance computing systems. In particular, supervised machine-learning tries to learn the relationship between the input and the output of an unknown response function by fitting a model from few representative examples. When the model is accurate enough, it can predict the output at new unseen inputs, which provides numerous benefits, in particular when the evaluation becomes expensive.

In this paper, we present a machine-learning-based approach for static load-balancing problems, and we apply it to find high-quality parameter configurations for load balancing the CICE component of the CESM on IBM Blue Gene/P (BG/P). The novelty of the proposed algorithm consists of iteratively using the model to choose configurations with shorter predicted runtime for evaluation on the target architecture. We emphasize, however, that the algorithm is general and not specific to the CESM and/or BG/P. The paper is structured as follows: (1) a machine-learning-based algorithm for static load-balancing problem, (2) deployment of a machine-learning method as a diagnostic tool for analyzing the sensitivity of the load-balancing parameters on the execution time, (3) empirical analysis of several state-of-the-art machine-learning methods for modeling the relationship between the load-balancing parameters and their corresponding execution time, and (4) 6x savings in core-hour usage for load balancing the CICE component of the CESM on BG/P.

2   The CICE Component on BG/P

For the CICE component, we need to find the optimal load-balancing parameter configuration x^* with the shortest the runtime (f^*) for task counts $\in \{80, 128, 160, 256, 320, 376, 512, 640, 800, 1024\}$. The task count corresponds to number of

Table 1. Decomposition strategies and their corresponding `block.x` and `block.y` sizes

decomp.set	decomp.typ	block.x	block.y
null	blkrobin, blkcart roundrobin, spacecurve	1, 2, 4, 8	24, 48, 96, 192, 3840
slenderX1 slenderX2	cartesian	4, 5, 8, 10	4 6 8 12

MPI tasks; the number of OpenMP threads per MPI task is set to four because of memory restrictions on BG/P. The CICE component comprises six parameters. Three integer parameters, namely, maximum number of CICE blocks, `max.block`; the size of a CICE block in the first and second horizontal dimensions `block.x` and `block.y` respectively. Two categorical parameters that determine the decomposition strategy, `decomp.typ` $\in$ {blkrobin, roundrobin, spacecurve, blkcart, cartesian} and `decomp.set` $\in$ {null, slenderX1, slenderX1}. A binary parameter `mask.h` $\in$ {0,1} that specifies to run the code with or without halo.

The constraints that define a feasible set $\mathcal{D}$ of configurations are as follows. The parameter `max.block` $\in$ {1, 2, 3, 4, 5, 6, 7, 8, 10, 11, 12, 13, 14, 15, 16, 20, 24, 26, 30, 32, 40, 48, 64} is determined by computing (CICE_X_Grid_Size × CICE_Y_Grid_Size) / (`block.x` × `block.y` × task count). The feasible values for `decomp.set`, `decomp.typ`, `block.x`, and `block.y` are constrained as shown in Table 1. The decomposition strategies have different rules, and not all combinations of block sizes are possible. The blkcart method must have a multiple of four blocks per compute core. The spacecurve method must have 2, 3, and 5 only in `max.block`. The slenderX1 method requires that the `block.x` multiplied by the task count divide evenly into the CICE X grid size. The value of `block.y` must also be divisible by the CICE Y grid size. The slenderX2 method requires that the `block.x` multiplied by the task count be divisible by the CICE X grid size multiplied by 2. The decomposition also requires that the `block.y` multiplied by 2 divide evenly into the CICE Y grid size.

3   Machine-Learning Based Load-Balancing Algorithm

Given a set of training data $\{(x_1, y_1)), \ldots, (x_l, y_l))\}$, where $x_i \in \mathcal{D}$ and $y_i = f(x_i) \in \mathbb{R}$ are the load-balancing parameter configuration and its corresponding runtime, respectively, the supervised machine-learning approach includes finding a surrogate function h for the expensive f such that the difference between $f(x_i)$ and $h(x_i)$ is minimal for $\forall i \in \{1, \ldots, l\}$. The function h, which is an empirical performance model, can be used to predict the runtimes of unevaluated $x' \in \mathcal{D}$. The key idea behind the machine-learning-based load-balancing algorithm is iteratively using the model to choose configurations with shorter predicted runtime for evaluation and retrain the model with the evaluated configurations.

The pseudo-code is shown in Algorithm 1. The symbols $\cup$ and $-$ denote set union and difference operators, respectively. Given a task count c, a pool $\mathcal{X}_p$ of unevaluated configurations of task count c, the maximum number n_{max} of

---

Algorithm 1. Pseudo-code for the machine-learning-based load-balancing algorithm

---

Input: task count c, configuration pool $\mathcal{X}_p$ of task count c, max evaluations $n_{\max}$, initial sample size n_s

1 $\mathcal{X}_{\mathrm{out}} \leftarrow$ sample $\min\{n_s, n_{\max}\}$ distinct configurations from $\mathcal{X}_p$
2 $\mathcal{Y}_{\mathrm{out}} \leftarrow$ Evaluate_Parallel(c, $\mathcal{X}_{\mathrm{out}}$)
3 $\mathcal{M} \leftarrow$ fit($\mathcal{X}_{\mathrm{out}}, \mathcal{Y}_{\mathrm{out}}$)
4 $\mathcal{X}_p \leftarrow \mathcal{X}_p - \mathcal{X}_{\mathrm{out}}$
5 **for** $i \leftarrow n_s + 1$ to $n_{\max}$ **do**
6      $\mathcal{Y}_p \leftarrow$ predict($\mathcal{M}, \mathcal{X}_p$)
7      $x_i \leftarrow x \in \mathcal{X}_p$ with the shortest runtime in $\mathcal{Y}_p$
8      $y_i \leftarrow$ Evaluate(c, x_i)
9   **retrain** $\mathcal{M}$ with (x_i, y_i)
10     $\mathcal{X}_{\mathrm{out}} \leftarrow \mathcal{X}_{\mathrm{out}} \cup x_i$; $\mathcal{Y}_{\mathrm{out}} \leftarrow \mathcal{Y}_{\mathrm{out}} \cup y_i$
11     $\mathcal{X}_p \leftarrow \mathcal{X}_p - x_i$
12 **end for**

Output: $x \in \mathcal{X}_{\mathrm{out}}$ with the shortest runtime in $\mathcal{Y}_{\mathrm{out}}$

---

allowed evaluations, and initial sample size n_s, the algorithm proceeds in two phases: parallel initialization phase and sequential iterative phase. In the initialization phase, the algorithm first samples n_s configurations at random and evaluates them in parallel to obtain their corresponding runtimes. A supervised learning method uses these points as a training set to build a predictive model. The sequential iterative phase consists of predicting the runtimes of all remaining unevaluated configurations using the model, evaluating the configuration with shortest predicted runtime, and retraining the model with the evaluation results. Without loss of generality, Algorithm 1 can be run in parallel for each task count $c \in C$. Because the best supervised learning algorithms depends on the relationship between the input and output, we test four state-of-the-art machine-learning algorithms as candidates for Algorithm 1: random forest (RF) [5], support vector machines (SVM) [6], Gaussian process regression (GP) [7], and neural networks (NN) [8].

RF belongs to the class of recursive partitioning methods [9]. They are widely used tools for predictive modeling in many scientific fields. These methods recursively partition the multi-dimensional input space $\mathcal{D}'$ of training points into a number of hyper rectangles. The partition is done in such a way that input configurations with similar outputs fall within the same rectangle. The partition gives rise to a set of if-else rules that can be represented as a decision tree. For each hyper rectangle, a constant value is assigned—typically this is an average over the output values that fall within the given hyper rectangle. An example tree which is obtained on the CICE component data is shown in Fig. 1. Given an unseen input $x^* \in \mathcal{D}^* \subset \mathcal{D}$, the algorithm uses the if-else rule to find the leaf and returns the corresponding constant value as the predicted value. RF uses a collection of regression trees, where each tree is obtained by the principle of recursive

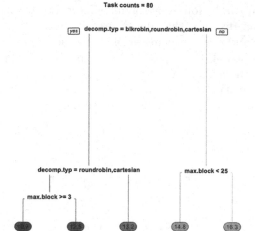

Fig. 1. Illustration of a decision tree obtained via recursive partitioning on CICE component data for the task count 80.

partitioning. For each tree generation, the algorithm takes a subsample of random points from the given training set. The subsample is either a bootstrap sample of the same size drawn with replacement or a subset of smaller size, drawn without replacement. Due to the randomness in the sampling, each subsample differs from each other. Given that each individual tree is build on the subsample, it can differ significantly from other trees. For a given x^*, each tree can make a prediction with respect to its own subsample. The power of RF comes from the aggregation of predicted output values from different trees and the natural way of handling the categorical parameters. Consequently, it can deal with large dimensional inputs even in the presence of complex interactions and non-linearity.

SVM for nonlinear regression consists of mapping the given $\mathcal{D}'$ of the training points into a high dimensional feature space and performing linear regression in the feature space:

$$g(\mathcal{D}') = \langle w \cdot \psi(\mathcal{D}') \rangle + b, \tag{1}$$

where $\psi : \mathbb{R}^n \to \mathcal{F}$ being the nonlinear transformation, b being the bias term, and $w \in \mathcal{F}$. Finding $g(\mathcal{D}')$ consists in specifying a loss function that need to be optimized and a kernel function $k(\cdot)$ for nonlinearity transformation ψ. For the former, we use ϵ intensive-loss function in which zero penalty is added to the loss function when predicted value of a training point is within ϵ from its observed value. For the latter, we use the widely used Gaussian radial basis function kernel. Now, Eq. 1 can be written as follows:

$$g(\mathcal{D}') = \sum_{i=1}^{l} \alpha_i \times [k(x_i, x_1), \dots, k(x_i, x_l)] + b, \tag{2}$$

$$k(x_i, x_j) = \exp\left(-\frac{||x_i - x_j||_2^2}{2\sigma^2}\right), \tag{3}$$

where coefficients α_i can be found by solving ϵ intensive-loss function, $||\mathbf{x} - \mathbf{x}'||_2^2$ is squared Euclidean distance that decreases with an increase in dissimilarity between x_i and x_j, and σ is a parameter of the kernel.

GP follows a probabilistic approach for regression. Given a training data of l points, GP assumes that $\mathbf{Y} = [y_1, \ldots, y_l]$ as a sample from a l-variate Gaussian distribution. For an unseen input x^*, the probability $p(y^*|\mathbf{Y})$ follows the Gaussian distribution $\mathcal{N}$ with a user defined kernel function $k(\cdot)$:

$$y^*|\mathbf{Y} \sim \mathcal{N}(\mathbf{K}_*\mathbf{K}^{-1}\mathbf{Y}, \mathbf{K}_{**} - \mathbf{K}_*\mathbf{K}^{-1}\mathbf{K}_*^{\mathbf{T}}), \tag{4}$$

where

$$\mathbf{K} = \begin{bmatrix} k(x_1, x_1) & \cdots & k(x_1, x_n) \\ \vdots & \ddots & \vdots \\ k(x_n, x_1) & \cdots & k(x_n, x_n), \end{bmatrix} \tag{5}$$

$$\mathbf{K}_* = [k(x_*, x_1), \ldots, k(x_*, x_n)],$$
$$\mathbf{K}_{**} = k(x_*, x_*).$$

Note that $\mathbf{T}$ represents matrix transpose operation. For $k(\cdot)$, we use the Gaussian radial basis function as in Eq. 3. The predicted value $\hat{y}_*$ and variance $var(y^*)$ of y^* are given by the parameters of $\mathcal{N}$:

$$\hat{y}^* = \mathbf{K}_*\mathbf{K}^{-1}\mathbf{Y},$$
$$var(y^*) = \mathbf{K}_{**} - \mathbf{K}_*\mathbf{K}^{-1}\mathbf{K}_*^{\mathbf{T}}. \tag{6}$$

NN is a classical and one of most widely used supervised learning approaches. We focus on a single-hidden-layer neural network, an effective variant that comprises one input layer, one hidden layers, and one output layer. The nonlinear regression performed by NN can be written as follows:

$$\mathbf{Y} = h(\mathcal{D}') = \mathbf{B}\varphi(\mathbf{A}\mathcal{D}' + \mathbf{a}) + \mathbf{b},$$

where $\mathbf{A}$ and is the matrix of weights and bias vector for the first layer (between input and hidden layer) and $\mathbf{B}$ and $\mathbf{b}$ are the weight matrix and the bias vector of the second layer (between hidden and output layer). The function φ denotes an element wise nonlinearity. The training in neural network consists in adapting all the weights and biases $\mathbf{A}$, $\mathbf{B}$, $\mathbf{a}$, and $\mathbf{b}$ to their optimal values for the training set $\{(x_1, y_1)), \ldots, (x_l, y_l))\}$. The optimization problem consists in minimizing the squared reconstruction error $\sum_{i=1}^{l} ||h(x_i) - y_i||^2$ and it can be solved effectively with back-propagation algorithm.

4   Experimental Results

We evaluated the effectiveness of the proposed load-balancing algorithm with the four machine-learning methods. In addition, we include two approaches in the comparison: Expert-knowledge-based enumeration (EE) and random search (RS). EE is the current practice for finding the optimal load-balancing configuration for the CICE component of the CESM. In addition to the application-specific constraints, expert knowledge of the code and the architecture were used to prune the feasible set of configurations $\mathcal{D}$ for the CICE component. As a result, for each task count c, there are 50 to 60 ($|\mathcal{D}_c|$) feasible configurations; in total, for all the 10 task counts, there are $|\mathcal{D}| = 653$ parameter configurations. This method evaluates all 653 parameter configurations. Moreover, we followed the current practice for defining the runtime $f(x)$ for x: the code was run twice with the same x and the shortest runtime was taken as $f(x)$. In RS, for each task count c, parameter configurations were sampled at random without replacement from $\mathcal{D}_c$ and were evaluated. To minimize the impact of randomness involved in the initialization procedure of Algorithm 1 and in the five approaches, we repeated all of them 10 times, each with a different random seed. Moreover, we stored the runtime of each configuration from EE in a lookup table and reused the results for running all other algorithms. For Algorithm 1, for each task count c, $\mathcal{D}_c$ obtained in the EE approach was given as the configuration pool $\mathcal{X}_p$, and the initial sample size n_s was set to 5. The approaches were implemented and run in the **R** programming language and environment [10] version 2.15.2 using the nnet (NN), kernlab (SVM, GP), and randomForest (RF) packages. The default parameter values were used for each method. Experiments were carried out on Intrepid, a BG/P supercomputer at Argonne.

Sensitivity Analysis: First, we present an empirical analysis to explain why the previously proposed analytical performance model fails to predict the runtime of the CICE component and why distinct models may be constructed for each task count. For this purpose, we used the RF method to analyze the impact of each load-balancing parameter on the resulting runtimes. For the training data, we randomly sampled 50 % of the data (parameter configuration and runtimes) obtained with EE approach. An RF model was fitted on this training set. The mean squared error (MSE) on the original training set is given by $\frac{\sum_{i=1}^{l}(f(x_i)-\hat{f}(x_i))^2}{l}$, where l is the number of training points, and $f(x_i)$ and $\hat{f}(x_i)$ are the original and predicted runtime value of parameter configuration x_i, respectively. In order to assess the impact of a parameter m, the values of m in the training set were randomly permuted. Again, an RF model was fitted on this imputed training set, and the mean squared error was computed. If a parameter m is important, then permuting the values of m should affect the prediction accuracy significantly and eventually increase the mean squared error. The results are shown in Fig. 2. We observe that the trend in the parameter importance is not the same over all the task counts. For task counts up to 320, decomp.set and/or decomp.type have a strong impact on the runtimes; for large task counts, they become relatively less important—max.block, block.x, and block.y have a strong impact on the runtime. For 1024, only max.block, block.x, and block.y have an impact on the runtime; the other three parameters have negative %IncMSE, suggesting that they do not affect the runtime. In summary, the impact of parameter values on the runtimes and the type of

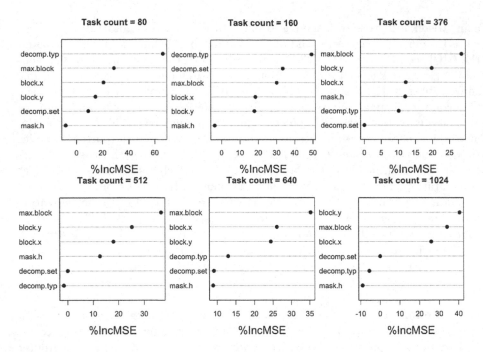

Fig. 2. Sensitivity analysis of the load-balancing parameters on the runtime of the CICE component for different task counts. For each parameter, the plot shows the percentage increase in mean squared error (%IncMSE) when the values of the corresponding parameter gets imputed.

nonlinear interactions between them change with an increase in the task counts. The previously developed analytical model does not take this effect into account for the CICE component, and consequently it falls short in runtime prediction. Moreover, if we build a single model for all task counts with task count being an input to the model, we might loose these task-count-specific interactions, thus affecting the runtime quality of the obtained configurations.

Comparison Between Variants: With EE as a baseline, we next examined the effectiveness of the five approaches in finding the optimal load-balancing configuration for the CICE component. As a measure of the effectiveness of each variant, we use the percentage deviation from the optimal runtime (%dev). Given a variant v and task count c, this is given by $\frac{f_v^c - f_{opt}^c}{f_{opt}^c} \times 100$, where f_v^c is the shortest runtime obtained by variant v and f_{opt}^c is the optimal runtime obtained from EE. Because we repeated each method 10 times to reduce the impact of randomness, we consider the mean percentage deviation from the optimal runtime of a variant as %dev averaged over 10 repetitions. We also used a statistical t-test to check whether the observed differences in the %dev of the variants are significant. Figure 3 shows the comparison between the approaches. The results show that RS requires almost the same number of evaluations as does EE for all task counts. These results indicate that the problem of finding high-quality configurations is not an easy task;

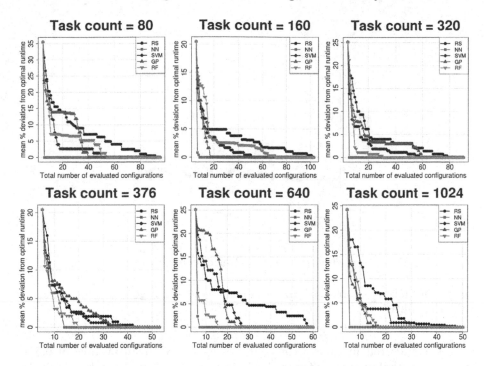

Fig. 3. Comparison between approaches for different task counts of the CICE component. The lines represent the mean percentage deviation from the optimal runtime as a function of the number of evaluated configurations.

clearly, we need more sophisticated approaches to find high-quality configurations within fewer evaluations. The variants of Algorithm 1 obtain optimal configurations with fewer evaluations, and they outperform RS. NN completely dominates all other variants and RS. The key advantage of NN comes from its requiring less than 10 evaluations to find the optimal parameter configuration on 9 out of 10 task counts—only on $c = 376$, does it require 15 evaluations.

In Table 2, we analyze $\%dev$ of each variant, when it is allowed to perform only 10 evaluations (for machine-learning variants this corresponds to five evaluations after the initialization). The results show that mean $\%dev$ of NN is zero and it lower than all other variants. For all but one task counts, the observed differences are significant in statistical sense. NN fails to find optimal runtime for $c = 376$, where it is 6 % away from the optimal runtime and it is comparable to other approaches.

As soon as a new evaluation becomes available, each machine-learning variants is retrained on all the available input-output pairs. This is the most computationally expensive part in the iterative phase of Algorithm 1. In Fig. 4, we analyze the retraining time required by the machine-learning variants after each evaluation. The reported time is an average time over all repetitions and task count. The results show that NN outperforms all other variants in retraining

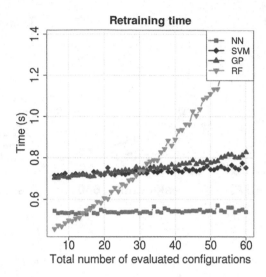

Fig. 4. Time taken by various machine-learning methods in Algorithm 1 for retraining after each evaluation.

time. The time remains fairly constant throughout with an average of 0.5 s. This can be attributed the effective back propagation algorithm adopted in the underlying optimization routine. For GP and SVM, there is a slight increase in retraining time. Nonetheless, the retraining time of RF increases linearly with an increase in the number of training points suggesting that it might not be suitable for sequential learning with large number of points. Note that there exists some advanced algorithm-specific techniques to avoid retraining from scratch, however, none of the machine-learning methods adopts such technique in our study. Furthermore, in all these algorithms, the time to predict an unseen input x^* is negligible (in the order of milli to micro seconds) because they belong to a class of eager learning algorithms as opposed to lazy learning algorithms where a model is built only when x^* needs to be predicted.

5   Related Work

Compared with dynamic strategies [11–16], static load-balancing approaches have received relatively less attention in the literature. The problem of static load-balancing can be formulated as a graph-partitioning problem that belongs to a class of $\mathcal{NP}$-hard problem for which finding optimal solution is computationally hard. Many efficient algorithms are developed to tackle this problem in operations research community and are used for static load-balancing. These algorithms can be grouped into geometry-based algorithms, graph-based algorithms, and partitioning algorithms [17]. In [18], the authors carried out an experimental comparison of eleven static load-balancing algorithms for heterogeneous distributed computing systems. They showed the relatively simple

Table 2. Mean percentage deviation from the optimal runtime averaged over 10 replications with the maximum budget of 10 evaluations

Task count	NN	RF	GP	SVM	RS
80	0.000	*12.668*	*15.032*	*18.089*	*20.246*
128	0.000	3.269	*7.620*	*5.177*	*12.846*
160	0.000	*12.649*	*8.050*	*6.989*	*8.563*
256	0.000	4.575	*8.468*	*7.340*	*10.024*
320	0.000	*2.208*	*8.818*	*6.709*	*13.105*
376	6.005	**3.186**	8.132	7.206	7.456
512	0.000	*10.269*	*11.794*	*6.472*	*9.090*
640	0.000	2.674	*20.058*	*14.072*	*10.292*
800	0.000	*7.435*	*5.996*	*6.182*	*8.770*
1024	0.000	*6.645*	*4.985*	5.966	*13.241*

Note: The value is typeset in *italics* (**bold**) when a variant is significantly worse (better) than NN according to a *t*-test with significance (alpha) level 0.05.

Min-Min heuristic performs well in comparison to the other techniques such as simulated annealing and genetic algorithms, and tabu search. However, the state-of-the-art high-performing algorithms comprises hybrid algorithms, multilevel approaches, and parallel implementations of the above algorithms [17]. We refer the reader to [17,19] for a survey for static load balancing approaches. Recently, in [20], a genetic algorithm was adopted for tasks scheduling and load balancing in heterogeneous parallel multiprocessor system. Nonetheless, the domain-specific constraints of the CICE component make the search problem hard and prevents the straightforward adoption of heuristic search algorithms [21]. In order to handle these constraints effectively, the search algorithms need a sophisticated constraint-handling mechanism; consequently they loose generality and become problem-specific.

The idea of using machine learning in load-balancing has received considerable attention for dynamic strategies. Examples include neural network [22], decision tree [23], and reinforcement learning approaches [24]. However, to the best of our knowledge, the adoption of machine-learning approaches for application and architecture specific static load-balancing has not been investigated before. Finally, this is the first work on the use of machine learning approaches for analyzing the sensitivity of the load-balancing parameters.

6   Summary and Outlook

We developed a machine-learning-based approach for static load-balancing problem and applied it for load balancing the CICE component of the CESM running on BG/P. We deployed a machine-learning method as a diagnostic tool for analyzing the sensitivity of the load-balancing parameters on the runtime and provided an explanation for inadequacy of the analytical performance model. The main contribution of the paper is the development and empirical analysis of the machine-learning-based algorithm that allowed us to load balance the CICE component of the CESM on BG/P with significant savings in core-hour usage. Compared to the current practice of expert-knowledge-based enumeration over feasible parameter configurations, we showed that the proposed algorithm requires 6x fewer evaluations to find the optimal load-balancing configurations.

A inherent limitation of our algorithm consists in the sequential evaluation of parameter configurations that will affect the wall clock time. To address this issue, we will develop unsupervised learning methods to partition the feasible set into a number of similar groups and learning those regions in parallel. To that end, we will investigate parallel machine-learning algorithms. Since the inefficient processor allocations of CICE component can affect overall scaling of the CESM, we will use the proposed approach and assess the overall performance of the CESM. Furthermore, two projects, Climate-Science Computational End Station Development and Attributing Changes in the Risk of Extreme Weather and Climate, granted computational time on ALCF's BG/P and Q supercomputers under the DOE INCITE program will directly benefit from this work. We are planning to investigate the effectiveness of the proposed algorithm for load-balancing various climate simulations in these projects.

Acknowledgments. This work was supported by the U.S. Department of Energy, Office of Science, Advanced Scientific Computing Research, under Contract DE-AC02-06CH11357. An award of computer time was provided by the Innovative and Novel Computational Impact on Theory and Experiment (INCITE) program. This research used resources of the Argonne Leadership Computing Facility at Argonne National Laboratory, which is supported by the Office of Science of the U.S. Department of Energy under contract DE-AC02-06CH11357.

References

1. Metz, B., Davidson, O., Bosch, P., Dave, R., Meyer, L.: Contribution of working group III to the fourth assessment report of the Intergovernmental Panel on Climate Change (2007)
2. MINOTAUR: a toolkit for MINLP. http://wiki.mcs.anl.gov/minotaur/index.php/Main_Page
3. 2013. http://www.cesm.ucar.edu/events/ws.2012/Presentations/SEWG2/craig.pdf
4. Bishop, C.M., et al.: Pattern Recognition And Machine Learning. Springer, New York (2006)
5. Breiman, L.: Random forests. Mach. Learn. **45**(1), 5–32 (2001)
6. Hearst, M.A., Dumais, S., Osman, E., Platt, J., Scholkopf, B.: Support vector machines. Intell. Syst. Appl. **13**(4), 18–28 (1998). IEEE
7. Rasmussen, C.E., Williams, C.K.: Gaussian Processes For Machine Learning. adaptive computation and machine learning. MIT Press, Cambridge (2005)

8. Haykin, S.: Neural Networks: A Comprehensive Foundation, 1st edn. Prentice Hall PTR, Upper Saddle River (1994)
9. Atkinson, E.J., Therneau, T.M.: An Introduction To Recursive Partitioning Using The Rpart Routines. Mayo Foundation, Rochester (2000)
10. R Core Team, R: A Language and Environment for Statistical Computing, R Foundation for Statistical Computing, Vienna, Austria (2013). http://www.r-project.org
11. Kale, L.V., Krishnan, S.: CHARM++: a portable concurrent object oriented system based on C++. ACM SIGPLAN Not. **28**(10), 91–108 (1993)
12. Barker, K., Chernikov, A., Chrisochoides, N., Pingali, K.: A load balancingframework for adaptive and asynchronous applications. IEEE Trans. Parallel Distrib. Syst. **15**(2), 183–192 (2004)
13. Barker, K.J., Chrisochoides, N.P.: An evaluation of a framework for the dynamic load balancing of highly adaptive and irregular parallel applications. In: Proceedings of the 2003 ACM/IEEE Conference on Supercomputing, p. 45. ACM (2003)
14. Huang, C., Zheng, G., Kalé, L., Kumar, S.: Performance evaluation of adaptive MPI. In: Proceedings of the Eleventh ACM SIGPLAN Symposium on Principles and Practice of Parallel Programming, pp. 12–21. ACM (2006)
15. Boneti, C., Gioiosa, R., Cazorla, F.J., Valero, M.: A dynamic scheduler for balancing HPC applications. In: Proceedings of the 2008 ACM/IEEE Conference on Supercomputing, p. 41. IEEE Press (2008)
16. Sharma, R., Kanungo, P.: Dynamic load balancing algorithm for heterogeneous multi-core processors cluster. In: 2014 Fourth International Conference on Communication Systems and Network Technologies (CSNT), pp. 288–292. IEEE (2014)
17. Hu, Y., Blake, R.: Load balancing for unstructured mesh applications. Parallel Distrib. Comput. Pract. **2**(3), 117–148 (1999)
18. Braun, T.D., et al.: A comparison of eleven static heuristics for mapping a class of independent tasks onto heterogeneous distributed computing systems. J. Parallel Distrib. Comput. **61**(6), 810–837 (2001)
19. Ichikawa, S., Yamashita, S.: Static load balancing of parallel PDE solver for distributed computing environment. In: Proceedings of the 13th International Conference on Parallel and Distributed Computing Systems, pp. 399–405 (2000)
20. Effatparvar, M., Garshasbi, M.: A genetic algorithm for static load balancing in parallel heterogeneous systems. Procedia Soc. Behav. Sci. **129**, 358–364 (2014)
21. Balaprakash, P., Wild, S.M., Hovland, P.D.: Can search algorithms save large-scale automatic performance tuning? In: International Conference on Computational Science (2011)
22. Jia, Y., Sun, J.-Z.: A load balance service based on probabilistic neural network. In: International Conference on Machine Learning and Cybernetics, vol. 3, pp. 1333–1336. IEEE (2003)
23. Dantas, M.A., Pinto, A.R.: A load balancing approach based on a geneticmachine learning algorithm. In: 19th International Symposium on HighPerformance Computing Systems and Applications (HPCS 2005), pp. 124–130. IEEE (2005)
24. Helmy, T., Shahab, S.A.: Machine learning-based adaptive load balancing framework for distributed object computing. In: Chung, Y.-C., Moreira, J.E. (eds.) GPC 2006. LNCS, vol. 3947, pp. 488–497. Springer, Heidelberg (2006)

Domain Decomposition for Heterojunction Problems in Semiconductors

Timothy Costa[1]($^{\boxtimes}$), David Foster[2], and Malgorzata Peszynska[1]

[1] Department of Mathematics, Oregon State University, Corvallis, OR 97330, USA
costat@math.oregonstate.edu
[2] Department of Physics, Oregon State University, Corvallis, OR 97330, USA

Abstract. We present a domain decomposition approach for the simulation of charge transport in heterojunction semiconductors. The problem is characterized by a large variation of primary variables across an interface region of a size much smaller than the device scale, and requires a multiscale approach in which that region is modeled as an internal boundary. The model combines drift diffusion equations on subdomains coupled by thermionic emission heterojunction model on the interface which involves a nonhomogeneous jump computed at fine scale with Density Functional Theory. Our full domain decomposition approach extends our previous work for the potential equation only, and we present perspectives on its HPC implementation. The model can be used, e.g., for the design of higher efficiency solar cells for which experimental results are not available. More generally, our algorithm is naturally parallelizable and is a new domain decomposition paradigm for problems with multiscale phenomena associated with internal interfaces and/or boundary layers.

Keywords: Domain decomposition · Drift-diffusion equations · Density Functional Theory · Heterojunction · Multiscale semiconductor modeling · Solar cells

1 Introduction

In this paper we present a multiscale approach for heterojunction interfaces in semiconductors, part of a larger interdisciplinary effort between computational mathematicians, physicists, and material scientists interested in building more efficient solar cells. The higher efficiency (may) arise from putting together different semiconductor materials, i.e., creating a *heterojunction*.

The computational challenge is that phenomena at heterojunctions must be resolved at the angstrom scale while the size of the device is on the scale of microns, thus it is difficult to simultaneously account for correct physics and keep the model computationally tractable. To model charge transport at the device scale we use the drift diffusion (D-D) system [14]. For interfaces, we follow the approach from [9] in which the interface region is shrunk to a low-dimensional internal boundary, and physics at this interface is approximated

© Springer International Publishing Switzerland 2015
M. Daydé et al. (Eds.): VECPAR 2014, LNCS 8969, pp. 92–101, 2015.
DOI: 10.1007/978-3-319-17353-5_8

by the thermionic emission model (TEM) which consists of unusual internal boundary conditions with jumps.

We determine the data for these jumps from an angstrom scale calculation using Density Functional Theory (DFT), and we model the physics away from the interface by the usual (D-D) equations coupled by TEM. The D-D model can be hard-coded as a monolithic approach which appears intractable and/or impractical in 2d/3d with complicated interface geometries. Our proposed alternative is to apply a domain decomposition (DDM) approach which allows the use of "black box" D-D solvers in subdomains, and enforces the TEM conditions at the level of the DDM driver. DDM have been applied to D-D, e.g., in [12,13], where the focus was on a multicore HPC implementation of efficiently implemented suite of linear and nonlinear solvers. Here we align the DDM with handling microscale physics at material interfaces. More importantly, fully decoupling the subdomains is a first step towards a true multiscale simulation where the behavior in the heterojunction region is treated simultaneously by a computational method at microscale.

The DDM approach we propose is non-standard because of the nonhomogeneous jumps arising from TEM. In [7] we presented the **DDP** algorithm for the potential equation. In this paper we report on the next nontrivial step which involves carrier transport equations. Here the interface model is an unusual Robin-like interface equation. The algorithm **DDC** works well and has promising properties.

This paper consists of the following. In Sect. 2 we describe the model. In Sect. 3 we present our domain decomposition algorithms, and in Sect. 4 we present numerical results for the simulation of two semiconductor heterojunctions. Finally in Sect. 5 we present conclusions, HPC context, and describe future work.

2   Computational Model for Coupled Scales

The continuum D-D model with TEM is described first, followed by the angstrom scale DFT model.

2.1   Device Scale Continuum Models: Drift Diffusion (D-D) System

Let $\Omega \in \mathbb{R}^N$, $N \in \{1, 2, 3\}$, be an open connected set with a Lipschitz boundary $\partial\Omega$. Let $\Omega_i \in \Omega$, $i = 1, 2$, be two non-overlapping subsets of Ω s.t. $\overline{\Omega}_1 \cup \overline{\Omega}_2 = \overline{\Omega}$, $\Omega_1 \cap \Omega_2 = \emptyset$, and denote $\Gamma := \overline{\Omega}_1 \cap \overline{\Omega}_2$. We assume Γ is a N-1 dimensional manifold, and $\Gamma \cap \partial\Omega = \emptyset$. Each subdomain Ω_i corresponds to a distinct semiconductor material, and Γ the interface between them. We adopt the following usual notation: $w_i = w|_{\Omega_i}$, $w_i^\Gamma = \lim_{x \to \Gamma} w_i$, and $[w]_\Gamma = w_2^\Gamma - w_1^\Gamma$ denotes the jump of w.

In the bulk semiconductor domains Ω_i, $i = 1, 2$, the charge transport is described by the D-D system: a potential equation solved for electrostatic potential ψ, and two continuity equations solved for the Slotboom variables u and v; these relate to the electron and hole densities n and p, respectively, via $n = \delta_n^2 e^\psi u$, $p = \delta_p^2 e^{-\psi} v$.

(The scaling parameters δ_n^2 and δ_p^2 depend on the material and the doping profile). We recall that in Slotboom variables the continuity equations are self-adjoint [14]. The stationary D-D model is

$$-\nabla \cdot (\epsilon_i \nabla \psi_i) = \frac{1}{\eta}(\delta_p^2 e^{-\psi_i} v - \delta_n^2 e^{\psi_i} u + N_T) := q(\psi_i, p_i, n_i), \qquad (1)$$

$$-\nabla \cdot (D_{n_i} \delta_n^2 e^{\psi_i} \nabla u_i) = R(\psi_i, u_i, v_i), \qquad (2)$$

$$-\nabla \cdot (D_{p_i} \delta_p^2 e^{-\psi_i} \nabla v_i) = -R(\psi_i, u_i, v_i). \qquad (3)$$

For background on the D-D model the reader is referred to [1,10,14,15,19,20]. In (1)–(3) we use data: the net doping profile N_T, a given expression for the electron-hole pair generation and recombination R, electrical permittivity ϵ, and electron and hole diffusivities D_n, D_p. Also, η is another scaling parameter [7].

The model (1)–(3) is completed with external boundary conditions; we impose Dirichlet conditions for the potential and recombination-velocity (Robin type) conditions for electron and hole densities. To this we add the TEM transmission conditions on the interface [9]

$$[\psi]_\Gamma = \psi_\triangle, \qquad \left[\epsilon \frac{\partial \psi}{\partial \nu}\right]_\Gamma = 0, \qquad (4)$$

$$J_{n_1} = a_2^n (e^\psi u)_2^\Gamma - a_1^n (e^\psi u)_1^\Gamma, \qquad [J_n]_\Gamma = 0, \qquad (5)$$

$$J_{p_1} = a_1^p (e^{-\psi} v)_1^\Gamma - a_2^p (e^{-\psi} v)_2^\Gamma, \qquad [J_p]_\Gamma = 0. \qquad (6)$$

Here J_n and J_p are the electron and hole currents

$$J_n = D_n \delta_n^2 e^\psi \nabla u, \qquad (7)$$

$$J_p = D_p \delta_p^2 e^{-\psi} \nabla v. \qquad (8)$$

Also, a_i^n and a_i^p are constants dependent on material properties and temperature, and $\psi_\triangle$ is a jump discontinuity in the electrostatic potential. These can be determined by a DFT calculation, see Fig. 1 for illustration.

The model (1)–(6) must be discretized. Here we use simple finite difference formulation following [14,20] with N nodal unknowns; we skip details for brevity. In what follows we refer to ψ, u, v, n, p meaning their discrete counterparts.

2.2   Density Functional Theory for Atomic Scale

Heterojunction parameters $a_i^n, a_i^p, \psi_\triangle$ in (4)–(6) are determined by quantum mechanics of electrons. The Schrödinger equation solved for wave function Ψ is fundamental for quantum behavior, but the problem of interacting N electrons is computationally intractable for large N.

DFT [4,5] provides an efficient method of determining material properties from first principles by shifting focus from wave functions Ψ to electron density, $n(r)$. The density sought in DFT is a function in $\mathbb{R}^3$, while the Schrödinger equation is solved for $\Psi \in \mathbb{C}^{3N}$. Finding n is possible via application of the

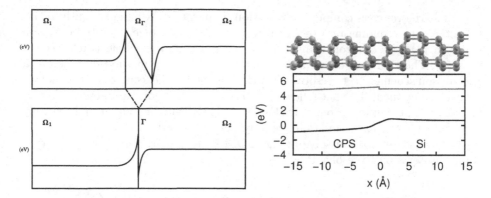

Fig. 1. Left top: schematic plot of potential across 1D interface region for Structure 1 (actual simulation in Fig. 2). Left bottom: schematic plot of potential with idealized heterojunction interface. Right top: interface atomic structure. Right bottom: smoothed local pseudopotential from the DFT calculation (black), and valence band jump construction (red), which determines a_i^n, a_i^p, and $\psi_\triangle$ [7] (Color figure online)

theory of Hohenberg and Kohn to a minimization problem in n, and is equivalent to the solution of the Schrödinger equation for the ground state. However, an energy functional $F[n]$ needed in the minimization principle in DFT is unknown, and DFT requires approximations to $F[n]$. The Kohn–Sham equations provide a basis for these approximations, and their solution can be found iteratively [4,5].

DFT is a widely used, low cost, first principles method which solves the zero temperature, zero current ground state of a system [4,5]. The local pseudopotential calculated by DFT is continuous at an interface (see Fig. 1), and can be used with known material properties to obtain the change in the continuous electrostatic potential ψ occurring close to a heterojunction. The potential jump (offset) $\psi_\triangle$ is a ground state property of the heterojunction structure, and DFT solution in the heterojunction region provides the data needed for TEM.

For the needs of this paper, we perform DFT calculations using the VASP code [11], with exchange-correlation treated using the Generalized Gradient Approximation and a $+U$ Hubbard term ($U = 6$ eV) for the Cu-d orbitals [4,5,7].

3   Domain Decomposition for Continuum Model

The procedure to solve (1)–(6) numerically is a set of nested iterations, with three levels of nesting.

First, when solving (1)–(6), we employ the usual Gummel Map [10,18], an iterative decoupling technique within which we solve each component equation of (1)–(3) independently. Note that each equation is still nonlinear in its primary variable, thus we must use Newton's iteration.

Furthermore, each component equation employs DDM independently to re-solve the corresponding part of TEM. In particular, we solve potential equation (1) with (4), the electron transport (2) with (5), and the hole transport (3) with (6). The DDM we use is an iterative substructuring method designed as a Richardson scheme [17] to resolve the TEM, defined and executed independently for each component. In what follows $\theta > 0$ is an acceleration parameter, different for each component equation. Since the DDM algorithm for p equation is entirely analogous to that for n equation, we only define the latter.

Last, each subdomain solve of the DDM is nonlinear, and we use Newton-Raphson iteration to resolve this.

3.1   Domain Decomposition for Potential Equation (1), (4)

Here we seek the interface value of λ with which (1), (4) is equivalent to

$$- \nabla \cdot (\epsilon_1 \nabla \psi_1) = q_1, \quad x \in \Omega_1; \quad \psi_1|_\Gamma = \lambda \tag{9}$$
$$-\nabla \cdot (\epsilon_2 \nabla \psi_2) = q_2, \quad x \in \Omega_2; \quad \psi_2|_\Gamma = \lambda + \psi_\triangle, \tag{10}$$

which requires $\left[\epsilon \frac{\partial \psi}{\partial \nu} \right]_\Gamma = 0$. The algorithm DDP we proposed in [7] is essentially a modification of the Neumann-Neumann algorithm [17].

Algorithm DDP to solve (1), (4): Given $\lambda^{(0)}$, for each $k \geq 0$,

1. Solve (9) and (10) for $\psi_i^{(k+1)}$, $i = 1, 2$.
2. Update λ by

$$\lambda^{(k+1)} = \lambda^{(k)} - \theta_\psi \left[\epsilon \frac{\partial \psi^{(k+1)}}{\partial \nu} \right]_\Gamma$$

3. Continue with (1) unless stopping criterium $\left\| \left[\epsilon \frac{\partial \psi^{(k+1)}}{\partial \nu} \right]_\Gamma \right\|$ holds.

3.2   Domain Decomposition for Continuity Equation (2), (5)

Here we seek to find data λ so that (2), (5) is equivalent to the problem:

$$- \nabla \cdot (D_{n_1} \delta_n^2 e^{\psi_1} \nabla u_1) = R_1, \quad x \in \Omega_1; \quad u_1|_\Gamma = \lambda \tag{11}$$
$$-\nabla \cdot (D_{n_2} \delta_n^2 e^{\psi_2} \nabla u_2) = R_2, \quad x \in \Omega_2; \tag{12}$$

$$u_2|_\Gamma = \frac{a_1^n (e^\psi)_1^\Gamma}{a_2^n (e^\psi)_2^\Gamma} \lambda + \frac{J_{n_1}}{a_2^n \delta^2 (e^\psi)_2^\Gamma} \tag{13}$$

which requires the homogeneous jump condition $[J_n]_\Gamma = 0$.

Algorithm **DDC** proposed in this paper is very different from **DDP** because it proceeds sequentially from domain Ω_1 to domain Ω_2. In addition, while it corrects λ in a manner similar to a Neumann-Neumann algorithm, in (13) it takes advantage of Neumann data from Ω_1 resulting from (11). An appropriate parallel algorithm for **DDC** which uses Neumann rather than Dirichlet data as in (11), (12) was promising for a synthetic example, but it has difficulties with convergence for realistic devices.

Algorithm DDC to solve (2), (5) *or* (3), (6): Given $\lambda^{(0)}$, for each $k \geq 0$,

1. Solve (11) for $u_1^{(k+1)}$ and then solve (12)–(13) for $u_2^{(k+1)}$.
2. Update λ by

$$\lambda^{(k+1)} = \lambda^{(k)} - \theta_n \left[D_n e^{\psi} \frac{\partial u^{(k+1)}}{\partial \nu} \right]_\Gamma$$

3. Continue with (1) unless stopping criterium $\|[J_n]_\Gamma\|$ holds.

While DDP and DDC are motivated by the multiphysics nature of the model, they may be viewed as extensions of Neumann-Neumann iterative substructuring methods to nonhomogeneous jumps and Robin-like transmission conditions. A scalable parallel implementation may be achieved in the future using two-level techniques [17, Sect. 3.3.2].

4   Heterojunction Semiconductor Simulation

Now we present numerical simulation results. Structure 1 is synthetic and solar cell-like, and is made of two hypothetical materials we call L1 and R1. Structure 2 is made of Si and Cu_3PSe_4 (CPS). In Table 2 we give details.

We use DFT to calculate $\psi_\triangle = -0.01$ eV for the $Cu_{0.75}P_{0.25}$-Si interface formed from CPS (001) and the Si (111) surfaces having normally oriented dangling bonds. Next we apply Domain Decomposition and specifically the algorithms **DDP, DDC**; see Fig. 2. For both structures we see the impact of heterojunction

Table 1. Number of iterations at each Gummel Iteration (GI) and parameters θ_n, θ_p for Structure 1 and algorithm **DDC**. **DDP** uses $\theta_\psi^1 = 0.0025$, $\theta_\psi^2 = 0.00025$. Also, we use $\theta_n^2 = 4\mathrm{e}11$, $\theta_p^2 = 1.4$

	DDC u, $\theta_n^1 = 2.5$				DDC v, $\theta_p^1 = 180$			
N	GI 1	GI 2	GI 3	GI 4	GI 1	GI 2	GI 3	GI 4
201	6	2	1	1	5	3	1	1
401	5	2	1	1	8	4	1	1
601	3	2	1	1	8	4	1	1
801	4	2	1	1	8	4	1	1

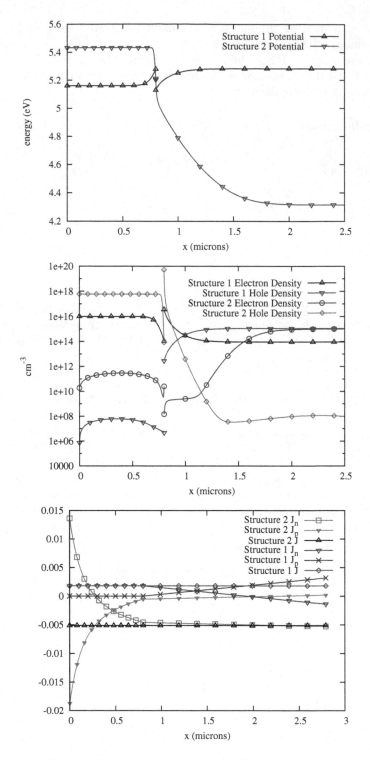

Fig. 2. Simulation of structure 1 and 2 with DDM

Table 2. Material and structure parameters

Property	L1	R1	CPS	Si
Permittivity ϵ	10.0	10.0	15.1 [6]	11.9 [21]
Electron affinity χ (eV)	5	5	4.05	4.05 [21]
Band gap E_g (eV)	1.0	0.5	1.4 [6]	1.12 [21]
Eff. electron density of states $\tilde{N}_C$ (cm^{-3})	5×10^{18}	5×10^{18}	3×10^{19}	2.8×10^{19}
Eff. hole density of states $\tilde{N}_V$ (cm^{-3})	5×10^{18}	5×10^{18}	1.2×10^{18}	1×10^{19}
Dopant charge density $\tilde{N}_T$ (cm^{-3})	1×10^{16}	-1×10^{15}	-6×10^{17} [6]	1×10^{15}
Electron diffusion constant $\tilde{D}_n$ (cm^2/s)	2.0	2.0	2.6	37.6 [21]
Hole diffusion constant $\tilde{D}_p$ (cm^2/s)	1.0	1.0	0.5	12.9 [21]
Constant photogeneration density G (cm^{-3}/s)	1×10^{17}	1×10^{20}	1×10^{21}	1×10^{18}
Direct recombination constant R_{dc} (cm^3/s)	1×10^{-10}	1×10^{-10}	1×10^{-10}	1×10^{-15}
Jump in potential ψ_Δ (eV)	-0.15		-0.01	

Table 3. Efficiency of DDM vs monolithic solvers. Column 4 estimates multicore efficiency

N	Monolithic time (sec)	DDM time (sec)	DDM parallel estimate	Current
501/501	0.5494	1.313	0.8	0.006637
751/751	0.8122	1.4537	0.9	0.006649
1001/1001	1.0231	1.4173	0.9	0.006655
1251/1251	failed	2.1665	1.3	0.0066592

and large jumps of ψ, n, p across the interface. The results are validated with a hard-coded monolithic solver.

As concerns solver's performance, in Table 1 we show that **DDC** is mesh independent, similarly to **DDP** [7]. Furthermore, the choice of θ is crucial. (In forthcoming work [3] we show how θ is determined from analysis of the jump data.)

5 Conclusions

The main contribution reported in this paper advances HPC methodology for solving problems with complex interface physics. We presented DDM for the simulation of charge transport in heterojunction semiconductors. Our method allows the coupling of "black-box" D-D (drift diffusion) solvers in subdomains corresponding to single semiconductor materials. We compared DDM to a monolithic solver, and the results are promising; see Table 3. As usual, DDM approach wins for large N. Also, it works when monolithic solver fails. In the model presented here DFT is used to determine heterojunction parameters but is currently entirely decoupled from D-D solvers in the bulk subdomains. Our approach is a first step towards a true multiscale simulation coupling the atomic and device scales.

At the current stage, the computational complexities of the microscale and macroscale simulations are vastly different. The microscale DFT simulations using VASP solver [11] for electronic structure simulations running on 4 machines with 12 cores with MPI2, take several days to complete. On the other hand, the D-D solver takes less than minutes at worst to complete; see Table 3. Thus, a true coupled multiscale approach is not feasible yet.

More broadly, problems with nonhomogeneous jump conditions across interfaces only begin to be investigated from mathematical and computational point of view. Our DDM approach is a new paradigm that applies elsewhere, e.g., for discrete fracture approximation models where nonhomogeneous jump conditions arise [8, 16].

Acknowledgements. This research was partially supported by the grant NSF-DMS 1035513 grant "SOLAR: Enhanced Photovoltaic Efficiency through Heterojunction assisted Impact Ionization." We would like to thank G. Schneider for useful discussions.

References

1. Bank, R.E., Rose, D.J., Fichtner, W.: Numerical methods for semiconductor device simulation. Siam J. Sci. Stat. Comput. **4**, 416–435 (1983)
2. Chang, K., Kwak, D.: Discontinuous bubble scheme for elliptic problems with jumps in the solution. Comput. Methods Appl. Mech. Eng. **200**, 494–508 (2011)
3. Costa, T., Foster, D.H., Peszynska, M.: Progress in modeling of semiconductor structures with heterojunctions, (2014, submitted)
4. Engel, E., Dreizler, R.M.: Density Functional Theory: An Advanced Course. Springer, Heidelberg (2011)
5. Fiolhais, C., Nogueira, F., Marques, M.A.: A Primer in Density Functional Theory. Springer, Heidelberg (2003)
6. Foster, D.H., Barras, F.L., Vielma, J.M., Schneider, G.: Defect physics and electronic properties of Cu3PSe4 from first principles. Phys. Rev. B. **88**, 195201 (2013)
7. Foster, D.H., Costa, T., Peszynska, M., Schneider, G.: Multiscale modeling of solar cells with interface phenomena. J. Coupled Syst. Multiscale Dyn. **1**, 179–204 (2013)
8. Frih, N., Roberts, J.E., Saada, A.: Modeling fractures as interfaces: A model for Forchheimer fractures. Comput. Geosci. **12**, 91–104 (2008)
9. Horio, K., Yanai, H.: Numerical modeling of heterojunctions including the thermionic emission mechanism at the heterojunction interface. IEEE Trans. Electron Devices **37**, 1093–1098 (1990)
10. Jerome, J.W.: Analysis of Charge Transport : A Mathematical Study of Semiconductor Devices. Springer-Verlag, Berlin (1996)

11. Kresse, G., Furthmüller, J.: Efficient iterative schemes for ab initio total-energy calculations using a plane-wave basis set. Phys. Rev. B **54**(16), 11169–11186 (1996)
12. Lin, P., Shadid, J., Sala, M., Tuminaro, R., Hennigan, G., Hoekstra, R.: Performance of a parallel algebraic multilevel preconditioner for stabilized finite element semiconductor device modeling. J. Comput. Phys. **228**(17), 6250–6267 (2009)
13. Lin, P., Shadid, J.: Towards large-scale multi-socket, multicore parallel simulations: Performance of an MPI-only semiconductor device simulator. J. Comput. Phys. **229**(19), 6804–6818 (2010)
14. Markowich, P.A.: The Stationary Semiconductor Device Equations. Computational Microelectronics. Springer-Verlag, Vienna (1986)
15. Markowich, P.A., Ringhofer, C.A., Schmeiser, C.: Semiconductor Equations. Springer-Verlag, Vienna (1990)
16. Martin, V., Jaffré, J., Roberts, J.E.: Modeling fractures and barriers as interfaces for flow in porous media. SISC **26**, 1667–1691 (2005)
17. Quarteroni, A., Valli, A.: Domain decomposition methods for partial differential equations. Oxford University Press, Oxford (1999)
18. Sacco, R., de Falco, C., Jerome, J.: Quantum-corrected drift-diffusion models: Solution fixed point map and finite element approximation. J. Comput. Phys. **228**, 770–1789 (2009)
19. Seeger, K.: Semiconductor Physics: An Introduction. Springer, Berlin (2010)
20. Selberherr, S.: Analysis and Simulation of Semiconductor Devices. Springer-Verlag, Heidelberg (1984)
21. Sze, S., Ng, K.: Physics of Semiconductor Devices. Wiley-Interscience, Berlin (2006)

A Hybrid Approach for Parallel Transistor-Level Full-Chip Circuit Simulation

Heidi K. Thornquist$^{(\boxtimes)}$ and Sivasankaran Rajamanickam

Sandia National Laboratories, Albuquerque, New Mexico, USA
hkthorn@sandia.gov

Abstract. The computer-aided design (CAD) applications that are fundamental to the electronic design automation industry need to harness the available hardware resources to be able to perform full-chip simulation for modern technology nodes (45 nm and below). We will present a hybrid (MPI+threads) approach for parallel transistor-level transient circuit simulation that achieves scalable performance for some challenging large-scale integrated circuits. This approach focuses on the computationally expensive part of the simulator: the linear system solve. Hybrid versions of two iterative linear solver strategies are presented, one takes advantage of block triangular form structure while the other uses a Schur complement technique. Results indicate up to a 27x improvement in total simulation time on 256 cores.

1 Introduction

Circuit simulation is a technique for checking and verifying the design of electrical and electronic circuits and systems prior to manufacturing and deployment. Circuit simulators use a detailed, transistor-level description of the circuit to achieve relatively accurate performance characteristics. For integrated circuit (IC) design, where probing the behavior of internal signals is extremely difficult, time-domain circuit simulation is an essential, yet expensive, part of the CAD process. Efficient, scalable simulation tools are even more important for simulation of modern technology nodes, where parasitic effects can increase the device count in an integrated circuit by an order of magnitude or more. Traditional transistor-level simulation, made popular by the Berkeley SPICE program [9], becomes impractical beyond tens of thousands of devices, due to the reliance on sparse direct linear solvers [3]. Many attempts have been made to allow for faster, large-scale circuit simulation with Fast-SPICE tools or hierarchical simulators. Unfortunately, the approximations inherent to these simulation approaches can break down under some circumstances, rendering such tools unreliable. With the transition to manycore processors, parallel transistor-level simulation has received more interest from the electronic design automation community.

Sandia National Laboratories is a multi-program laboratory managed and operated by Sandia Corporation, a wholly owned subsidiary of Lockheed Martin Corporation, for the U.S. Department of Energy's National Nuclear Security Administration under contract DE-AC04-94AL85000.

© Springer International Publishing Switzerland 2015
M. Daydé et al. (Eds.): VECPAR 2014, LNCS 8969, pp. 102–111, 2015.
DOI: 10.1007/978-3-319-17353-5_9

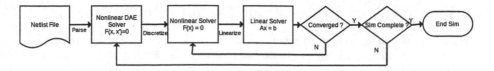

Fig. 1. General circuit simulation flow

Our contributions in this paper are: a *scalable and robust* transistor-level circuit simulation approach for large and challenging problems that uses a hybrid version of the BTF-based preconditioner [11] or a multithreaded Schur complement computation within a hybrid (direct+iterative) solver [10], integration of these linear solvers to a distributed memory parallel circuit simulator, Xyce [6], and a thorough comparison with other serial, multithreaded and distributed solvers for the full simulation. Our hybrid solvers can achieve a parallel speedup of up to *27x* when compared with fastest third party solver. This paper improves upon the MPI-only Schur complement solver that was presented before [1,12].

2   Xyce Framework

Xyce is a transistor-level simulator and adheres to a general flow, as shown in Fig. 1. The circuit is described by a netlist file, which lists the individual components and how they are connected together. This list of devices and interconnectivity is transformed via modified nodal analysis (MNA) into a set of nonlinear differential algebraic equations (DAEs)

$$\frac{dq(x(t))}{dt} + f(x(t)) = b(t), \tag{1}$$

where $x(t) \in \mathbb{R}^N$ is the vector of circuit unknowns, q and f are functions representing the dynamic and static circuit elements (respectively), and $b(t) \in \mathbb{R}^M$ is the input vector.

Time-domain simulation, or transient analysis, solves the nonlinear DAEs (1) implicitly through numerical integration methods, resulting in the nested solver loop in Fig. 1. Any numerical integration method requires the solution to a sequence of nonlinear equations, $F(x) = 0$. Typically, Newton's method is used to solve these nonlinear equations, which generates a sequence of linear systems

$$Ax = b \tag{2}$$

with conductance, $G(t) = \frac{df}{dx}(x(t))$, and capacitance, $C(t) = \frac{dq}{dx}(x(t))$, matrices.

The computational expense in large-scale circuit simulation is dominated by repeatedly solving linear systems of equations, which are at the center of the nested solver loop (Fig. 1). This requires their assembly, which depends upon device evaluations for the whole circuit and the Jacobian matrix and residual vector load. So the dominant computational expense includes both the device loads and the numerical method used to solve the linear systems. For smaller problems, the device loads dominate the total simulation time. However, as the circuit size increases, the linear solve phase starts to dominate.

3   Linear Solver Strategies for Circuit Simulation

Circuit simulation generates some of the most challenging sparse linear systems for both direct and iterative methods because of their heterogeneous structure and ill-conditioning. Direct sparse linear solvers [3] are the industry standard approach because of their robustness, but the are less practical when the linear system has hundreds of thousands of unknowns or more. Iterative methods have more potential for scalability, but their efficacy is deeply reliant upon the preconditioner. Standard preconditioners, such as multigrid, domain decomposition and incomplete factorizations, do not generally work well for circuit problems and have scaling issues, since the number of iterations to solve the linear system will increase with the number of MPI processes. A hybrid approach, combining iterative and direct methods, controls the number of iterations and Schur complement approaches have been shown to work well as preconditioners [1, 2].

This section describes two hybrid linear solvers, one that takes advantage of block triangular form (BTF) structure to create a preconditioned iterative method and another that uses a Schur complement approach to combine iterative and direct techniques. The discussion will introduce a BTF-based preconditioner and Schur complement solver that is effective for circuit problems, then discuss their limitations in a distributed memory only implementation and hybrid techniques that improve scalability. For full-chip circuit simulation, where the number of devices can reach into the millions, these types of scalable hybrid linear solver strategies are imperative.

3.1   KLU

KLU is the only open-source sparse direct linear solver developed for matrices from circuit simulation [3]. At this time, KLU is a serial direct linear solver. It leverages the often reducible property of circuit matrices, permuting them to block triangular form (equivalently the Dulmage Mendelsohn decomposition) before performing an AMD ordering of each diagonal block. The block triangular form is determined in two steps: first a maximum matching permutation to generate a matrix with a zero-free diagonal, and second a topological sort which finds the strongly-connected components of the associated directed graph. Circuit matrices are very sparse and the permutations performed by KLU retain this property, so there are no dense substructures for BLAS to be useful. Because it is an efficient and reliable serial direct solver for this application, it is used as the block diagonal solver for both hybrid linear solvers presented in this paper. For the Schur complement solver, we introduced multithreaded triangular solves in KLU for block columns and leveraged the sparsity structure of the right-hand sides.

3.2   BTF-based Preconditioned Iterative Method

In general, a good preconditioner for the linear system (2) is inexpensive to apply and approximates the coefficient matrix A well. Unfortunately, these two properties often conflict. So, like with many applications, domain-specific structure must be leveraged to develop an effective preconditioner for circuit simulation. The motivation behind the BTF-based preconditioning technique is the observation that the conductance matrix $G(t)$ is often reducible when $t = 0$, and sometimes may be permuted to a block triangular form with small diagonal blocks [3,11].

This linear solution strategy has several steps that result in a block Jacobi preconditioner for the Generalized Minimal Residual (GMRES) method. The first step, *singleton removal*, removes the dense rows and columns that typically result from ideal power supplies and ground nodes, which are common to circuits [2]. These dense rows (or columns) correspond to singleton columns (or singleton rows) with one and only one non-zero entry. These matrix features have the potential to increase communication costs dramatically and can easily be removed from the linear system in a pre-processing (singleton rows) or post-processing (singleton columns) step.

The second step of this linear solution strategy is the permutation of the matrix resulting from the singleton removal to *block lower (or upper) triangular form*. We leverage the fact that circuits often give many small diagonal blocks to use the BTF structure in a novel way. A condensed (block) graph is constructed by contracting all the vertices within each diagonal block into a single vertex. This results in a coarse representation of the original graph that is often much smaller.

The *matrix partitioning* is the third step in this linear solution strategy. We partition the condensed (block) graph into parts that are only loosely connected using hypergraph partitioning [4]. These three steps produce a global matrix reordering, illustrated in Fig. 2, that is used to generate a block Jacobi preconditioner for GMRES. The number of MPI processes determines the number of diagonal blocks in the preconditioner. Since a block Jacobi preconditioner only applies the inverse of these diagonal blocks, no parallel communication is required to perform the factorization and solve, which makes it a scalable preconditioning technique. However, the number of GMRES iterations needed to solve the linear system to a given tolerance will increase with the number of subdomains (MPI ranks). This effectively limits the number of MPI processes that can be used for any given problem. Therefore, it is necessary to take advantage of intra-node parallelism for accelerating local computations. We use the multithreaded (MPI+threads) kernels in the Epetra package of Trilinos for the multithreaded sparse matrix-vector multiplication and vector operations.

Epetra MPI+threads Support. The Epetra package [5] in Trilinos provides fundamental data classes for application codes, like Xyce. Epetra supports piecewise construction of distributed sparse and dense matrices, vectors and graphs, and executes in parallel using either MPI or OpenMP, or both (with OpenMP running underneath each MPI process). Threaded execution via OpenMP is supported for all vector and multivector operations and for sparse matrix multiplication. Work and data distribution are managed dynamically through the OpenMP runtime system. The performance limits of Epetra computations are

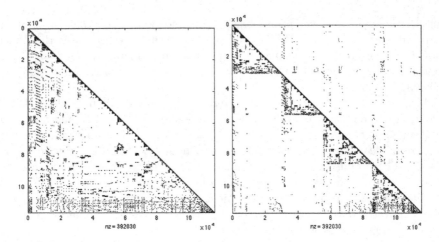

Fig. 2. Example matrix structure after the second (left) and third (right) step of the BTF-based preconditioning method.

almost always determined by the memory system performance of the computational nodes, since the operations-per-memory-reference ratio is very low. As a result–especially on non-uniform memory access (NUMA) architectures–proper data placement is extremely important for obtaining optimal performance. The more closely aligned the memory pages are with the processing core that uses the data most frequently, the better the effective bandwidth we expect to see.

3.3   ShyLU

ShyLU is a hybrid linear solver designed to be a black-box algebraic solver [10]. It is hybrid in both the parallel programming sense - using MPI and threads - and in the mathematical sense - using features from direct and iterative methods. ShyLU was originally designed to be a subdomain solver in a domain decomposition framework within Trilinos [10]. However, it can also be used as a global Schur complement solver, as we do in this paper. We introduce hybrid parallelism in the Schur complement computation of ShyLU for it to be more scalable for large circuits.

Let $Ax = b$ be the system of interest. Suppose A has the form

$$A = \begin{pmatrix} D\ C \\ R\ G \end{pmatrix},\tag{3}$$

where D and G are square and D is non-singular. The Schur complement is $S = G - R * D^{-1} * C$. Solving $Ax = b$,

$$\begin{pmatrix} D\ C \\ R\ G \end{pmatrix} \times \begin{pmatrix} x_1 \\ x_2 \end{pmatrix} = \begin{pmatrix} b_1 \\ b_2 \end{pmatrix},\tag{4}$$

consists of factoring D and solving

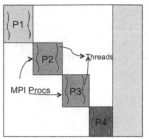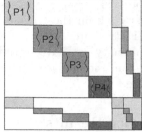

Fig. 3. Graph/Hypergraph based ordering of the sparse linear system for parallelism in ShyLU with unsymmetric ordering (left) and symmetric ordering (right).

1. $Dz = b_1$.
2. $Sx_2 = b_2 - Rz$.
3. $Dx_1 = b_1 - Cx_2$.

ShyLU uses hypergraph partitioning to permute the matrix into the bordered block diagonal form shown in Fig. 3. Each block diagonal in the permuted matrix corresponds to a MPI rank and is factored using a direct solver. An approximate Schur complement is used to compute a preconditioner for an iterative method to solve for the Schur complement. The approximation is computed using either dropping or a probing method for a fixed pattern. We will use the former in this paper.

The first expensive step in ShyLU is the factorization of block diagonals. For all the problems in this paper we use KLU [3] to factor the block diagonals. The second expensive step is computing the approximate Schur complement. Computing the approximate Schur complement is completely local within an MPI rank in ShyLU and is a good candidate for hybrid parallelism. It involves triangular solves in $D^{-1} * C$ computation of the Schur complement and a matrix-vector multiply. While it can also be formulated as a matrix-matrix multiply, ShyLU uses the matrix-vector formulation. The triangular solve in this particular case has multiple right-hand sides, where the right-hand sides are themselves sparse columns of C. We have modified KLU in order to more efficiently compute the approximate Schur complement as described below.

The hybrid Schur approximation uses a triangular solve with multiple right-hand sides to compute a block column of the Schur complement in parallel. KLU's triangular solve was optimized for multiple right-hand sides using vectorization. We have introduced the multithreaded triangular solves for block columns in addition to the existing vectorization. The change is to exploit the sparsity in the right-hand side of the triangular solve, since C is a sparse matrix. This is accomplished by avoiding the additional floating point operations in the triangular solve. The importance of exploiting the sparsity in the triangular solves has been observed before [13]. However, it is also important to exploit the BTF structure in the factorization step for circuit problems. The BTF here is within the direct solver and different from BTF-based preconditioner in Sect. 3.2.

As KLU uses the BTF structure in its factorization and triangular solve we exploit the sparsity within the triangular solve corresponding to the diagonal blocks of the BTF structure. Note that the numeric factorization of KLU is still sequential. In summary, we have introduced a multithreaded triangular solve with block right-hand sides that exploits sparsity in the right-hand side within the BTF structure.

4   Performance Results

This section presents results for the proposed hybrid approaches for parallel simulation of challenging problems. Results presented in this paper are generated using Xyce (post release 5.2.1) and Trilinos(10.10.1). The test machine is a capacity cluster, with 272 compute nodes, where each node has a 2.2 GHz AMD quad socket/quad core processor and 32GB RAM. Xyce, Trilinos, SuperLU v4.3 [7], and SuperLU_DIST v2.5 [8] are compiled using Intel 11.1 compilers, where the Intel MKL 11.1 provides the BLAS/LAPACK and PARDISO libraries. The integrated circuits selected for these tests are of varying scales and the simulation challenges even the sparse direct linear solvers. Table 1 partially describes the circuits used in the numerical experiments. All three of these are proprietary application-specific integrated circuits (ASICs).

Table 1. Circuits: matrix size(N), capacitors(C), MOSFETs(M), resistors(R), voltage sources(V), diodes (D).

Circuit	N	C	M	R	V	D
ckt1	116247	52552	69085	76079	137	0
ckt2	688838	93	222481	176	75	291761
ckt3	1944792	400234	211486	795827	36100	199992

4.1   Sparse Direct Solver Performance

The circuits selected for these experiments are small enough that sparse direct solvers are still practical. Therefore, we will start by looking at performance results from the state-of-the-art sparse solvers KLU, PARDISO, SuperLU, and SuperLU_DIST in Table 2. KLU and SuperLU are serial, while the parallel codes are run on 16 cores. The results illustrate the difficulty of these simulations. The only solver that consistently enables a transient simulation to complete is KLU. PARDISO performs well, beating KLU on ckt1, and ckt2, but fails to complete the DC analysis phase for ckt3. SuperLU_DIST is designed for problems with supernodal structure which is not present in any of our test cases. We believe its static pivoting choice causes the problems in completing the simulation. Note that these are representative simulations. Real simulations could be order of magnitude longer. We will compare our approaches to KLU in the rest of the paper.

Table 2. Total linear solve time (sec.) for various sparse direct solvers; "-" indicates simulation failed to complete; parentheses contain the # of threads/MPI processes.

	ckt1	ckt2	ckt3
KLU	9381.3	7060.8	**14222.7**
PARDISO (16)	**715.0**	**6690.5**	-
SuperLU	-	-	72176.8
SuperLU_Dist (16)	-	-	-

Table 3. Comparison of total linear solve time for ckt1 when the number of MPI processes per node (ppn) is varied with one thread.

MPI processes	4 ppn	8 ppn	16 ppn
4	253.8	-	-
8	125.9	136.7	-
16	77.5	83.5	94.7
32	74.8	84.5	100.4

4.2 Hybrid Linear Solver Performance

The linear solver dominates the simulation time for circuits ckt1 and ckt3 ($> 90\,\%$), while it is about half the total simulation time for ckt2. Thus, these circuits are the most useful in determining the effectiveness of the hybrid linear solvers. From past experience [11,12] with MPI-only simulations the BTF-based preconditioner will be used for ckt1 and ckt2. ShyLU will provide a much more robust solver strategy for ckt3, as it has a large irreducible conductance matrix. The BTF-based preconditioner is paired with the Epetra MPI+threads implementation in these experiments. For both linear solver strategies, KLU is chosen as the block diagonal solver.

A scaling study is performed using ckt1 to determine the number of MPI processes per node resulting in the best linear solver performance for the BTF-based preconditioner. The simulations are run for various numbers of MPI processes per node (ppn), 4, 8, or 16, from 1 to 8 nodes. The results in Table 3 indicate that using 4 processes per node enables the simulator to achieve a faster linear solver time than with 8 or 16 processes per node. In fact, both 8 and 16 processes per node achieve their peak linear solver performance with 16 MPI processes. The number of MPI processes is the same as the number of subdomains, as a result the preconditioner is more effective with fewer MPI ranks. The hybrid approach allows us to use the available cores effectively with fewer subdomains and a better preconditioner. For comparison the 4 ppn configuration (with 32 MPI processes) results in 25 % speedup over 16 ppn configuration and results in a 9.5x speedup over PARDISO's time (Table 2).

A larger scaling study is performed for ckt2, which generates a much larger linear system (Table 1). The strong scaling results for the linear solver time and

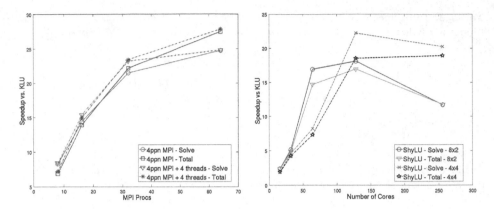

Fig. 4. Speedup of Xyce's simulation time and linear solve time for strong scaling experiments with different configurations of MPI tasks x threads per node using BTF (ckt2, left) and ShyLU (ckt3, right).

total simulation time are presented in Fig. 4(left). They indicate that the BTF-based preconditioning technique achieves scalable performance up to 64 MPI processes, or 256 cores. At 32 MPI processes, the Epetra MPI+threads implementation provides an additional $2x$ speedup over KLU. Overall the total simulation time is $27x$ faster on 64 MPI processes for this circuit.

The largest and most challenging test case is ckt3. For single node runs, only KLU and SuperLU can finish this simulation and take 4 hours and 20 hours, respectively, for a transient time of 1 ns. Typical simulations require a transient time of 20 ns or longer, which would result in the simulation taking more than a week. The hybrid linear solver - ShyLU - is used to simulate ckt3 on up to 256 cores. The results, shown in Fig. 4 (right), indicate a significant speedup in the linear solve time, up to $22x$, and total simulation time, up to $19x$.

The importance of hybrid parallelism is illustrated in Fig. 4(right) by comparing different MPI tasks x threads per node 8×2 vs 4×4). The 4×4 configuration clearly wins in the larger core counts (128 and 256). At 256 cores, the number of MPI processes for the 8x2 configuration is 128, which is equal to the number of parts for ShyLU's partitioning (see Fig. 3). Experiments indicate that the ideal part size for ckt3 is 64. Using 128 parts results in a matrix that is imbalanced in the direct factorization phase. This results in the 8x2 case having the best performance with 128 cores (or 64 MPI processes). The 4x4 case achieves its best performance with 64 MPI processes for 256 cores. Thus, using four threads instead of two threads allowed ShyLU to speedup the simulation by an additional $4x$ (from $15x$ to $19x$) for higher core counts. An MPI-only version [12] peaks at just 64 cores because of the limited inherent parallelism in the linear problem.

5   Conclusion

This paper proposes hybrid techniques for enabling fast, parallel circuit simulation of large-scale ASICs on modern multicore platforms. These techniques are implemented in a MPI-based parallel circuit simulator, Xyce, and tested on a set of challenging integrated circuits. The results presented indicate that the hybrid linear solver strategies provide a significant improvement to Xyce's scalability. For a 500 K device ASIC, the BTF-based preconditioned iterative method enables Xyce to achieve a $27x$ speedup on 256 cores. While, ShyLU, the Schur complement based hybrid linear solver, enables Xyce to achieve a $19x$ speedup on 256 cores for a 2 million device ASIC.

References

1. Baker, C., Boman, E., Heroux, M., Keiter, E., Rajamanickam, S., Schiek, R., Thornquist, H.: Enabling next-generation parallel circuit simulation with trilinos. In: Alexander, M., D'Ambra, P., Belloum, A., Bosilca, G., Cannataro, M., Danelutto, M., Di Martino, B., Gerndt, M., Jeannot, E., Namyst, R., Roman, J., Scott, S.L., Traff, J.L., Vallée, G., Weidendorfer, J. (eds.) Euro-Par 2011, Part I. LNCS, vol. 7155, pp. 315–323. Springer, Heidelberg (2012)
2. Basermann, A., Jaekel, U., Nordhausen, M., Hachiya, K.: Parallel iterative solvers for sparse linear systems in circuit simulation. Future Gener. Comp. Sys. **21**(8), 1275–1284 (2005)
3. Davis, T.A., Natarajan, E.P.: Algorithm 907: KLU, a direct sparse solver for circuit simulation problems. ACM Trans. Math. Softw. **37**(3), 36:1–36:17 (2010)
4. Devine, K.D., Boman, E.G., Heaphy, R.T., Bisseling, R.H., Çatalyürek, Ü.V.: Parallel hypergraph partitioning for scientific computing. In: Proceedings of 20th International Parallel and Distributed Processing Symposium (IPDPS'06). IEEE (2006)
5. Heroux, M.: Epetra performance optimization guide. Technical report SAND2005-1668, Sandia National Laboratories, March 2009
6. Keiter, E.R., Thornquist, H.K., Hoekstra, R.J., Russo, T.V., Schiek, R.L., Rankin, E.L.: Parallel transistor-level circuit simulation. In: Li, P., Silveira, L.M., Feldmann, P. (eds.) Adv. Simul. Verification Electron. Biol. Syst., pp. 1–21. Springer, Dordrecht (2011)
7. Li, X.S.: An overview of superLU: algorithms, implementation, and user interface. ACM Trans. Math. Softw. **31**, 302–325 (2005)
8. Li, X.S., Demmel, J.W.: SuperLU_DIST: a scalable distributed-memory sparse direct solver for unsymmetric linear systems. ACM Trans. Math. Soft. **29**(2), 110–140 (2003)
9. Nagel, L.W.: SPICE2, a computer program to simulate semiconductor circuits. Technical report ERL-M250, University of California, Berkeley, 1975
10. Rajamanickam, S., Boman, E.G., Heroux, M.A.: ShyLU: A hybrid-hybrid solver for multicore platforms. In: IEEE 26th International Parallel Distributed Processing Symposium (IPDPS), pp. 631–643, May 2012
11. Thornquist, H.K., Keiter, E.R., Hoekstra, R.J., Day, D.M., Boman, E.G.: A parallel preconditioning strategy for efficient transistor-level circuit simulation. In: Proceedings of the 2009 (ICCAD). ACM, November 2009
12. Thornquist, H.K., Keiter, E.R., Rajamanickam, S.: Electrical modeling and simulation for stockpile stewardship. XRDS **19**(3), 18–22 (2013)
13. Yamazaki, I., Li, X.S.: On techniques to improve robustness and scalability of a parallel hybrid linear solver. In: Palma, J.M.L.M., Daydé, M., Marques, O., Lopes, J.C. (eds.) VECPAR 2010. LNCS, vol. 6449, pp. 421–434. Springer, Heidelberg (2011)

Numerical Algorithms

Self-adaptive Multiprecision Preconditioners
on Multicore and Manycore Architectures

Hartwig Anzt[1]([✉]), Dimitar Lukarski[2], Stanimire Tomov[1], and Jack Dongarra[1]

[1] Innovative Computing Lab, University of Tennessee, Knoxville, USA
hanzt@icl.utk.edu, tomov@cs.utk.edu, dongarra@eecs.utk.edu
[2] Department of Information Technology, Uppsala University, Uppsala, Sweden
dimitar.lukarski@it.uu.se

Abstract. Based on the premise that preconditioners needed for scientific computing are not only required to be robust in the numerical sense, but also scalable for up to thousands of light-weight cores, we argue that this two-fold goal is achieved for the recently developed self-adaptive multi-elimination preconditioner. For this purpose, we revise the underlying idea and analyze the performance of implementations realized in the PARALUTION and MAGMA open-source software libraries on GPU architectures (using either CUDA or OpenCL), Intel's Many Integrated Core Architecture, and Intel's Sandy Bridge processor. The comparison with other well-established preconditioners like multi-coloured Gauss-Seidel, ILU(0) and multi-colored ILU(0), shows that the twofold goal of a numerically stable cross-platform performant algorithm is achieved.

1   Introduction

When solving sparse linear systems iteratively, e.g., via Krylov subspace solvers, using preconditioners is often the key to reducing the time needed to obtain a sufficiently accurate solution approximation. For this reason, significant effort is spent on the development of efficient preconditioners, usually optimized for one particular problem. However, the theoretical derivation of methods improving the convergence characteristics is often not sufficient, as the algorithms have to be implemented and parallelized on the respectively used hardware platform. The use of accelerator technology, like graphics processing units (GPUs) or Intel Xeon Phi Coprocessors (known also as Many Integrated Core Architectures, or MIC), in scientific computing centers requires a combination of deep mathematical background knowledge and software engineering skills to develop suitable methods. The challenge is to combine the robustness and efficiency of the preconditioner scheme with the scalability of the implementation up to hundreds and thousands of light-weight computing cores. The non-uniformity of the high-performance computing landscape introduces additional complexity to this endeavor, and complex sparse linear algebra algorithms that are designed to efficiently exploit one specific architecture often fail to leverage the computing power of other technologies. In this paper we show that, for the recently developed self-adapting and multi-precision preconditioner [10], the two-fold goal of deriving a

© Springer International Publishing Switzerland 2015
M. Daydé et al. (Eds.): VECPAR 2014, LNCS 8969, pp. 115–123, 2015.
DOI: 10.1007/978-3-319-17353-5_10

numerically robust method featuring cross-platform scalability is achieved. While the use of different floating point precision formats, and the combination of dense and sparse linear algebra operations, may challenge cross-platform suitability, we show that the self-adaptive mixed precision multi-elimination method can efficiently exploit different hardware architectures and is highly competitive to some of the most commonly used preconditioners. While the implementation of the algorithm is realized using the PARALUTION [8] and MAGMA [5] open source software libraries, both known to be able to efficiently exploit the computing power of accelerators, the hardware systems used in our experiments represent some of the most popular technologies used in current HPC platforms. The rest of the paper is structured as follows. First, we provide some details about the self-adaptive mixed precision multi-elimination preconditioner and the implementation we use. Next, we summarize some characteristics of the many-core accelerators we target in our experiments and introduce the test matrices we use for benchmarking. We then evaluate the performance of the mixed precision multi-elimination preconditioner, embedded in a Conjugate Gradient solver on the different hardware systems, and compare against other well-known preconditioners. Finally, we summarize some key findings and provide ideas for future research.

2   Self-adaptive Multi-elimination Preconditioner

Among the most popular preconditioners is the class based on the incomplete LU factorization (ILU) [15]. Although using ILU without fill-ins can lead to appealing convergence improvement to the top-level iterative method, it may also fail due to its rather rough approximation properties, e.g., when solving linear systems arising from complex applications like computational fluid dynamics [14]. To enhance the accuracy of the preconditioner, one can allow for additional fill-in in the preconditioning matrix, resulting in the (ILU(m) scheme, see [15]). Additional fill-in usually reduces the amount of parallelism in ILU(m) compared to ILU(0), but there are a number of techniques designed to retain it, such as the level-scheduling techniques [11,15] or the multi-coloring algorithms for the ILU factorization with levels based on the power(q)-pattern method [9]. Another workaround is given by the idea of multi-elimination [14,16], which is based on successive independent set coloring [6]. The motivation is that in a step of the Gaussian elimination, there usually exists a large set of rows that can be processed in parallel. This set is called the independent set. For multi-elimination, the idea is to determine this set, and then eliminate the unknowns in the respective rows simultaneously, to obtain a smaller reduced system. To control the sparsity of the factors, multi-elimination uses an approximate reduction based on a standard threshold strategy. Recursively applying this step, one obtains a sequence of linear systems with decreasing dimension and increasing fill-in. On the lowest level, the system must be solved, e.g., either by an iterative method, or by a direct solver based on an LU factorization. Recently, a multi-elimination preconditioner, using an adaptive level depth in combination with a direct solver based on LU factorization, was proposed in [10]. The advantage of this approach is that the once computed LU factorization for the bottom-level system can be reused in every iteration step, and the ability to utilize a lower precision format in the triangular solves allows for leveraging the often superior single precision performance of accelerators like GPUs. While we only shortly recall the central ideas of the multi-elimination concept, a detailed derivation can be found in [14]. The underlying scheme is to use permutations P to bring the original matrix A, of the system $Ax = b$ that we want to solve, into the form

$$PAP^T \equiv \begin{pmatrix} D & F \\ E & C \end{pmatrix},$$

where D is preferably a diagonal or at least an easy to invert matrix, so that

$$PAP^T \equiv \begin{pmatrix} D & F \\ E & C \end{pmatrix} = \begin{pmatrix} I & 0 \\ ED^{-1} & I \end{pmatrix} \times \begin{pmatrix} D & F \\ 0 & \hat{A} \end{pmatrix} \text{ with } \hat{A} = C - ED^{-1}F \quad (1)$$

is easy to compute [10]. One way to achieve this is by using an independent set ordering [6,7,13,18], where non-adjacent unknowns of the original matrix A are determined. Recursively applying this idea and using some threshold strategy to control the fill-in one obtains a sequence of successively smaller problems. To control the increasing density of $\hat{A}$, we propose a self-adapting algorithm which determines an appropriate sequence depth and a fill-in threshold based on the average of all non-zero entries of $\hat{A}$. In the iteration phase (see Fig. 1) the sequence of transformations must also be applied to the right-hand side and to the solution approximation. This is achieved by applying the decomposition [14]

$$x := \begin{pmatrix} \hat{y} \\ \hat{x} \end{pmatrix}$$

and computing, according to the partitioning in (1), the forward sweep as [14]: $\hat{x} := \hat{x} - ED^{-1}y$. Consequently, backward solution for y_j hence becomes $y := D^{-1}(y - F\hat{x})$. On the lowest level the linear system must be solved, either again via an iterative method, or, like suggested in [10] via triangular solves (in single precision), using a beforehand computed factorization. Algorithmic details, as well as a comparison between single and double precision triangular solves, can be found in [10]. As the level-depth is not preset but determined during the recursive factorization sequence using thresholds for drop-off and the direct solve size, the algorithm is self-adapting to a specific problem.

3 Hardware and Software Issues

Target Platforms. The trend to introduce accelerator technology into high performance computers is reflected in the top-ranked computer systems in both the performance-oriented TOP500, and the resource-aware Green500 list (see [3] and [1], respectively). While in recent years the usage of GPUs from different vendors drew attention, Intel responded with the development of the MIC

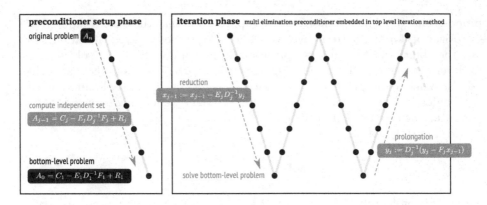

Fig. 1. Visualization of the multi-elimination scheme denoting the system matrix of the original problem A_n and a sequence of successively smaller problems down to the bottom-level system matrix A_0.

architecture (and in the November 2013 Top500 list, the number one ranked supercomputer was based on MICs). For the future, even more diversity may be expected as precise plans for building systems based on the low-power ARM technology already exist [2]. Despite attempts like OpenCL [17] and OpenACC [4], unfortunately no cross-platform language that allows for efficient usage of the different accelerator architectures currently exists. Therefore, it usually remains a burden to the software developer to implement algorithms for a specific target architecture using a suitable programming language for the respective hardware. Especially for numerical linear algebra algorithms, the algorithm-specific properties often make the implementation on different architectures challenging. To determine whether the challenge of deriving a cross-platform performant method is achieved for the recently developed self-adaptive multi-elimination preconditioner we introduced in the last section, we benchmark it on different multi- and many-core systems listed along with some key characteristics in Table 1.

The implementation of the preconditioner, as well as the other methods we compare against in Sect. 4, is realized using the PARALUTION [8] (version 0.4.0) and MAGMA [5] (version 1.4) open-source software libraries. The framework and the CPU solver implementations are based on C/C++, while the

Table 1. Key characteristics of the target architectures.

Acronym	System	Performance Peak	Memory	Bandwidth
ISB	2 × Intel Xeon E5-2670 (Sandy Bridge)	333 GFlop/s	65 GB	2 × 25.5 GB/s
K40	NVIDIA Tesla K40c	1,682 GFlop/s	12 GB	288 GB/s
AMD	AMD Radeon HD 7970 (Tahiti)	947 GFlop/s	3 GB	264 GB/s
MIC	Intel Xeon Phi 7110P	1,238 GFlop/s	16 GB	352 GB/s

Table 2. Description and properties of the test matrices.

Matrix	#nonzeros (nnz)	Size (n)	nnz/n
APACHE	4,817,870	715,176	6.74
ECOLOGY	4,995,991	999,999	5.00
G2_ CIRC	726,674	150,102	4.83
G3_ CIRC	7,660,826	1,585,478	4.83
LAPLACE	4,996,000	1,000,000	4.99
OFFSHORE	4,242,673	259,789	16.33
STocF	21,005,389	1,465,137	14.34
THERMAL	8,580,313	1,228,045	6.99

GPU-accelerated implementations use either CUDA [12] version 5.5 for the NVIDIA GPUs, or OpenCL [17], version 1.2 and clAmdBlas 1.11.314 for AMD GPUs. The MIC implementation, similar to GPU's, treats the MIC as an accelerator/coprocessor and is based on OpenMP and the BLAS operations provided in Intel's MKL 11.0, update 5.

Solver Parameters. All experiments solve the linear system $Ax = b$ where we set the initial right-hand-side to $b \equiv 1$, start with the initial guess $x \equiv 0$ and run the iteration process until we achieve a relative residual accuracy of $1e - 6$. In the preprocessing phase of the multi-elimination, the identification of an independent set via a graph algorithm is handled by the CPU of the host system; the factorization process itself, including the permutation and the generation of the lower-level systems via a sparse matrix-matrix multiplication is implemented on the GPU.

Test Matrices. For the experiments, we use a set of symmetric, positive definite (SPD) test matrices taken either from the University of Florida matrix collection (UFMC)[1], Matrix Market[2], or generated as finite difference discretization (LAPLACE). The test matrices are listed along with some key characteristics in Table 2. Although we target only SPD systems, we use ME-ILU factorization due to the fact that the IC requires non-zero diagonal elements. Positive diagonal entries for the IC can be obtained with non-symmetric permutation. This is not applicable because the multi-elimination uses maximal independent set (MIS) algorithm which produces a symmetric permutation.

4 Performance on Emerging Hardware Architectures

In Table 3 we list the runtime of the iteration phase of the self-adaptive mixed precision multi-elimination implementation on different hardware platforms. With the number of iterations constant over the architectures, the performance is determined by the available computing power and the efficiency of the programming

---

[1] UFMC; see http://www.cise.ufl.edu/research/sparse/matrices/.
[2] see http://math.nist.gov/MatrixMarket/.

Table 3. Iteration count and runtime (in seconds) of the Conjugate Gradient solver preconditioned with the self-adaptive mixed precision multi-elimination (MPME) preconditioner for different test matrices and hardware architectures.

Matrix	#iters	ISB	K40	AMD	MIC
APACHE	293	15.43	3.04	15.46	8.35
ECOLOGY	799	63.57	10.98	-	23.17
G2_ CIRC	359	11.11	2.49	15.99	5.29
G3_ CIRC	512	20.30	5.42	18.23	16.18
LAPLACE	338	9.13	3.41	14.31	10.01
OFFSHORE	1314	93.67	9.59	58.23	14.88
StocF	4388	178.56	52.06	-	115.05
THERMAL	916	57.41	13.93	59.20	35.73

model to exploit it. The results reveal that the best performance is achieved using the CUDA implementation on the NVIDIA Kepler architecture. The MIC implementation fails to achieve the K40 performance, but is in most cases superior to ISB . Switching from the CPU to the OpenCL programming model on the AMD platform accelerates the solver execution only for some problems, and even for those, the performance is significantly lower than on the NVIDIA GPU. Furthermore,

Table 4. Iteration count and runtime (in seconds) of the unpreconditioned Conjugate Gradient solver (labelled CG) and the implementations using a multi-coloured Gauss-Seidel preconditioner (labelled MCGS-CG), a ILU-0 and a multi-colored ILU-0 preconditioner (labelled ILU0-CG and MCILU0-C, respectively) for different test matrices and hardware architectures.

	CG					MCGS-CG			
matrix	#iters	ISB	K40	AMD	MIC	#iters	ISB	K40	AMD	MIC
APACHE	3971	16.60	5.02	15.39	10.12	1677	15.45	5.22	14.90	12.56
ECOLOGY	5392	24.17	8.20	19.74	15.35	2784	27.50	8.94	19.83	19.17
G2_CIRC	8911	5.32	3.76	13.83	10.27	907	1.61	1.29	5.47	4.80
G3_CIRC	12658	107.56	29.67	60.86	77.15	1329	28.55	9.01	15.82	22.35
LAPLACE	1633	8.03	2.53	5.87	4.73	817	9.00	2.63	5.76	5.60
OFFSHORE		– no convergence –				628	10.19	4.92	15.03	16.79
StocF		– no convergence –				66042	2200.46	1187.59	2679.99	2678.97
THERMAL	4589	53.06	9.63	28.44	30.52	2151	39.27	18.33	36.28	52.68

	ILU0-CG					MCILU0-CG			
matrix	#iters	ISB	K40	AMD	MIC	#iters	ISB	K40	AMD	MIC
APACHE	643	25.56	9.63	-	-	1438	16.37	4.07	11.55	9.80
ECOLOGY	1700	74.86	64.03	-	-	2854	38.18	8.35	18.72	18.09
G2_CIRC	481	3.28	6.15	-	-	857	1.54	1.12	4.37	3.94
G3_CIRC	680	51.73	33.77	-	-	1242	25.06	7.71	13.20	19.24
LAPLACE	537	23.18	19.30	-	-	817	8.49	2.37	5.29	5.29
OFFSHORE	365	13.83	23.22	-	-	487	6.88	3.57	8.54	11.72
StocF	2364	368.36	158.37	-	-	16740	544.91	290.38	634.35	624.89
THERMAL	1945	188.58	54.13	-	-	2095	42.33	16.79	30.57	49.53

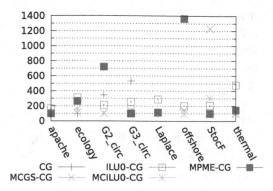

Fig. 2. Runtime of the different implementations normalized to the runtime of the best method on the Intel Sandy Bridge CPU.

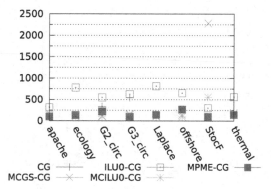

Fig. 3. Runtime of the different implementations normalized to the runtime of the best method on the Kepler K40 GPU.

the smaller memory size of the AMD architecture prevents it from handling all problems. While this performance drop may suggest that mixed precision multi-elimination is not suitable for OpenCL on AMD architectures, the runtime results for other preconditioner choices in Table 4 indicate that this behavior is not a singularity. None of the implementations using the OpenCL-AMD framework achieves performance competitive to the CUDA results on the Kepler K40. Finally, we want a comparison between the different preconditioners. In Figs. (2, 3, 4, and 5) we compare for the different architectures the performance of the plain CG with the implementations preconditioned by multi-colored Gauss-Seidel, ILU(0), multi-colored ILU(0), and the developed mixed precision multi-elimination with the runtime normalized to the respective best implementation.

From the results we can determine that the mixed precision multi-elimination is not suitable for the small G2_ CIRC problem, but reduces the runtime significantly in the STOCF case. Furthermore, it shows very good performance on the Kepler K40 GPU and Intel's Manycore Architecture. Overall, the developed self-adaptive preconditioner is competitive compared to the well-established methods.

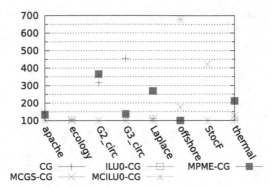

Fig. 4. Runtime of the different implementations normalized to the runtime of the best method on the AMD Radeon 7900.

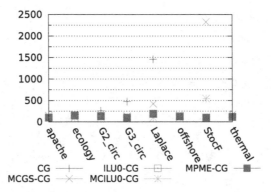

Fig. 5. Runtime of the different implementations normalized to the runtime of the best method on Intel's Many Integrated Core Architecture.

5   Summary and Future Research

In this paper we have analyzed the cross-platform suitability of the recently developed mixed precision multi-elimination preconditioner using self-adaptive level depth. We have analyzed the method's performance characteristics using different hardware platforms and compared the runtime with some of the most popular preconditioners. The numerical robustness combined with platform-independent scalability makes the method a competitive candidate when choosing a preconditioner for solving linear problems in scientific computing. Future research will target the question of how to leverage the computing power of platforms equipped with multiple, not necessarily uniform, accelerators.

Acknowledgments. This work has been supported by the Linnaeus centre of excellence UPMARC, Uppsala Programming for Multicore Architectures Research Center, the Russian Scientific Fund (Agreement N14-11-00190), DOE grant #DE-SC0010042, NVIDIA, and the NSF grant # ACI-1339822.

References

1. The green 500 list. http://www.green500.org/
2. The Mont Blanc Project. http://montblanc-project.eu
3. The top 500 list. http://www.top.org/
4. O. Corp. Openacc 2.0a spec - revised august 2013, June 2013
5. I. C. Lab. Software distribution of MAGMA version 1.4 (2013). http://icl.cs.utk. edu/magma/
6. Leuze, M.R.: Independent set orderings for parallel matrix factorization by gaussian elimination. Parallel Comput. **10**(2), 177–191 (1989)
7. Luby, M.: A simple parallel algorithm for the maximal independent set problem. SIAM J. Comput. **15**(4), 1036–1053 (1986)
8. Lukarski, D.: PARALUTION project. http://www.paralution.com/
9. Lukarski, D.: Parallel Sparse Linear Algebra for Multi-core and Many-core Platforms - Parallel Solvers and Preconditioners. Ph.D. thesis, Karlsruhe Institute of Technology (KIT), Germany (2012)
10. Lukarski, D., Anzt, H., Tomov, S., Dongarra, J.: Multi-Elimination ILU Preconditioners on GPUs. Technical report UT-CS-14-723, Innovative Computing Laboratory, University of Tennessee (2014)
11. Naumov, M.: Parallel solution of sparse triangular linear systems in the preconditioned iterative methods on the GPU. Technical report, NVIDIA (2011)
12. NVIDIA Corporation. NVIDIA CUDA Compute Unified Device Architecture Programming Guide, 2.3.1 edition, August 2009
13. Robson, J.: Algorithms for maximum independent sets. J. Algorithms **7**(3), 425–440 (1986)
14. Saad, Y.: Ilum: a multi-elimination ilu preconditioner for general sparse matrices. SIAM J. Sci. Comput **17**, 830–847 (1999)
15. Saad, Y.: Iterative Methods for Sparse Linear Systems. Society for Industrial and Applied Mathematics, Philadelphia (2003)
16. Saad, Y., Zhang, J.: Bilum: block versions of multi-elimination and multi-level ilu preconditioner for general sparse linear systems. SIAM J. Sci. Comput. **20**, 2103–2121 (1997)
17. Stone, J.E., Gohara, D., Shi, G.: Opencl: a parallel programming standard for heterogeneous computing systems. IEEE Des. Test **12**(3), 66–73 (2010)
18. Yao, L., Cao, W., Li, Z., Wang, Y., Wang, Z.: An improved independent set ordering algorithm for solving large-scale sparse linear systems. In: 2010 2nd International Conference on Intelligent Human-Machine Systems and Cybernetics (IHMSC), vol. 1, pp. 178–181 (2010)

Fault Tolerance in an Inner-Outer Solver: A GVR-Enabled Case Study

Ziming Zheng[1], Andrew A. Chien[1]([⊠]), and Keita Teranishi[2]

[1] University of Chicago, Chicago, IL 60637, USA
achien@cs.uchicago.edu
[2] Sandia National Laboratories, Livermore, CA 94551, USA

Abstract. Resilience is a major challenge for large-scale systems. It is particularly important for iterative linear solvers, since they take much of the time of many scientific applications. We show that single bit flip errors in the Flexible GMRES iterative linear solver can lead to high computational overhead or even failure to converge to the right answer. Informed by these results, we design and evaluate several strategies for fault tolerance in both inner and outer solvers appropriate across a range of error rates. We implement them, extending Trilinos' solver library with the Global View Resilience (GVR) programming model, which provides multi-stream snapshots, multi-version data structures with portable and rich error checking/recovery. Experimental results validate correct execution with low performance overhead under varied error conditions.

Keywords: Resilience · Numerical solver · High performance computing

1 Introduction

The scaling of semiconductor technology and increasing power concerns combined with system scale make fault management a growing concern in high performance computing systems [1,4,11,13]. Soft errors and higher error rates all expected. Just as they played an important role in achieving scalable, high performance, we expect that widely-used numeric solvers such as Flexible Generalized Minimal Residual Method (FGMRES) will play an important key role in achieving resilience and performance for large-scale applications in future "exa" scale systems.

Flexible GMRES with restarting (see Fig. 1 [2,17]) is robust to soft errors due to three aspects. First, the inner solver in Step 3 is inexact, and the outer solver can tolerate large changes to inner solver. Second, the minimal residual procedure can reduce the impact of error on inner solver and keep the residual decreasing (see Step 11). Third, FGMRES restarts the computation after m outer iterations (see Step 17). While the major purpose of restarting is to address the performance and memory usage, restarting can also eliminate errors in outer solver data structures. However in our experiments some bit-errors are still problematic. Errors in inner solver can incur high computational overhead for convergence. Errors in outer solver can even lead to divergence failure. Restarting may lead to stagnation of convergence.

© Springer International Publishing Switzerland 2015
M. Daydé et al. (Eds.): VECPAR 2014, LNCS 8969, pp. 124–132, 2015.
DOI: 10.1007/978-3-319-17353-5_11

With these insights, we design and evaluate error checking and recovery strategies. For inner solver, residual based checking is deployed to identify significant error; recomputing and multi-versioning are exploited for recovery in different cost and granularity. For outer solver, double modular redundancy and data reloading strategies are utilized for error checking and recovery. Our experiments employ the Trilinos library [12], extending FGMRES inner-outer solver with the Global View Resilience (GVR) framework [10], use 5 matrices from the Florida sparse collection [7], running on up to 128 processes. Experimental results illustrate that our GVR-enabled FGMRES solver successfully tolerates the bit flip errors and significantly reduces the impact on performance. Specific contributions include:

- Characterizing situations where bit-errors cause resilience problems for both inner and outer solvers in FGMRES.
- Employ GVR programming model with Trilinos library for portable and rich error checking/recovery strategies in inner-outer solver.
- Evaluate each recovery method, empirically validating that they are efficient and that each is best for regime of error rates.

The rest of the paper is organized as follows. Section 2 introduces the background of GVR and Trilinos for our implementation. Sections 3 and 4 explore the error impact error checking and recovery methods for inner solver and outer solver respectively. Section 5 discusses experimental results, and Sect. 6 surveys related work. Finally, we summarize and discuss future directions in Sect. 7.

Input: Linear system $Ax = b$ and initial guess x_0.
Output: Approximate solution x_m.
1: $r_0 := Ax - b, \beta := ||r_0||_2, q_1 := r_0/\beta$
2: **for** $j = 1, \ldots, m$ **do**
3:      Inner solver for inexact solution z_j in $q_j = Az_j$
4:      $v_{j+1} := Az_j$
5:      **for** $i = 1, \ldots, j$ **do**
6:          $H(i, j) := (v_{j+1}, q_i)$
7:          $v_{j+1} := v_{j+1} - q_i H(i, j)$
8:      **end for**
9:      $H(j + 1, j) := ||v_{j+1}||_2$
10:     $q_{j+1} := v_{j+1}/H(j + 1, j)$
11:     $y_j := argmin_y ||H(1 : j + 1, 1 : j)y - \beta e_1|||_2$
12:     $x_j := x_0 + [z_1, \ldots, z_j]y_j$
13: **end for**
14: **if** converged **then**
15:     Return x_m
16: **else**
17:     $x_0 := x_m,$ go to 1
18: **end if**

Fig. 1. Flexible GMRES with restarting

2   GVR and Trilinos

Our implementation of fault tolerance inner-outer solver is based Global View Resilience (GVR) [10] and Trilinos [12]. Trilinos is an object-oriented software framework for solving big complex science and engineering problems. Kernel classes of Trilinos include vector, matrix, and map. It provides common abstract solvers, such as iterative linear solvers and preconditioners. Based on the kernel class and solvers, Trilinos provides comprehensive algorithmic packages such as stochastic PDEs.

GVR is a novel programming model to enable sophisticated, application-specific fault tolerance in parallel computing. It enables the application to create global data store (GDS) objects for flexible, portable and efficient fault management. We extend the kernel classes of Trilinos using GVR APIs, including the GDS object creation, put/get operations, and GDS versioning. Based on the extended kernel classes, we implement GVR enabled inner-outer solver package, which can be directly used for other Trilinos applications. Especially, GVR facilitates our inner-outer solver in the following aspects.

1. GVR provides multi-stream scheme to create multiple GDS objects for distributed basis vectors and solution vectors. Each GDS object can periodically take snapshots at application specified stable point such as the end of iteration. GVR explores the benefits of local and hierarchy storage to reduce the runtime overhead of snapshot.
2. Multiple older versions of the GDS object remain available for access. The multi-version scheme is motivated for latent error, i.e., errors that retain for some iterations. We use it for recovery inside of the inner solver.
3. It is flexible to configure different versioning, error-checking, and error-recovery schemes to each GDS object. It is helpful to customize the explored strategies thus adapting to different error rates.
4. GVR provides erasure code based on resilience mechanisms for the multi-version snapshots. Since the snapshot is used only for recovery, the overhead is negligible. It is also configurable to explore NVRAM with low error rate for snapshot resilience.
5. The application can provide each GDS object with specific callback routines for error checking and error-recovery in a uniform framework. Error-recovery routines can respond to errors raised by either the application or by the underlying system, such as uncorrectable ECC signal from operating system. Combining with multi-version, GVR can recover the application from catastrophic memory failures.

In this paper, we only use 1–4 GVR features to address soft errors. We will explore using more features in the future.

3   Inner Solver

In this study, we presume that the inner solves takes most of execution time and arbitrarily set 30 iterations inner solver. In this scenario, the inner solver takes more than 90 % execution time, which is a key factor to make trade-off between system reliability and inner solve reliability. We will study other scenarios as a future work.

3.1   Error Impact

To study the impact of errors on inner solver, we randomly inject the error during SpMV or vector dot product operation as the most error-prone, or inject the result vector z_j directly as the most important data visible to the outer solver. In this study, we focus on double precision floating-point data, which consists of 1 sign bits, 11 exponent bits, and 52 bits for mantissa. Bit-flips not in the first 2 bytes only introduce a relative error $<= 2^{-4}$ [9], thus having little impact on execution correctness and convergence.

As inner solver result is approximate, if error occurs not in the first 2 bytes which only introduces a relative error $<= 2^{-4}$ [9], error impact is minimal on execution correctness and convergence. However, as shown in Fig. 2, if a bit error significantly increase the residual of a significant inner solver comparing with previous inner-outer iteration, it generally incurred 2 or 3 additional inner-outer solver iterations, which is consistent with the study in [9]. In extreme cases, as many as 48 additional inner-outer solver iterations can be required. Further, the error impact can accumulate. As the increasing of errors, we observe 8× number of inner-outer iterations in extreme cases.

3.2   Error Check and Recovery: Outside

First, we study outside error checking and recovery; such coarse-grained recovery is relevant even in current-day error environments, and applies to many inner solvers such as GMRES and CG. We exploit two symptoms to identify significant error: (1) residual increase (vs previous iteration) and (2) the matrix $H(1:j, 1:j)$ is not full rank [2]. For these methods, checking overhead is low. Explicit residual checks can be calculated by outer solver, as well as checks for errors in A and q_j. In our experiments, the explicit residual check incurs only take 0.2 % overhead per iteration. Further, checking rank deficiency of matrix $H(1:j, 1:j)$ is essentially free as the SVD-based method to calculate step-11 (see Fig. 1) computes the its rank directly.

There are two simple strategies for recovery outside of inner solver. The first is recomputing the inner solver, incurring high overhead since the inner solver is 90 % of the computation. Despite that, recomputing is still viable as the significant inner solver errors generally introduce 2–3 more inner-outer solver iterations (see Fig. 2). The second is restarting the whole computation as step-17 in Fig. 1 [2]. Restarting may lead to stagnation of convergence, so it is employed only if recomputing fails.

3.3   Error Recovery: Inside

For higher error rates, it is necessary to handle the errors inside of the inner solver rather than recomputing the whole inner solver. In this study, we keep

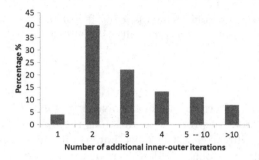

Fig. 2. Distribution of additional inner-outer solver iterations incurred by significant inner solver errors.

one snapshot of q_j at the beginning and multi-version snapshot of z_j during the inner solver iterations. If a significant error is detected outside of inner solver, we check the versions in descend order. If any version of intermediate result has significant lower residual than the final result, inner solver rolls back to that point, reload q_j and A, and executes the rest iterations. Otherwise, inner solver is recomputed from the scratch.

An alternative solution is to check and handle the error during the inner solver iterations. In this study, we do not adopt it due to two reasons. First, it is difficult to identify the error with low overhead and high coverage. Second, it is hard to predict the impact of the error on the inner-outer iterations. We will study this solution as a future work.

The crossover between outside and inside error handling happens when the overhead of error recovery within the solver is less than later recomputing. In this study, we define the *error probability* as the ratio between the number of iterations with errors and the total iterations. Suppose the probability of inner solver error is P, the error-free execution is Φ longer due to the overhead of snapshot, and the error handling inside reduce the recomputing time by Θ shorter. So the error handling inside inner solver become beneficial when $1-P+2P > (1-P)\Phi+(1+\Theta)P$. We validate these tradeoffs in our experiments in Sect. 5.

4   Outer Solver

The outer solver typically consumes less execution time, but errors in outer solver are more critical for correctness and performance. In most cases, if significant errors occur in the basis vectors or Hessenberg matrix H, the residual may increase or stay constant. Even a single bit-flip may lead to divergence no matter at which iteration the bit-flip occurs.

To tolerate the error in outer solver, we adopt simple double modular redundancy (DMR) [15]. It executes the outer solver twice and compare the results. DMR based method may fail to tolerate the memory error staying in both executions. To address this problem, at each error-free iteration, we take snapshots

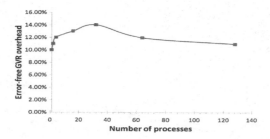

Fig. 3. GVR overhead in error-free execution. Here one snapshot of related vectors are taken at each inner-outer iteration.

of subspace basis $[v_1, v_2, \ldots, v_{j-1}]$, $[z_1, z_2, \ldots, z_{j-1}]$, matrix H, and result vector x_j. Notice that GVR provides resilience mechanism for these snapshots. Reloading A, b, and these snapshots in previous iteration from GDS objects before the second execution, can detect memory errors in the original execution. If any inconsistency between two outer solver executions, the third one will be triggered to identify the correct execution. This approach has low overhead, high error checking accuracy and prevents error propagation to the next iteration.

5   Experiments

Based on our implementation from GVR and Trilinos, we run 128 processes with 5 matrices from the Florida sparse collection. The data is the average result of 1,000 error trials for each error probability and matrix. As the error impact studies, we mainly focus on significant error in the first 2 bytes of double precision data.

5.1   GVR Overhead in Error-Free Execution

In our GVR enabled FGMRES, we create GDS objects on the solution and basis vectors, put the data into GDS objects and make the versions. We vary the number of processes to study the overhead of GVR in error-free execution. Here we take one snapshot of z_j, $[v_1, v_2, \ldots, v_{j-1}]$, $[z_1, z_2, \ldots, z_{j-1}]$, H, and x_j at each inner iteration.

As shown in Fig. 3, the overhead is less than 15 % and keeps stable with the increasing of processes. The major overhead is on versioning since we use collective call to get consistent snapshot. We plan to bundle these vectors into one GDS object thus reducing the number of versioning operation. Note that the overhead is much lower for a realistic error rate, which is unnecessary for versioning at each inner iteration.

5.2   Inner Solver

To explore a range of error rates, we vary probability of significant error inside of inner solver computation or the result vector z_j. The recovery process is

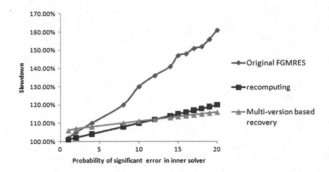

Fig. 4. Execution slowdown - original FGMRES, recomputing based recovery, and multi-version based recovery.

triggered only if the inner solver residual becomes 100× larger than the previous iteration. We compare the total execution time without recovery and with our inside/outside resilience method. We calculate the *slowdown* as the ratio between the total execution time with errors and the failure-free execution time.

Our results of slowdown (see Fig. 4) show that both recomputing and multi-versioning based recovery outperforms the original FGMRES with restarting. When error probability is < 11 %, recomputing has lower slowdown than multi-versioning based recovery due to the cost of snapshot. As the growth of error probability, multi-versioning based recovery becomes more beneficial by reducing the work loss.

5.3   Outer Solver

We vary probability of significant error in the basis vectors and solution vectors of outer solver and present the slowdown in Fig. 5. The significant error in outer solver always leads to divergence for FGMRES without restarting. When the error probability is low, restarting can tolerate the error but introduce extreme high overhead, e.g. 1682.5 % slowdown on 10 % of error probability. As the increasing of error probability, it also becomes divergence. Our GVR enabled FGMRES successfully addresses the high error probability with relatively small overhead because it can isolate the errors in each iteration.

6   Related Work

In large-scale system, traditional studies have focused on system level checkpoint/restart to tolerate fail-stop process failures [16]. As the growing concern around soft errors, more recent studies have focused on application level and cross layer solutions, especially for numeric solvers. Huang and Abraham developed the checksums based algorithm-based fault tolerance (ABFT) technique for matrix operations [14]. In [6], Chen developed theoretical conditions based error checking for

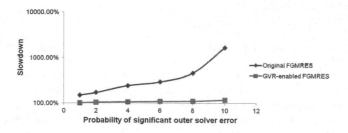

Fig. 5. Execution slowdown - original FGMRES, recomputing based, and multi-version based recovery.

Krylov subspace iterative methods. In [3], Bronevetsky analyzed soft error vulnerability for linear solvers. In [18], fault tolerant PCG solver is presented for sparse linear systems. Du presented encoding strategy for LU factorization based dense liner systems [8]. Unlike these works, this study is focusing on inner-outer solver.

The studies on fault tolerant inner-outer solver are limited. In [5], Chen analyzed flexible BiCGStab to bound the inner solver error for convergence. In [9], Elliott studied the impact of inner solver error in FGMRES. In [2] FGMRES solver was extended to tolerate inner solver error. Distinguished from these studies, this paper presents comprehensive error analysis for FGMRES and develops GVR-enabled methods for both inner and outer solvers under various error rate.

7 Summary and Future Work

We analyze the impact of bit-flip errors on the FGMRES inner-outer solver, which can lead to divergence failure or extreme high computation overhead. Based on the analysis results, we design the error checking/recovery strategies for inner solver and outer solver. We implement it by extending Trilinos solver library with our Global View Resilience (GVR) system. Our experiments show that our GVR-enabled inner-outer solver successfully tolerate the bit flip errors for execution convergence with low overhead.

Interesting future directions include, studying a wider range of inner-outer solver configurations, error checking and recovery methods, and employing additional GVR features. Another direction would include study of errors from other sources – other hardware elements, or even intentional inaccuracies such as reduced coverage ECC or probabilistic CMOS.

Acknowledgments. We thank Mark Hoemmen from Sandia National Laboratories for his advice. This work supported by the Office of Advanced Scientific Computing Research, Office of Science, U.S. Department of Energy, under Award DE-SC0008603 and Contract DE-AC02-06CH11357. Also under the DOE National Nuclear Security Administration (NNSA) Advanced Simulation and Computing (ASC) program. Sandia National Laboratories is a multi-program laboratory managed and operated by Sandia Corporation, a wholly owned subsidiary of Lockheed Martin Corporation, for the U.S. Department of Energy National Nuclear Security Administration under contract DE-AC04-94AL85000.

References

1. Borkar, S., Chien, A.A.: The future of microprocessors. Commun. ACM **54**(5), 67–77 (2011)
2. Bridges, P.G., Ferreira, K. B., Heroux, M. A., Hoemmen, M.: Fault-tolerant linear solvers via selective reliability. ArXiv e-prints, June 2012. Provided by the SAO/NASA Astrophysics Data System
3. Bronevetsky, G., de Supinski, B.: Soft error vulnerability of iterative linear algebra methods. In: Proceedings of ICS (2008)
4. Cappello, F., Geist, A., Gropp, W., Kale, L., Kramer, W., Snir, M.: Towards exascale resilience. Int. J. High Perform. Comput. Appl. **23**(4), 374–388 (2009)
5. Chen, J., McInnes, L.C., Zhang, H.: Analysis and practical use of flexible BiCGStab. Technical report ANL/MCS-P3039-0912, Argonne National Laboratory (2012)
6. Chen, Z.: Online-ABFT: an online algorithm based fault tolerance scheme for soft error detection in iterative methods. In: Proceedings of PPoPP (2013)
7. Davis, T.A., Hu, Y.: The University of Florida sparse matrix collection. ACM Trans. Math. Softw. **38**(1), 1–25 (2011)
8. Du, P., Luszczek, P., Dongarra, J.: High performance dense linear system solver with resilience to multiple soft errors. In: Proceedings of ICCS (2012)
9. Elliott, J., Hoemmen, M., Mueller, F.: Evaluating the impact of SDC on the GMRES iterative solver. In: Proceedings of IPDPS (2014)
10. Chien, A., et al.: Global View Resilience Project (GVR). http://gvr.cs.uchicago.edu
11. Elnozahy, M., et al.: System resilience at extreme scale (2009). White Paper written for the Defense Advanced Research Project Agency (DARPA), with Ricardo Bianchini et al.
12. Heroux, M., et al.: An overview of the trilinos project. ACM Trans. Math. Softw. **31**(3), 397–423 (2005)
13. Kogge, P., et al.: Exascale computing study: Technology challenges in achieving exascale systems. Technical report TR-2008-13, University of Notre Dame CSE Department (2008)
14. Huang, K., Abraham, J.: Algorithm-based fault tolerance for matrix operations. IEEE Trans. Comput. **C–33**(6), 518–528 (1984)
15. Lidman, J., Quinlan, D. J., Liao, C., McKee, S.A.: ROSEFTTransform - a source-to-source translation framework for exascale fault-tolerance research. In: DSN-W (2012)
16. Moody, A., Bronevetsky, G., Mohror, K., Supinski, B.: Design, modeling, and evaluation of a scalable multi-level checkpointing system. In: Proceedings of Supercomputing (2010)
17. Saad, Y.: Iterative Methods for Sparse Linear Systems, 2nd edn. SIAM, Philadelphia (2003)
18. Shantharam, M., Srinivasmurthy, S., Raghavan, P.: Fault tolerant preconditioned conjugate gradient for sparse linear system solution. In: Proceedings of ICS (2012)

Direct/Hybrid Methods for Solving Sparse Matrices

Using Random Butterfly Transformations to Avoid Pivoting in Sparse Direct Methods

Marc Baboulin[1], Xiaoye S. Li[2], and François-Henry Rouet[2]([✉])

[1] Inria Saclay, University of Paris-Sud, Orsay, France
[2] Lawrence Berkeley National Laboratory, Berkeley, CA, USA
fhrouet@lbl.gov

Abstract. We consider the solution of sparse linear systems using direct methods via LU factorization. Unless the matrix is positive definite, numerical pivoting is usually needed to ensure stability, which is costly to implement especially in the sparse case. The Random Butterfly Transformations (RBT) technique provides an alternative to pivoting and is easily parallelizable. The RBT transforms the original matrix into another one that can be factorized without pivoting with probability one. This approach has been successful for dense matrices; in this work, we investigate the sparse case. In particular, we address the issue of fill-in in the transformed system.

1 Introduction

When solving the linear systems using the LU or the LDL^T factorizations, numerical pivoting is often needed to ensure stability. Pivoting prevents division by zero or by small quantities by permuting on the fly the rows and/or columns of the matrix so that the pivotal element is relatively large in magnitude. Pivoting involves irregular data movement and can significantly impact the speed of the factorization, especially on large parallel machines. This issue arises in both unsymmetric and symmetric cases, and for both dense and sparse factorizations. The ScaLAPACK [7], MAGMA [20] and PLASMA [17] dense linear algebra libraries contain a Cholesky factorization for positive definite matrices, for which no pivoting is required, but they do not contain an LDL^T factorization. They contain an LU factorization with partial pivoting (i.e. $PA = LU$, where P is a permutation matrix), but partial pivoting can significantly slow down the speed. For example, on a hybrid CPU/GPU system, the LU algorithm in the MAGMA library spends over 20 % of the factorization time in pivoting even for a large random matrix of size $10,000 \times 10,000$.

Pivoting poses additional problems in sparse factorizations because of the *fill-in*, which corresponds to the new nonzeros generated in the factored matrices L and U. For sparse Cholesky, where pivots can be chosen on the diagonal, we often use a sparsity-preserving ordering algorithm, such as minimum degree or nested dissection, to reorder the matrix first so that the Cholesky factor of the permuted matrix PAP^T has less fill-in than that of A. For sparse LU, we often factorize

© Springer International Publishing Switzerland 2015
M. Daydé et al. (Eds.): VECPAR 2014, LNCS 8969, pp. 135–144, 2015.
DOI: 10.1007/978-3-319-17353-5_12

PAQ^T with both row and column permutation matrices P and Q. The purpose is to preserve sparsity as well as to maintain numerical stability. There are complex interplays between ordering (for sparsity) and pivoting (for stability). Often, the two objectives cannot be well achieved simultaneously. Several relaxed pivoting schemes, other than partial pivoting, have been developed to trade off stability and sparsity, which allow larger pivot growth while maintaining better sparsity. These include *threshold pivoting* [10], *restricted pivoting* [19], and *static pivoting* [14].

One difficulty with dynamic pivoting, either partial pivoting or threshold pivoting, is that the fill-ins are produced on the fly depending on the permutation at each step. It is thus not possible to have the separate ordering and symbolic preprocessing algorithms that precisely minimize the number of fill-ins and forecast the fill-in positions. A good ordering strategy to accommodate dynamic row pivoting is to apply any ordering algorithm to the graph of the symmetrized matrix $A^T A$ which gives a fill-reducing permutation Q. Then, Q is applied to the columns of A *before* performing the LU factorization with row pivoting: $P(AQ^T) = LU$. The rationale behind this is that the nonzero structure of the Cholesky factor R of $A^T A = R^T R$ upper bounds the nonzero structures of L^T and U of $PA = LU$, for *any* row permutation P [12]. That is, the Cholesky factor R_q of $(AQ^T)^T(AQ) = R_q^T R_q$ upper bounds the L_q^T and U_q of $P(AQ^T) = L_q U_q$, and R_q contains smaller amount of fill than that of R. In essence, the column ordering Q tends to minimize an upper bound on the actual fill-ins in the LU factors, taking into account all the possible row pivotings. This strategy can be pessimistic when most pivots happen to be on the diagonal (e.g. diagonally dominant matrices). The sequential SuperLU library uses this ordering strategy together with partial pivoting [9]. This is our comparison baseline to be used in Sect. 3 about the numerical results.

The cost of dynamic pivoting in parallel is even more dramatic than in the dense case. For example, for matrix nlpkkt80 of a KKT system from nonlinear optimization, the parallel factorization with threshold pivoting using MUMPS [2] took 639 s with 128 processes. After the matrix is modified to be diagonally dominant with the same sparsity structure, the parallel factorization without pivoting took only 87 s, even though the size of the LU factors and the flop count are roughly the same in both cases.

In the parallel direct solver SuperLU_DIST [14], a static pivoting strategy is used to enhance scalability. Here, P is chosen *before* factorization based solely on the values of the original A. A maximum weighted matching algorithm and the code MC64 [11] is currently employed. The algorithm chooses P to maximize the magnitude of the diagonal entries of PA. During factorization, the pivots are chosen on the diagonal and the tiny ones are replaced by a fixed value. Since this does not involve dynamic row permutation, a sparsity-reducing algorithm can be applied to the graph of another symmetrized matrix $PA + (PA)^T$, producing the permutation matrix Q. This tends to minimize the amount of fill in the L and U of $Q(PA)Q^T = LU$. The static pivoting improves speed and scalability but it might fail for very challenging problems. MC64 is sequential in nature and there is no good parallel algorithm yet. Riedy [18] suggests a parallel auction algorithm but concludes that parallel performance is too unpredictable to make it a black-box tool. Therefore, the pre-pivoting phase will be a severe obstacle for solving larger problems on extreme-scale parallel machines.

In 1995, Parker introduced a randomization algorithm to *eliminate the need for pivoting* [16]. In this approach, the Random Butterfly Transformation (RBT) is used to transform the original system into an "easier" one such that, with probability one, the LU factorization of the transformed matrix can be performed without pivoting. This technique was successfully applied and implemented into the dense libraries for LU and LDL^T factorizations [3,6]. In this work, we investigate the potential of the RBT method for sparse cases.

2   Random Butterfly Transformations

In this section we recall the main concepts and definitions related to RBT where the randomization of the matrix is based on a technique initially described by Parker [16] and recently revisited by Baboulin et al. [3] for general dense systems. The procedure to solve $Ax = b$, where A is a general matrix, using a random transformation and the LU factorization is:

1. Compute $A_r = U^T A V$, with U, V random matrices,
2. Factorize $A_r = LU$ (without pivoting),
3. Solve $A_r y = U^T b$ and compute $x = V y$.

The random matrices U and V are chosen among a particular class of matrices called *recursive butterfly matrices*. A *butterfly matrix* is an $n \times n$ matrix of the form

$$B^{<n>} = \frac{1}{\sqrt{2}} \begin{bmatrix} R_0 & R_1 \\ R_0 & -R_1 \end{bmatrix}$$

where R_0 and R_1 are random diagonal $\frac{n}{2} \times \frac{n}{2}$ matrices. A *recursive butterfly matrix* of size n and depth d is defined recursively as

$$W^{<n,d>} = \begin{bmatrix} B_1^{<n/2^{d-1}>} & & \\ & \ddots & \\ & & B_{2^{d-1}}^{<n/2^{d-1}>} \end{bmatrix} \cdot W^{<n,d-1>}, \text{ with } W^{<n,1>} = B^{<n>}$$

where the $B_i^{<n/2^{d-1}>}$ are butterflies of size $n/2^{d-1}$, and $B^{<n>}$ is a butterfly of size n.

In the original work by Parker, $d = \log_2 n$; he shows that, given two recursive butterfly matrices U and V, the matrix $U^T A V$, where A is the original matrix of the system to be solved, can be factored into LU without pivoting with probability 1 in exact arithmetic, or with probability $1 - O(2^{-t})$ using t-bit floating point numbers. For symmetric problems, $V = U$ and the same result holds with LDL^T. Baboulin et al. studied extensively the use of RBT for dense matrices and showed that in practice, $d = 1$ or 2 is enough; in most cases a few steps of iterative refinement can recover the digits that have been lost. They also showed that random butterfly matrices are cheap to store and to apply ($O(nd)$ and $O(dn^2)$ respectively) and they proposed implementations on hybrid multicore/GPU systems for the unsymmetric [3] case. For the symmetric case, they proposed a tiled algorithm for multicore architectures [4] and more recently a distributed solver [5] combined with a runtime system [8]. As was demonstrated, the preprocessing by RBT can be easily parallelized with good scalability.

3   Using RBT in Sparse Direct Solvers

We first describe and compare different strategies and parameters when applying RBT to the sparse LU factorization. We carry out the experiments on a large set of sparse matrices in order to identify the best practical strategy.

3.1   Influence of the Degree d

In the dense case, the use of RBT incurs small amount of extra operation and memory. The cost is limited to storing and applying RBT *prior to* the factorization. However, in the sparse case, applying RBT modifies the *nonzero structure* of the transformed matrix. The number of nonzeros in the transformed matrix $U^T A V$ can be up to 4^d times the number of nonzeros in A in the worst case. This increase in nonzeros may lead to an even larger increase in the size of the LU factors and thus to prohibitive costs. We therefore limit our investigation to small degrees: $d = 1$ or 2, which correspond to the practical setting used by Baboulin et al. in the dense case [3–5].

3.2   Combining RBT and Fill-Reducing Permutations

Fill-reducing ordering is critical to preserve sparsity. This operation is usually performed after all the preprocessings that modify the sparsity pattern of the input matrix (e.g., MC64). At first glance, it seems that the most natural way of combining RBT with a fill-reducing permutation is:

1. transform the original matrix A into $U^T A V$,
2. permute with a fill-reducing algorithm (then factorize).

However, one can show that the matrix resulting from steps 1 and 2 is not guaranteed to be factorizable without pivoting. We provide an example here. Let A be a 4×4 matrix; A can be written in a 2×2 form as

$$A = \begin{bmatrix} A_{11} & A_{12} \\ A_{21} & A_{22} \end{bmatrix}$$

Let U and V two recursive butterflies of size 4 and degree 2. By Parker's theorem, if A is non-singular then $U^T A V$ is factorizable without pivoting. Let p be the permutation vector [1 3 2 4] and P the associated permutation matrix. We consider $B = P U^T A V P^T$. One can show that if $\sum A_{11} = \sum A_{22} = \sum A_{21} = \sum A_{12}$

then the leading submatrix $B_{1:2,1:2}$ is singular, regardless of the random values in U and V. Therefore B is not factorizable without pivoting (B_{22} becomes 0 after eliminating the first pivot, in the absence of roundoff errors). One can easily build a non-singular matrix satisfying this property, e.g., $A_{11} = \begin{bmatrix} 2 & 0 \\ 0 & 2 \end{bmatrix}$, $A_{12} = A_{21} = \begin{bmatrix} 1 & 1 \\ 1 & 1 \end{bmatrix}$, $A_{22} = \begin{bmatrix} 3 & 0 \\ 0 & 1 \end{bmatrix}$ (leading to $\det(A) = -4$, i.e., A non-singular).

As a consequence, the strategy consisting in permuting for sparsity after the transformation may not work in theory, but we still wish to investigate its practical performance. We compare the following two strategies:

Strategy 1: the matrix is permuted using a fill-reducing (or bandwidth minimization) heuristic then transformed with RBT. This guarantees that the factorization would succeed for $d = \log_2 n$ but it might yield large number of fill-ins in the factors. The first step is an attempt to minimize the nonzeros in the transformed matrix and the fill-ins.

Strategy 2: the matrix is transformed with RBT then permuted using a fill-reducing heuristic. This might fail even for $d = \log_2 n$ but it provides a much better control of fill-in.

3.3   Evaluation of the Different Strategies and Parameters

The experiments were carried out on 90 non-singular matrices with size $n \le 10,000$. Table 1 shows the success rate of the factorization, the increase in nonzeros and the increase in the size of the LU factors with respect to partial pivoting. We use the partial pivoting code SuperLU [9]; for RBT, pivoting is disabled and the factorization is stopped whenever a zero diagonal pivot is found (although a possibility could be to replace it by a small perturbation such as $\varepsilon\|A\|$). The random values we use are $e^{r/10}$ with r randomly chosen in $[-\frac{1}{2}, \frac{1}{2}]$ from a uniform distribution. This guarantees a small condition number for U and V [3].

Table 1. Influence of the different strategies and parameters for 90 matrices with size $n \le 10,000$. "Success rate" is the percentage of matrices for which the factorization completes. "Increase in nonzeros" is the ratio $nnz(U^T AV)/nnz(A)$ and "Increase in factors" is the ratio $nnz(LU(U^T AV))/nnz(LU(A))$; we report the minimum, geometric mean, arithmetic mean, and maximum.

Strategy and degree		Success rate	Increase in nonzeros				Increase in factors			
			min	geo	avg	max	min	geo	avg	max
Strategy 1	$d = 1$	81.1 %	1.00	2.97	3.14	3.99	1.12	9.92	21.07	362.32
	$d = 2$	92.2 %	2.01	9.53	10.52	15.79	1.14	19.35	45.41	635.84
Strategy 2	$d = 1$	82.2 %	1.00	2.02	2.25	4.00	0.03	1.55	2.62	20.42
	$d = 2$	80.0 %	1.50	4.95	5.98	15.01	0.06	2.96	6.78	144.49

We make the following observations: (1) Strategy 1 and Strategy 2 have similar success rates. Although both strategies lead to an increase in the size of the LU factors (with respect to partial pivoting), this increase is much more limited with Strategy 2. Therefore, Strategy 2 will be our method of choice. (2) Similar to what was observed in the dense case, most matrices succeed with $d = 1$. With Strategy 1, $d = 2$ yields a near-perfect success rate at the price of a large increase in the size of factors; the effect is less clear with Strategy 2.

Using Strategy 2 with $d = 1$ seems to be the most practical setting. Figure 1 shows how this approach compares with partial pivoting. Figure 1(a) shows how the size of the factors varies when RBT is used. 37 out of 90 matrices have a smaller size of LU factors; as explained in the introduction, this is due to the fact that partial pivoting relies on a fill-reducing permutation that can only aim at minimizing an upper bound of the fill-in, since the order in which variables are eliminated is not known in advance. On the other hand, not doing pivoting allows the fill-reducing permutation to focus on the right problem (minimizing the actual fill-in). For 30 matrices, the increase (due to the larger structure of the transformed matrix) is moderate (larger than one but less than two). Although this means that the number of operations with RBT might be larger than with partial pivoting, RBT may catch up since doing no pivoting yields better flop rate and scalability. For 23 matrices, the increase is large (between 2 and 20), which means it is more unlikely that RBT will yield better runtime. Figure 1(b) shows the ratio between the forward error $||x - x_{\text{true}}||/||x_{\text{true}}||$ with RBT and that with partial pivoting. For 69 out of 90 problems, the ratio is less than 10^2 i.e. at most 2 digits are lost when using RBT instead of partial pivoting. The loss in accuracy found for some matrices is due to a larger growth factor with RBT, meaning that some elements found during the factorization become very large (relative to the elements in the matrix to be factored) and lead to inaccuracies. In most cases, a few steps of iterative refinement recover the lost digits. Overall, we found that 48 out of 90, i.e. 53.3 % have both a moderate increase in the factors size (less than twice) and a moderate loss in accuracy (less than 2 digits).

3.4   One-Sided Transformation

The original approach proposed by Parker relies on a two-sided transformation $U^T AV$. We showed that a one-sided transformation is sufficient to maintain the main numerical property, i.e., $U^T A$ can be factorized without pivoting when U is a recursive butterfly matrix with degree $d = \log_2 n$. The benefit is that the number of nonzeros in $U^T A$ (and the LU factor size) can be less than the number of nonzeros in $U^T AV$. Through private communication, Parker mentions that it is analogous to using partial pivoting rather than complete pivoting, i.e., although no zero pivot appears, the growth factor may be larger.

We experimented this one-sided approach, and found that, with $d = 1$, the success rate of the one-sided and two-sided approaches are similar. For $d = 2$, the success rate is marginally higher with the two-sided approach. Figure 2 illustrates how the two approaches influence the size of the transformed matrix and the size

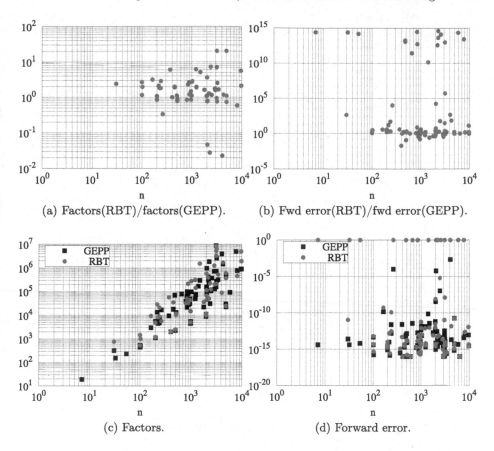

(a) Factors(RBT)/factors(GEPP).   (b) Fwd error(RBT)/fwd error(GEPP).

(c) Factors.   (d) Forward error.

Fig. 1. RBT (Strategy 2, $d = 1$) vs partial pivoting for 90 matrices sorted by size.

of the factors. We observed that the one-sided approach marginally decreases the size of the factors on average, but the results are problem-dependent.

3.5   Source of Failures

Parker's theorem states that, using $d = \log_n$, no zero pivot can be found during the factorization (with probability one). However, in [15], there is no proof about the *magnitude* of the pivots. Small (with respect to machine precision) pivots lead to large element growth in the U factor, and elements of smaller size are lost.

Consider a matrix M and its LU factorization $M = L_M U_M$. The standard metric used in error analysis is the *growth factor* [13]

$$\rho = \frac{\max_{i,j,k} |m_{ij}^{(k)}|}{\max_{i,j} |m_{ij}|}$$

where $a_{ij}^{(k)}$ is the element met at the k-th step of Gaussian Elimination on M. ρ is expensive to compute, and, in practice, solvers report the *pivot growth*, which

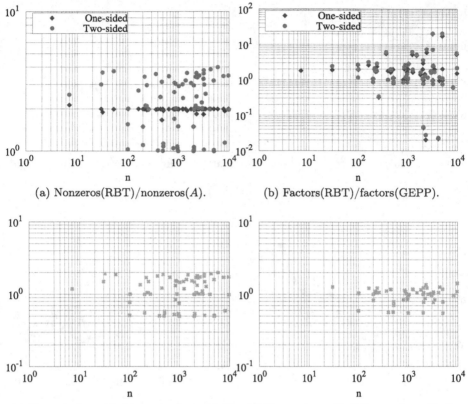

(a) Nonzeros(RBT)/nonzeros(A).   (b) Factors(RBT)/factors(GEPP).

(c) Nonzeros(two-sided)/nonzeros(one-sided).   (d) Factors(two-sided)/factors(one-sided).

Fig. 2. One-sided vs two-sided (Strategy 2, $d = 1$) for 90 matrices sorted by size.

is either $\frac{||U_M||_\infty}{||M||_\infty}$ (e.g., in LAPACK [1]), or $\max_j \frac{||U_M(:,j)||_\infty}{||M(:,j)||_\infty}$ (e.g., in SuperLU). These definitions assume that partial pivoting is used; in this case, elements in the L factor are bounded by 1. To accomodate the fact that we do not use pivoting when we use RBT, we propose the following definition:

$$\rho = \max \left(\max_j \frac{||L_M(:,j)||_\infty}{||M(:,j)||_\infty} , \max_j \frac{||U_M(:,j)||_\infty}{||M(:,j)||_\infty} \right)$$

In Fig. 3, we report the pivot growth for our collection of test matrices, using partial pivoting (i.e., growth factor for the original matrix A) and using RBT (i.e., pivot growth for the factorization without pivoting of $U^T AV$). We observe that for most matrices the pivot growth is slightly larger (but reasonable) when using RBT instead of partial pivoting. However, a few matrices have very large pivot growth, leading to instability in the factorization.

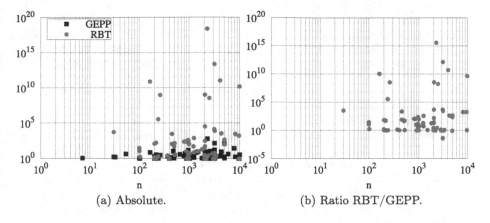

(a) Absolute.   (b) Ratio RBT/GEPP.

Fig. 3. Pivot growth for 90 matrices sorted by size.

4   Conclusion and Perspectives

For sparse direct solvers using LU factorization, a serious scalability bottle-neck is numerical pivoting. A number of relaxed pivoting algorithms have been developed, but none of them have shown promise of scalable implementation. In this exploratory work, through large number (90) of real-world test matrices, we demonstrated that the Random Butterfly Transformation is a good alternative to pivoting, especially with properly chosen ordering strategies and transformation parameters. RBT is particularly appealing for extreme-scale systems because it is highly parallelizable. This opens the possibilities of several avenues of new research, such as application of RBT to the LDL^T factorization, classification of the problems according to various RBT strategies, and investigation of RBT's impact on the scalability of existing parallel direct solvers.

Acknowledgement. We would like to thank Stott Parker for insightful discussions about the one-sided transformation. Partial support for this work was provided through Scientific Discovery through Advanced Computing (SciDAC) program funded by U.S. Department of Energy, Office of Science, Advanced Scientific Computing Research (and Basic Energy Sciences/Biological and Environmental Research/High Energy Physics/ Fusion Energy Sciences/Nuclear Physics). We used resources of the National Energy Research Scientific Computing Center, which is supported by the Office of Science of the U.S. Department of Energy under Contract No. DE-AC02-05CH11231.

References

1. Anderson, E., Bai, Z., Dongarra, J.J., Greenbaum, A., McKenney, A., Du Croz, J., Hammarling, S., Demmel, J.W., Bischof, C., Sorensen, D.: LAPACK: a portable linear algebra library for high-performance computers. In: Proceedings of the 1990 ACM/IEEE Conference on Supercomputing (1990)

2. Amestoy, P.R., Guermouche, A., L'Excellent, J.-Y., Pralet, S.: Hybrid scheduling for the parallel solution of linear systems. Parallel Comput. **32**(2), 136–156 (2006)
3. Baboulin, M., Dongarra, J.J., Hermann, J., Tomov, S.: Accelerating linear system solutions using randomization techniques. ACM Trans. Math. Softw. **39**(2), 1–13 (2013)
4. Baboulin, M., Becker, D., Dongarra, J.J.: A parallel tiled solver for dense symmetric indefinite systems on multicore architectures. In: Parallel & Distributed Processing Symposium (IPDPS) (2012)
5. Baboulin, M., Becker, D., Bosilca, G., Danalis, A., Dongarra, J.J.: An efficient distributed randomized algorithm for solving large dense symmetric indefinite linear systems. Parallel Comput. **40**(7), 213–223 (2014)
6. Becker, D., Baboulin, M., Dongarra, J.: Reducing the amount of pivoting in symmetric indefinite systems. In: Wyrzykowski, R., Dongarra, J., Karczewski, K., Waśniewski, J. (eds.) PPAM 2011. LNCS, vol. 7203, pp. 133–142. Springer, Heidelberg (2012)
7. Blackford, L., Choi, J., Cleary, A., D'Azevedo, E., Demmel, J.W., Dhillon, I., Dongarra, J.J., Hammarling, S., Henry, G., Petitet, A., Stanley, K., Walker, D., Whaley, R.: ScaLAPACK Users' Guide. SIAM, Philadelphia (1997)
8. Bosilca, G., Bouteiller, A., Danalis, A., Herault, T., Lemarinier, P., Dongarra, J.J.: DAGuE: a generic distributed DAG engine for high performance computing. Parallel Comput. **38**(1&2), 37–51 (2011)
9. Demmel, J.W., Eisenstat, S.C., Gilbert, J.R., Li, X.S., Liu, J.W.H.: A supernodal approach to sparse partial pivoting. SIAM J. Matrix Anal. Appl. **20**(3), 720–755 (1999)
10. Duff, I.S., Erisman, I.M., Reid, J.K.: Direct Methods for Sparse Matrices. Oxford University Press, London (1986)
11. Duff, I.S., Koster, J.: The design and use of algorithms for permuting large entries to the diagonal of sparse matrices. SIAM J. Matrix Anal. Appl. **20**(4), 889–901 (1999)
12. George, A., Ng, E.: Symbolic factorization for sparse Gaussian elimination with partial pivoting. SIAM J. Sci. Stat. Comput. **8**(6), 877–898 (1987)
13. Higham, N.J.: Accuracy and Stability of Numerical Algorithms. SIAM, Philadelphia (2002)
14. Li, X.S., Demmel, J.W.: SuperLU_DIST: a scalable distributed-memory sparse direct solver for unsymmetric linear systems. ACM Trans. Math. Softw. **29**(9), 110–140 (2003)
15. Parker, D.S.: Explicit formulas for the results of Gaussian elimination, Technical report CSD-950025, UCLA Computer Science Department (1995)
16. Parker, D.S.: Random butterfly transformations with applications in computational linear algebra, Technical report CSD-950023, UCLA Computer Science Department (1995)
17. PLASMA users' guide, parallel linear algebra software for multicore architectures, Version 2.3 (2010). University of Tennessee
18. Riedy, E.J.: Making static pivoting scalable and dependable, Technical report, UC Berkeley, EECS-2010-172 (2010)
19. Schenk, O., Gärtner, K.: Solving unsymmetric sparse systems of linear equations with PARDISO. Future Gener. Comput. Syst. **20**, 476–487 (2004)
20. Tomov, S., Dongarra, J.J., Baboulin, M.: Towards dense linear algebra for hybrid GPU accelerated manycore systems. Parallel Comput. **36**(5&6), 232–240 (2010)

Hybrid Sparse Linear Solutions with Substituted Factorization

Joshua Dennis Booth[1]([✉]) and Padma Raghavan[2]

[1] Sandia National Laboratories, Albuquerque, NM 87185, USA
jdbooth@sandia.gov
[2] The Pennsylvania State University, University Park, PA 16802, USA
raghavan@cse.psu.edu

Abstract. We develop a computationally less expensive alternative to the direct solution of a large sparse symmetric positive definite system arising from the numerical solution of elliptic partial differential equation models. Our method, *substituted factorization*, replaces the computationally expensive factorization of certain dense submatrices that arise in the course of direct solution with sparse Cholesky factorization with one or more solutions of triangular systems using substitution. These substitutions fit into the tree-structure commonly used by parallel sparse Cholesky, and reduce the initial factorization cost at the expense of a slight increase cost in solving for a right-hand side vector. Our analysis shows that substituted factorization reduces the number of floating-point operations for the model $k \times k$ 5-point finite-difference problem by 10 % and empirical tests show execution time reduction on average of 24.4 %. On a test suite of three-dimensional problems we observe execution time reduction as high as 51.7 % and 43.1 % on average.

1 Introduction

The solution of sparse linear systems arising from finite discretization of second-order elliptic partial differential equations (PDEs) dominates the execution time of scientific codes [1]. These systems of linear equations may be solved by two broad categories of solvers, namely direct or iterative [2–5]. In application, a third hybrid category may appear, which tries to balance trade-offs between direct and iterative solvers [6,7]. They make no claim on the number of iterations needed to solve the system, and therefore hybrid methods can be seen as a preconditioned iterative method. We provide a new formulation used as a direct method by making substitutions in place of factorization, and demonstrate how this formulation results in savings over the direct method sparse Cholesky while providing a robust solution.

This paper provides a detailed understanding and analysis of our new *substituted factorization* (*SF*), and demonstrates our method as an alternative direct

---

J.D. Booth—This work was completed while attending The Pennsylvania State University.

M. Daydé et al. (Eds.): VECPAR 2014, LNCS 8969, pp. 145–155, 2015.
DOI: 10.1007/978-3-319-17353-5_13

method formulation for solving systems arising from finite discretization of a second-order elliptic PDE on quasi-uniform grids. Although other methods exist, the choice of the best method depends on the number of solves, robustness, and memory constraints. Direct methods provide a desired robustness and low cost for multiple solves that many iterative solvers cannot provide. However, the high computational cost of factorization in direct methods makes iterative method more desirable unless this cost can be amortized with multiple solves. Our method provides a middle ground in terms of the number of solves needed to amortize factorization cost while still providing the desirable robustness.

2   Background and Related Work

A number of methods exist to solve systems arising from second-order elliptic PDE models. Sparse direct solvers [2,4] use sparse factorization to compute robust solutions. Iterative solvers, such as Krylov space based Conjugate Gradients (CG) [3,5], attempt to solve systems faster by iteratively converging to a solution. Many elliptic problems can be efficiently solved using domain decomposition methods (DDM) by splitting the problem into smaller domains. DDM may solve the arising systems or be used to precondition Krylov space iterative methods, and the effectiveness varies with problem and decomposition.

When using an iterative method similar to CG, the number of iterations needed to solve for a single right-hand side vector (RHS) depends on the condition number of the matrix. The condition number of a matrix is defined as: $\delta(A) = \|A\|_2\|A^{-1}\|_2$ [3]. In particular, the number of iterations needed for CG to converge is $O(\sqrt{\delta(A)})$ [8].

For symmetric systems of equations arising from finite discretization of a second-order elliptic PDE on regular grids, the condition of the matrix is known to be greater than $O(h^{-2})$. Generally, h^{-1} equals $n^{1/2}$ for two-dimensional (2D), and $n^{1/3}$ for three-dimensional (3D) grids. However, the condition number for submatrices used may be better. Mansfield [9] demonstrates the condition number for the Schur complement is bounded by $O(h^{-1})$ for the broad class of linear systems arising from finite discretization of a second-order elliptic PDE on regular grids. SF uses this bound to fix the number of iterations for CG.

Hybrid solvers [6,7] use various combinations of direct and iterative methods. These hybrid solvers partition the matrix into submatrices linked by other submatrices that correspond to separators in the graph representation, and the resulting form may be similar to:

$$\begin{bmatrix} A_{11} & 0 & A_{31}^T \\ 0 & A_{22} & A_{32}^T \\ A_{31} & A_{32} & A_{33} \end{bmatrix} \begin{bmatrix} x_1 = x_1^s + x_1^d \\ x_2 = x_2^s + x_2^d \\ x_3 \end{bmatrix} = \begin{bmatrix} f_1 \\ f_2 \\ f_3 \end{bmatrix}. \tag{1}$$

Sparse factorization or incomplete factorization of these submatrices are used to compute a solution vector.

SF is related to our earlier work, Booth, Chatterjee, Raghavan, & Frasca. [6], which uses Eq. 1, and we refer to this method as DI. In DI, sparse Cholesky is applied to A_{11} and A_{22}, incomplete Cholesky to the submatrix corresponding to the separator, and A_{31} and A_{32} remain unfactored. After factorization, DI applies direct forward/backwards solves to find x_1^d and x_2^d, and uses preconditioned CG to solve for x_3. Finally, DI uses forward/backward solves to find x_1^s and x_2^s. SF uses DI's framework, and provides a new formulation by expanding A_{33} and using factorization on A_{31} and A_{32}.

3   Substituted Factorization Formulation

In this section, we present our alternative direct method known as *substituted factorization* (*SF*). For our method, we consider a sparse linear system $Ax = f$, where A is a $n \times n$ sparse symmetric positive definite matrix ($A \in \mathbb{R}^{n \times n}$), $f \in \mathbb{R}^{n \times 1}$, and $x \in \mathbb{R}^{n \times 1}$. We wish to solve for x, and we rely on a graph representation of the sparse matrix A to construct a tree-structure. In this graph representation, each row/column represents a node and an undirected edge exists iff $A(i,j) \neq 0$. Additionally, self edges, i.e., $A(i,i) \neq 0$, are removed.

Nodes in the graph are split into two disconnected sets of nodes (Ω_0 and Ω_1) and a small set of nodes (Λ_0) that separates them. Λ_0 is commonly called a separator and Ω_is the domains. In a tree-structure, each separator is the parent of two domains. These domains can be split recursively. When done recursively, the resulting tree has leaf nodes of disconnected domains and internal parent nodes of separators. This tree-structure may impose a new ordering for the sparse matrix A by numbering the nodes corresponding to separators after those of their children. This ordering is known as nested dissection (ND) [10,11].

By splitting the graph twice and reordering, we rewrite A and its associated Cholesky factorization as follows:

$$
\begin{pmatrix}
A_{11} & A_{31}^T & & & & & A_{71}^T \\
& A_{22} & A_{32}^T & & & & A_{72}^T \\
A_{31} & A_{32} & A_{33} & & & & A_{73}^T \\
& & & A_{44} & A_{64}^T & A_{74}^T \\
& & & & A_{55} & A_{65}^T & A_{75}^T \\
& & & A_{64} & A_{65} & A_{66} & A_{76}^T \\
A_{71} & A_{72} & A_{73} & A_{74} & A_{75} & A_{76} & A_{77}
\end{pmatrix}
\quad \& \quad
\begin{pmatrix}
L_{11} \\
& L_{22} \\
L_{31} & L_{32} & L_{33} \\
& & & L_{44} \\
& & & & L_{55} \\
& & & L_{64} & L_{65} & L_{66} \\
L_{71} & L_{72} & L_{73} & L_{74} & L_{75} & L_{76} & L_{77}
\end{pmatrix},
$$

where $A = LL^T$. We divide the submatrices of the sparse Cholesky into two groups, namely domains and separators.

The domain ($i \in \{1, 2, 4, 5\}$) and their corresponding off-diagonal entries can be written as $A_{ii} = L_{ii}L_{ii}^T$ and $A_{ji} = L_{ji}L_{ii}^T$ where j is an index of a separator such that $j \in \{3, 6, 7\}$. Additionally, the diagonal entries corresponding to our domains will be relatively sparse compared to our separators after factorization. Therefore, we perform sparse Cholesky on both the diagonal and off-diagonal domain submatrices.

The separators ($i \in \{3, 6, 7\}$) submatrices have the following equations

$$A_{73} = L_{71}L_{31}^T + L_{72}L_{32}^T + L_{73}L_{33}^T, \tag{2}$$
$$A_{76} = L_{74}L_{64}^T + L_{75}L_{65}^T + L_{76}L_{66}^T, \tag{3}$$
$$A_{77} = L_{71}L_{71}^T + L_{72}L_{72}^T + L_{73}L_{73}^T + L_{74}L_{74}^T + L_{75}L_{75}^T + L_{76}L_{76}^T + L_{77}L_{77}^T. \tag{4}$$

Because these submatrices become dense during factorization, we avoid computing their factorization. The matrices, $L_{jj}L_{jj}^T$, for the top levels are commonly referred to as the Schur complement, and denote them as $\overline{A_{jj}}$. To avoid factorization, we solve each $\overline{A_{jj}}$ with associated right-hand side using a fixed number of iterations of CG guarantee to converge, thus making our method direct.

Traditional direct methods that implement sparse Cholesky have a forward and backward stage when solving for a given right-hand side (f). During the forward and backward stages, an updated right-hand side must be formed for each of the separating pieces, i.e., $j \in \{3,5,7\}$. These new right-hand sides (denoted $\hat{f}_j$) need updates by all $L_{j,i}$ where i corresponds to all node underneath j in the tree-structure, e.g., $\hat{f}_3 = f_3 - L_{31}L_{11}^{-1}f_1 - L_{32}L_{22}^{-1}f_2$. However, we save operations by not finding the factorization $L_{jj}L_{jj}^T$ and its corresponding off-diagonal entries for the top levels in our tree-structure. Instead, SF keeps the following equations from (2) and (3) as:

$$E_{73} = L_{73}L_{33}^T = A_{73} - L_{71}L_{31}^T - L_{72}L_{32}^T, \tag{5}$$
$$E_{76} = L_{76}L_{66}^T = A_{76} - L_{74}L_{64}^T - L_{75}L_{65}^T, \tag{6}$$

and substitutes them when forming the associated right-hand side for $\overline{A_{77}}$.

Using substitution, we can write the solution to the set of linear equations in two sets, namely forward and backward. The forward set is defined as:

$$x_1^f = L_{11}^{-1}f_1, \qquad x_4^f = L_{44}^{-1}f_4, \qquad x_3^f = \overline{A_{33}}^{-1}\hat{f}_3,$$
$$x_2^f = L_{22}^{-1}f_2, \qquad x_5^f = L_{55}^{-1}f_5, \qquad x_6^f = \overline{A_{66}}^{-1}\hat{f}_6,$$
$$x_7 = \overline{A_{77}}^{-1}\hat{f}_7,$$

where

$$\hat{f}_3 = f_3 - L_{31}x_1^f - L_{32}x_2^f,$$
$$\hat{f}_6 = f_6 - L_{64}x_4^f - L_{65}x_5^f,$$
$$\hat{f}_7 = f_7 - L_{71}x_1^f - L_{72}x_2^f - L_{74}x_4^f - L_{75}x_5^f - E_{73}x_3^f - E_{76}x_6^f.$$

Using this forward set, the backward set is defined as:

$$x_1 = L_{11}^{-T}(x_1^f - L_{31}^T x_3 - L_{71}^T x_7), \qquad x_5 = L_{55}^{-T}(x_5^f - L_{65}^T x_6 - L_{75}^T x_7),$$
$$x_2 = L_{22}^{-T}(x_2^f - L_{32}^T x_3 - L_{72}^T x_7), \qquad x_3 = x_3^f - \overline{A_{33}}^{-1}E_{73}^T x_7,$$
$$x_4 = L_{44}^{-T}(x_4^f - L_{64}^T x_6 - L_{74}^T x_7), \qquad x_6 = x_6^f - \overline{A_{66}}^{-1}E_{76}^T x_7.$$

4   Cost Analysis

In this section, we provide the cost analysis in terms of the number of multiplication operations (Ops) and in terms of number of nonzeros (NNZ) needed by SF. This analysis focuses on applying SF on the model $k \times k$ finite-difference 5-point stencil problem. We will examine the setup cost, i.e., the cost of factorization, and solve cost for SF compared to sparse Cholesky with ND ordering ($Chol$). The setup and solve cost are denoted as Γ^i_{setup} and Γ^i_{solve} where i is either SF or $Chol$. Additionally, we provide a comparison of the cost to solve multiple RHS, and provide a break even point between the two methods in terms of the number of RHS. Lastly, we provide the number of nonzeros needed to store the partially factorization in SF (Ψ).

Setup Cost. We first consider the setup cost of $Chol$ and SF. The setup cost is defined as the number of Ops during factorization. The setup cost to apply sparse Cholesky to a matrix A approximately equals $1/2 \sum_{i=1}^{n} nnz(i)^2$ where $nnz(i)$ is the number of nonzeros in the ith column of L, such that $A = LL^T$ [2]. When A is reordered with ND, George [10] provides that $\Gamma^{Chol}_{setup} \approx \frac{267}{28}k^3 - 17k^2 \log_2 k + \frac{847}{28}n^2 + O(k \log_2 k)$.

   SF has the same number of Ops as the previous analysis minus the cost to factor the top level "+" separator which accounts for A_{33}, A_{66}, and A_{77}. The cost to factor the "+" separator is $\sum_{i=k}^{3k/2} i^2 - \frac{1}{2} \sum_{i=1}^{k} i^2 = \frac{23}{24}k^3 + \frac{7}{8}k^2 + \frac{1}{6}k$.

   Therefore, our setup cost equals:

$$\Gamma^{SF}_{setup} \approx \frac{1441}{168}k^3 - 17k^2 \log_2 k + \frac{235}{8}k^2 + O(k \log_2 k). \tag{7}$$

As our result, setup cost for SF has approximately **10 %** fewer Ops than $Chol$.

Solve Cost. When solving using $Chol$, the number of Ops is on the order of the number of nonzeros. CG with no precondition requires $O(n^{1.5})$ Ops to solve the model problem [8]. George and Liu [2] state $\Gamma^{Chol}_{solve} = 31/4k^2 \log_2 k + O(k^2)$.

   SF performs sparse Cholesky on the leaf nodes and a fixed number of CG steps on the parent nodes. These parent nodes correspond to the Schur complements, and these nodes have a bounded condition number less than that of the whole system. For the model problem, the condition number is $O(h^{-2})$, and the condition number for our Schur complement will be less than $O(h^{-1}) \approx k$ [9]. Additionally, we know that CG will theoretically converge to a solution in $\sqrt{\delta(A)}$ with exact arithmetic [8]. Using these two facts, we set our fixed number of iterations to $\gamma = \beta\sqrt{k}$ where $\beta > 0$ is some unknown constant. For SF, the cost for a solve is

$$\Gamma^{SF}_{solve} = \frac{31}{4}k^2 \log_2 k + (c-2)k^2 + 2\beta k^{5/2} \tag{8}$$

where c is the constant from Γ^{Chol}_{solve}.

Number of Solves. Despite slightly more Ops during solving, SF has fewer Ops for factorization and solving some m RHS. The m is easily seen to be $\frac{23}{24}k^3 \approx 2\beta k^{5/2}m$, which translates to $m \approx \frac{23}{48\beta}k^{1/2}$. In Fig. 1, the number of right-hand side vectors solved is plotted that can be solved while costing less than or equal to a direct method using sparse Cholesky. Each line represents the variation due to the number of iterations needed.

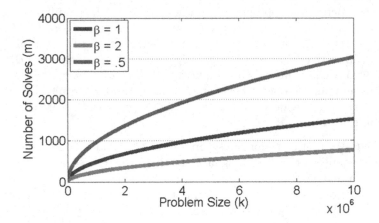

Fig. 1. Number of right-hand sides solved

Number of Nonzeros. *SF* has the same number of nonzeros as full sparse Cholesky minus the nonzeros to store L_{33}, L_{66}, and L_{77}. The number of nonzeros in a full sparse Cholesky is $\Psi^{Chol} \approx \frac{32}{4}k^2 \log_2 k - \frac{73}{3}k^2 + 24k \log_2 k + O(k)$ [10]. The reduction of nonzeros would be equal $\frac{7}{4}k^2$ resulting in

$$\Psi^{SF} \approx \frac{32}{4}k^2 \log_2 k - \frac{271}{12}k^2 + 24k \log_2 k + O(k). \tag{9}$$

For $k = 1000$, the number of nonzeros would reduce by approximately 5.9 %.

5 Experiments and Evaluation

In this section, we present numerical experimentation of *SF*. *SF* is constructed using DSCPACK [12] for factorization, and using WSMP [13] for CG. The code is constructed using *C* and *FORTRAN*, and compiled using *gnu 4.7.2* compilers with *O3* optimization. GotoBLAS2 [14] is used by all solvers. We compare Ops to DSCPACK (DSC), the direct symmetric WSMP, and the hybrid solver from Booth et al. [6] (DI). However, we only compare times against DSC and WSMP, since DI is coded in MATLAB. Additionally, we compare *SF* against WSMP for preconditioned CG.

All measurements are taken on a Linux SUSE 12 system. This system contains a 2nd generation Intel i7 processor and 2 GB of DDR3 DRAM. All runs are done in serial with hyperthreading and Turbo-Boost turned off.

Test Suites. Our experimentation evaluates *SF* over a collection of 2D and 3D problems. Test matrices are divided into two test suites. *Test suite A* comprises

Table 1. *Test suite A*

Matrix	n	nonzeros
2DL300	90,000	448,800
2DL500	250,000	1,248,000
2DL700	490,000	2,447,200

Table 2. *Test suite B*

Matrix	n	nonzeros
3DL40	64,000	401,896
3DL50	125,000	802,376
Brick20	13,860	1,912,944
Brick30	44,640	6,500,784

of $k \times k$ 5-point stencil Laplacian problems (*2DL*) of varying size. *Test suite A* evaluates the analysis of the previous section using perfect separators similar to those in the analysis. *Test suite B* contains 3D problems, and uses separators found using a multi-level partitioning scheme from WSMP. *Test suite B* comprises of 3D regular grid Laplacian problems (*3DL*) and stiffness matrices for a brick in 3D using 20 node serendipity elements and the equations of linear elasticity (*Brick*). Brick is generated using *ex10.c* from PETSc [15]. These matrices are found in Tables 1 and 2.

Metrics for Evaluation. We evaluate methods in terms of the number of Ops and time to factorand solve (T). When comparing Ops, we will commonly use relative improvement ($ROps(x)$) to compare method x to WSMP, and compare the time for setup and solving m RHS for method x to WSMP as $RTime(x)$, i.e., $ROps(x) = \dfrac{\text{Ops(WSMP)}-\text{Ops(x)}}{\text{Ops(WSMP)}}$, $RTime(x) = \dfrac{\text{T(WSMP)}-\text{T(x)}}{\text{T(WSMP)}}$.

Evaluation of Test Suite A: 2D Laplacian Problems. We evaluate our cost analysis of the 2D stencil problem in Figs. 2 and 3 using 1 and 20 solves. For this evaluation, we assume that β is $1/2$ and therefore use the fixed number of $1/2\sqrt{k}$ CG iterations. For all RHS, the solution found with *SF* has error on the same order of magnitude as WSMP. We notice in Fig. 2 that *SF* reduces the number of Ops similar to predicted in the analysis section by approximately 10 %. The slight increase of reduced Ops with size comes from Ops reduction during solve. This growing reduction with problem size demonstrates the usefulness of *SF* for large problems, and this is further demonstrated in Fig. 3. Figure 3 demonstrates the Ops for solving 20 RHS will be better for *SF* on larger problems than *Chol*. Note from our analysis and assumption of β, the number of RHS while costing less is $\approx .96\sqrt{k}$.

In further investigation, we examine the execution time for solving 1 and 20 RHS in Figs. 4 and 5. We first notice *SF*'s improvement over other methods when performing one solve and how *SF*'s improvement increases with the size of the problem from 21 % to 31 %. Even when solving 20 RHS, our method outperforms sparse Cholesky when $k > 500$.

Additionally, we examine the memory reduction for using *SF* compared to sparse Cholesky. In particular, we consider the metric

$$RNNZ(x) = \frac{NNZ(WSMP) - NNZ(x)}{NNZ(WSMP)} \tag{10}$$

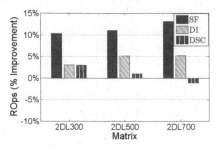

Fig. 2. ROps 1RHS

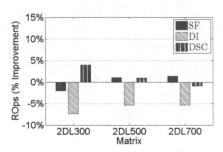

Fig. 3. ROps 20RHS

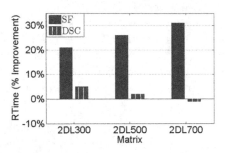

Fig. 4. RTime 1RHS

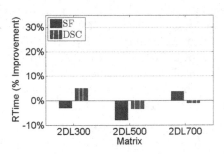

Fig. 5. RTime 20RHS

in order to validate our memory analysis. The resulting $RNNZ$ for *test suite A* is found in Table 3. We observe that the number of nonzeros reduced decreases as the grid size increases.

Table 3. The relative decrease in the number of nonzeros need by *SF* compared to sparse Cholesky.

$Matrix$	2DL300	2DL500	2DL700
$RNNZ$	7.80 %	6.70 %	6.10 %

Evaluation of Test Suite B: 3D Laplacian and Brick. Our second set of problems represents a key area, because they aremore difficult for iterative methods, e.g., CG, and ND using multi-level partitioning schemes may have larger separators. For this set, the fix number of CG iterations is set to 1000 based on observed condition number, however we allow CG to stop if the relative error of the solution is below $O(10^{-12})$. All the solutions have error similar in magnitude to *Chol*.

Figures 6 and 7 present the relative time for *SF* and DSC relative to WSMP for 1 and 50 RHS. We first notice the execution times are better than those from *test suite A*. The execution time is greater than 40 % for all matrices solving once. This better performance is in part due to the larger Schur complement in

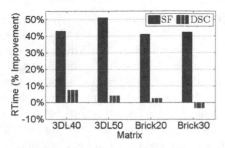

Fig. 6. RTime 1 RHS

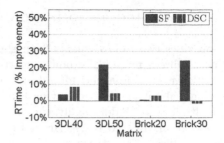

Fig. 7. RTime 50 RHS

the top levels of our tree. For Brick30, the top level separator dimension is 3960 where the largest 2D top level separator dimension is 700.

Additionally, we observe that 3D problems allow *SF* to solve a greater number of RHS while costing less than sparse Cholesky. In Fig. 7, *SF* solves 50 RHS before costing more than sparse Cholesky, and demonstrates a trade-off between the number of solves and the number of nonzeros.

Experimentation with Iterative Solvers. *SF* outperforms DI which is shown to be better for multiplesolves than precondition CG in Booth et al. [6]. Here we demonstrate that *SF* does better than preconditioned conjugate gradients with level-0 incomplete (PCG). We demonstrate *SF* outperforms PCG for both test suites reordered with reverse Cuthill-McKee (RCM) by examining the relative time improvement over PCG, i.e.,

$$RTimeI(x) = \frac{T(PCG) - T(x)}{T(PCG).} \tag{11}$$

In Fig. 8, the RTimeI for *test suite A* is given for 1 RHS, and we notice that the savings can be as high as 57.9 %. In Fig. 9, the RTimeI for *test suite B* is given for 1 RHS, and we notice that the savings are not as large as only 14.9 %. For both *test suite A and B*, the relative time improvement decreases as the size of the problem increases.

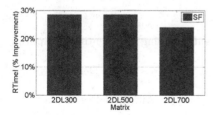

Fig. 8. RTimeI 1 RHS Test Suite A

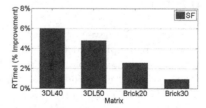

Fig. 9. RTimeI 1 RHS Test Suite B

6   Substituted Factorization Contributions

SF provides two improvements from the work Booth et al. [6]. First, we construct a direct method by bounding the number of iterations and using CG. Second, we provide a new formulation that allows for the factorization of the off-diagonal entries, thus removing the need for both forward/backward solves at multiple steps. Additionally, the off-diagonal entries allow for our method to expand terms related to A_{33} from Eq. 1 into submatrices that will be solved using our derived substitutions. This expansion reduces Ops and provides better matrix conditioning as observed in the experimental section.

7   Conclusions

Finding the most efficient solutions of sparse linear systems arising from finite discretization of second-order elliptic PDEs is very important to the performance of scientific codes. We have developed *substituted factorization* (SF) that finds solutions with the robustness of direct methods in less time than a direct solution based on sparse Cholesky factorization for a given number of solves. Specifically, the number of floating-point operations during factorization will reduce by 10 % for the $k \times k$ finite-difference problem. Additionally, experiments on our test suites demonstrate that execution time may reduce by as much as 48.9 % on 3D problems and 31.7 % on our 2D problems. *SF* method provides a needed method between sparse Cholesky and CG in terms of the number of solves needed to amortize factorization or preconditioning cost.

Acknowledgment. We would like to thank Anshul Gupta for all his assistance with WSMP that made our experiments possible. Additionally, we acknowledge the support of NSF grants CCF-0830679, CCF-1319448, CNS-1017882.

Sandia National Laboratories is a multi-program laboratory managed and operated by Sandia Corporation, a wholly owned subsidiary of Lockheed Martin Corporation, for the U.S. Department of Energy's National Nuclear Security Administration under contract DE-AC04-94AL85000.

References

1. Heroux, M., Raghavan, P., Simon, H.: Parallel Processing for Scientific Computing. SIAM, Philadelphia (2006)
2. George, A., Liu, J.: Computer Solution of Large Sparse Positive Definite Systems. Prentice-Hall, Englewood Cliffs, NJ, USA (1981)
3. Golub, G.H., Van Loan, C.F.: Matrix computations, 3rd edn. Johns Hopkins University Press, Baltimore, MD, USA (1996)
4. Gupta, A., Kumar, V.: A scalable parallel algorithm for sparse cholesky factorization. In: Proceedings of Supercomputing 1994, pp. 793–802 (1994)
5. Saad, Y.: Iterative Methods for Sparse Linear Systems, 2nd edn. Society for Industrial and Applied Mathematics, Philadelphia, PA, USA (2003)

6. Booth, J.D., Chatterjee, A., Raghavan, P., Frasca, M.: A multilevel cholesky conjugate gradients hybrid solver for linear systems with multiple right-hand sides. Procedia Comput. Sci. **4**, 2307–2316 (2011)
7. Gaidamour, J., Henon, P.: A parallel direct/iterative solver based on a schur complement approach. In: 11th IEEE International Conference on Computational Science and Engineering, CSE 2008, pp. 98–105 (2008)
8. Bern, M., Gilbert, J., Hendrickson, B., Nguyen, N., Toledo, S.: Support-graph preconditioners. SIAM J. Matrix Anal. and Appl. **27**(4), 930–951 (2006)
9. Mansfield, L.: On the conjugate gradient solution of the schur complement system obtained from domain decomposition. SIAM J. Numer. Anal. **27**(6), 1612–1620 (1990)
10. George, A.: Nested dissection of a regular finite element mesh. SIAM J. Numer. Anal. **2**, 24–45 (1973)
11. Heath, M.T., Raghavan, P.: A cartesian parallel nested dissection algorithm. SIAM J. Matrix Anal. Appl. **16**(1), 235–253 (1995). doi:10.1137/S0895479892238270
12. Raghavan, P.: DSCPACK. Technical report CSE-02-004, Penn State (2002)
13. Gupta, A., Joshi, M.: WSMP: A high-performance shared- and distributed-memory parallel sparse linear equation solver (2001)
14. Goto, K., van de Geijn, R.A.: High-performance implementation of the level-3 BLAS. ACM Trans. Math. Softw. **35**(1), 16:1–16:11 (2008)
15. Balay, S., Brown, J., Buschelman, K., Eijkhout, V., Gropp, W.D., Kaushik, D., Knepley, M.G., McInnes, L.C., Smith, B.F., Zhang, H.: PETSc users manual. Technical report ANL-95/11 - Revision 3.4, Argonne National Laboratory (2013)

Modeling 1D Distributed-Memory Dense Kernels for an Asynchronous Multifrontal Sparse Solver

Patrick R. Amestoy[1], Jean-Yves L'Excellent[2], François-Henry Rouet[3], and Wissam M. Sid-Lakhdar[4($\boxtimes$)]

[1] University of Toulouse, INPT(ENSEEIHT)-IRIT, Toulouse, France
[2] University of Lyon, Inria and LIP (CNRS, ENS Lyon, Inria, UCBL), Lyon, France
[3] Lawrence Berkeley National Laboratory, Berkeley, USA
[4] University of Lyon, ENS Lyon and LIP (CNRS, ENS Lyon, Inria, UCBL), Lyon, France
mohamed.sid_lakhdar@ens-lyon.fr

Abstract. To solve sparse systems of linear equations, multifrontal methods rely on dense partial LU decompositions of so-called frontal matrices; we consider a parallel asynchronous setting in which several frontal matrices can be factored simultaneously. In this context, to address performance and scalability issues of acyclic pipelined asynchronous factorization kernels, we study models to revisit properties of left and right-looking variants of partial LU decompositions, study the use of several levels of blocking, before focusing on communication issues. The general purpose sparse solver MUMPS has been modified to implement the proposed algorithms and confirm the properties demonstrated by the models.

1 Introduction

Multifrontal methods [2] are widely used to solve sparse systems of equations of the form $Ax = b$, where A is a sparse matrix, b is the right-hand side and x the unknown. They cast the factorization of the sparse matrix A into a series of partial factorizations of smaller dense matrices, called *fronts*, or *frontal matrices*. The dependency graph between those partial dense factorizations is a tree (the *assembly tree*), processed from the leaves to the root, such that the Schur complement so called *contribution block (CB)* produced after the partial factorization of a front is used at the parent node to build the front of the parent in a so-called *assembly operation*, before the parent node is in turn partially factored.

In this paper, we focus on the dense factorization kernels used in multifrontal methods for unsymmetric matrices where an LU decomposition is applied. For more information on multifrontal methods, we refer the reader to [11,15]. Much work has been done and is being done by the dense linear algebra community on LU factorizations, using for example static 2D block-cyclic data distributions [8], sometimes 2.5D communications [19], or DAG-based tiled algorithms

© Springer International Publishing Switzerland 2015
M. Daydé et al. (Eds.): VECPAR 2014, LNCS 8969, pp. 156–169, 2015.
DOI: 10.1007/978-3-319-17353-5_14

in both shared memory [1,7] and distributed-memory environments [5]. Recent asynchronous approaches often rely on a task scheduling engine [4,6] and on fine-grain parallelism for an efficient utilization of the computing resources. Most often, the choice of using an asynchronous approach with fine-grain parallelism in both directions (2D) implies relaxed pivoting strategies (such as *tournament pivoting*, typically used in communication-avoiding algorithms [13]). This is because neither full rows nor full columns are available to test for pivots stability. This is especially the case in distributed-memory environments, with the exception of the (synchronous) ScaLAPACK library [8].

In multifrontal-based, asynchronous, distributed-memory sparse factorization methods, many dense frontal matrices may be factorized simultaneously. Processes might thus be involved in more than one dense factorization, depending on dynamic scheduling decisions based on current CPU load and memory usage of each process and this is thus quite difficult to predict. We are also concerned with numerical accuracy and thus want to maintain standard numerical threshold pivoting [10] even in a distributed-memory context, which is quite a unique feature for a general purpose distributed-memory solver. In this context, a one-dimensional distribution of the dense factorization of fronts makes sense and has been adapted [3]. We are thus interested in analyzing and pushing the limits of this one-dimensional distribution. As we will show, analytical models can be complex because of the discrete nature of the phenomena. We have therefore also developed simulator that models parallel executions for standard blocked variants (so-called left and right-looking [12]) of the dense factorization of multifrontal fronts. We note that our objective here is to model the individual dense multifrontal kernels and not an entire sparse multifrontal factorization, although the findings will have an impact on the overall peformance of a sparse multifrontal factorization. Although cyclic pipelined factorizations have been modeled in the past [9], we are not aware of a clear illustration of the natural and intuitive properties of left-looking and right-looking approaches in a context comparable to ours, with acyclic factorizations, and where the process in charge of factorizing rows of the matrix does it either in an LL or RL way whereas other processes always perform their updates as soon as possible in a RL manner. For ScaLAPACK which relies on a 2D block cyclic distribution, right-looking is preferred over left-looking [8]; however, with our 1D technique, our conclusion is different, as will be illustrated in this paper.

The paper is organized as follows. In Sect. 2, we first study the theoretical behaviour of left-looking and right-looking variants for both one and two levels blocked algorithms. In order to better reveal and illustrate some of the intrinsic properties of those algorithms, we first consider a network with infinite bandwith. Communication models are then studied in more detail in Sect. 3 where we also analyse buffer memory requirements, cost of asynchronous one-to-many communications and impact on the blocked variants. This analysis has been used to modify the general purpose distributed-memory solver MUMPS [3] and to illustrate in Sect. 4 the benefits of the proposed approach in a distributed-memory environment.

2   Modeling Left-Looking and Right-Looking Computations

We consider a distributed-memory dense partial factorization relying on a dynamic asynchronous pipelined algorithm. A one-dimensional (1D) data distribution is used to allow for efficient pivot searches without synchronization between processes. In order to partially factorize the first *npiv* rows/columns of a front of order *nfront* using *nproc* MPI processes, one process designated as the *master* will handle the factorization of the *npiv* rows and the *nproc*-1 other processes (called *workers*) will manage the update of the so called *CB rows* of size *ncb* = *nfront* − *npiv* (see Fig. 1). The master uses a blocked *LU* algorithm with threshold partial pivoting: pivots are checked against the magnitude of the row but pivots can only be chosen within the first *npiv* × *npiv* block. After factorizing a panel of size *npan*, the master sends it to the workers in a non-blocking way, along with pivoting information. The master can immediately update its remaining non-factored rows (right-looking approach) or postpone this to when the next panel will start (left-looking approach). In parallel, the workers update all their rows at each panel reception. Thus, the behaviour of the workers always follows a right-looking scheme. The factorization operations rely on BLAS1 and BLAS2 routines inside panels, whereas update operations (both on master and workers) rely on BLAS3 routines, where TRSM is used to update columns of newly eliminated pivots and GEMM is used to update the remaining columns. For the sake of clarity, we consider that CB rows are uniformly distributed over the workers.

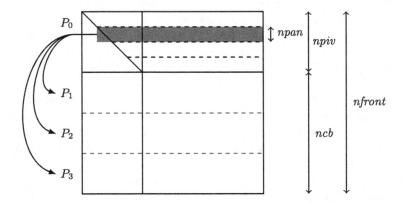

Fig. 1. Partial factorization of a front of order *nfront*, with *npiv* variables to eliminate by panels of *npan* rows, and *ncb* = *nfront* − *npiv* rows to be updated.

We have first modeled the factorizations analytically. Figure 2 shows the context and main notations. We have used the *MAPLE* software to help in this task, due to the complexity of the equations arising when solving problems such as finding optimal parameters of the factorizations.

Equation 1 represents the number of floating-point operations necessary for the factorization of a panel of k rows and $k + n$ columns.

$$W\!f(k,n) \rightarrow \left(\frac{2}{3}\right) k^3 + \left(n - \frac{1}{2}\right) k^2 - \left(n + \frac{1}{6}\right) k \qquad (1)$$

This is the result of the sum of the floating-point operations of the factorization of each row:

$$W\!f(k,n) \rightarrow \sum_{i=k-1}^{0} i + 2 * i * (i + n) \qquad (2)$$

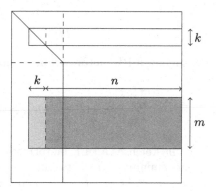

Fig. 2. Illustration of the factorization of a panel of size $k \times (k + n)$ on the master and the corresponding update on a worker. The light and dark gray areas represent the pieces of the front on a worker on which a TRSM and GEMM are applied respectively.

Equation. 3 represents the number of floating-point operations necessary for the update of a block (factorization of the L factors and update of the contribution part) of m rows and $k + n$ columns by a panel of k rows and $k + n$ columns (we thus assume a right-looking algorithm).

$$Wu(m,n,k) \rightarrow W_{TRSM}(m,k) + W_{GEMM}(m,n,k) \qquad (3)$$

with

$$W_{TRSM}(m,k) \rightarrow mk^2 \qquad (4)$$

and

$$W_{GEMM}(m,n,k) \rightarrow 2\,mnk \qquad (5)$$

– Given a GFlops/s rate β for update operations (TRSM and GEMM), MU_i, the time of update related to the ith panel by the master is given by:

$$MU_i = \beta \times Wu\,(npiv - \min\,(npiv, i * npan)\,, npiv + ncb-\,,$$
$$\min\,(npiv, i * npan)\,\min\,(npan, npiv - (i - 1)\,npan)) \quad (6)$$

– SU_i, the time of update related to the ith panel by a worker is given by:

$$SU_i = \beta \times Wu \left(\frac{ncb}{nslave}, npiv + ncb - \min(npiv, i * npan), \right.$$

$$\left. \min(npan, npiv - (i-1)\,npan) \right) \qquad (7)$$

– Given a GFlops rate α for the panel factorization (including some BLAS2 operations), MF_{i+1}, the time of factorization of the $(i+1)^{th}$ panel (if it exists) by the master is given by:

$$MF_{i+1} = \alpha \times Wf \left(\min \left(npan, npiv - npan \min \left(i, floor \left(\frac{npiv}{npan} \right) \right) \right), \right.$$

$$\left. npiv + ncb - (i+1) * npan \right) \qquad (8)$$

The total factorization time of a RL factorization is then given by Eq. 9:

$$TRight = MF_1 + \sum_{i=1}^{ceil\left(\frac{npiv}{npan}\right)} \max(SU_i, MU_i + MF_{i+1}) \qquad (9)$$

An algorithm where the master uses an LL factorization can be modeled in a similar way. Furthermore, communication costs can also be taken into account in Formula 9 in a simple manner. At the price of complicated formulas it is then possible with the help of Maple to build analytical formulas to express some properties (efficiency, speed-up, ...). However, we have preferred to consider the implementation of a simpler Python simulator for distributed-memory factorizations. Our simulator is naturally able to take into account varying communication and computation models and to produce Gannt-charts of the factorization. In order to illustrate some intrinsic properties of the algorithms that do not depend on the network bandwidth, we consider, to start with, that communications take place on a network with infinite bandwidth γ and that computations take place at a constant GFlops rate ($\alpha = \beta$). Because the messages sent are always reasonably large, we consider that the network latency is always negligible.

Right-Looking and Left-Looking Algorithms. In order to better characterize the main properties of our algorithms, we consider here a situation where the number of floating-point operations (flops) on the master is equal to that of each worker. Figure 3 represents the Gantt charts for $nfront = 10000$ and $nproc = 8$ (in this case $npiv = 2155$ to equilibrate flops) using both right-looking (RL) and left-looking (LL) blocked factorizations on the master, while workers perform their updates at each received panel, in a right-looking way. In each subfigure, the Gantt chart on the top represents the activity of the master and the bottom one that of a single worker. Because all workers theoretically behave the same way, only one worker is represented in the Gantt chart. Figure 3(a)

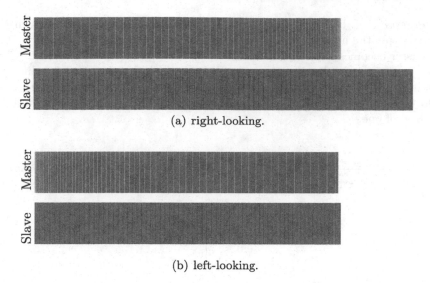

(a) right-looking.

(b) left-looking.

Fig. 3. Gantt chart of the RL and LL algorithms. Factorization in green, updates in blue and idle times in red (Color figure online).

clearly illustrates the weakness of the RL approach. Given that *npiv* balances the total amount of work (flops) between master and workers, one would expect all processes to finish at the same time. However, the workers finish much later because they have idle phases that sum up to the gap between master and workers completion times. When computing the first panels, the master process performs more update operations than the workers, which makes them become idle. The amount of update operations relative to each panel decreases faster on the master process than on the workers, and idle times decrease. When panels get smaller, the master process performs less operations than the workers and sends panels to the workers quicker than the workers manage to perform the corresponding updates; the workers then work continuously, desperately trying to catch up with their delay. As the consumption of factored panels is critical on the workers, the master should produce panels as soon as possible, delaying its own updates as much as possible. A solution consists in applying on the master a left-looking algorithm instead, resulting in the perfect Gantt chart of Fig. 3(b). In the following subsections, we compare the behavior of both variants.

Load Balance and Scalability. Although the ratio between *npiv* and *nfront* is mainly defined by the sparsity pattern of the matrix to be factored, we will show at the end of this section that we have some leeway to modify this ratio; in Fig. 4(a), we study the influence of *npiv* for a fixed *nfront*. We distinguish three parts, depending on *npiv*. In the first part, for *npiv* under a certain value $npiv_0$ ($npiv_0 \approx 5000$), LL and RL algorithms behave exactly the same: workers are the bottleneck because they have much more work than the master. For

$npiv > npiv_0$, LL becomes better than RL: $npiv_0$ is the value above which the time to apply the update (RL) of the first panel and factorize the second one on the master becomes bigger than time to apply the update associated to the first panel on the worker. Both variants reach their peak speed-up but for different values of $npiv$. Then, for large values of $npiv$, the master has much more work to do than the workers and becomes the bottleneck, leading to an asymptotic speed-up of one.

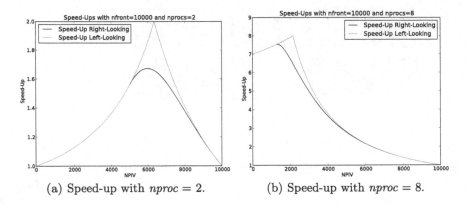

(a) Speed-up with $nproc = 2$.   (b) Speed-up with $nproc = 8$.

Fig. 4. Influence of $npiv$ on LL and RL algorithms with 2 (left) and 4 (right) processes: speed-ups with respect to the serial version ($nfront = 10000$).

When $nproc$ is larger — Fig. 4(b), the maximum speed-ups of RL and LL tend to get closer. LL reaches its maximum speed-up when all processes (master and worker) get the same amount of computations *Flops equilibrium* (*eqFlops*), so that neither the master nor the workers are bottlenecks to each other. On the other hand, RL reaches its maximum speed-up when all processes (master and worker) are roughly assigned the same number of rows *Rows equilibrium* (*eqRows*). This latter approximation relies on the fact that this keeps workers always busy, leading to a speed-up at least equal to $nproc - 1$.

The previous theoretical model showed interesting results. However, in order to benefit from them, we must first ensure that some fundamental hypotheses hold true in practice. We show here the observed discrepancies and the algorithms and techniques we applied to fix or reduce them.

Generalization to Multiple Levels of Panels and to Arbitrary Front Shapes. The previous models showed that front factorizations are efficient when the ratio $\frac{npiv}{nfront}$ respects *eqRows* and *eqFlops* for RL and LL, respectively. In order to improve locality and BLAS3 effects on the master, recursive algorithms can be used [20]. However, at the first level of recursion, the update of the second block with the first one would take a significant amount of time, possibly making the workers idle for a huge period. The adopted solution consists in using multiple levels of blocking (in our case, two levels), which means computing an external panel using internal ones. Because the GFlops rate on the master may still be slightly lower than on the workers, corresponding to a smaller value of α than β in the models, one must slightly modify the *eqFlops* ideal $\frac{npiv}{nfront}$ ratio (for LL) in order to obtain $\frac{flops_{master}}{GFlops\ rate_{master}} = \frac{flops_{worker}}{GFlops\ rate_{worker}}$.

Another issue is that in practice, the multifrontal method results in frontal matrices that often have an $\frac{npiv}{nfront}$ ratio larger than the ideal one, especially for large *nproc*. Fortunately, assembly trees are not rigid entities and can be reshaped, for example using two standard operations known as *amalgamation* and *splitting*. Amalgamation consists in merging two related fronts into a single one (a child and its parent, usually). It has the advantage of generating larger fronts, which increases factorizations efficiency, sometimes at the cost of extra fill-ins, inducing more computations and memory requirements. Contrarily, splitting consists in cutting a front into a so called *split chain* of fronts such that in the chain, the Schur complement of a child is considered as a new, parent, front. We note that remapping may have to be done between two successive fronts in a chain, and that, although we consider *nproc* constant in our models and experiments, dynamic scheduling strategies may imply variations of *nproc* between two successive pipelined factorizations. Lost processes can be assigned to other fronts in other subtrees; vice versa, new processes can be assigned to parent fronts in the chain. In both cases, the shape of the fronts and the length of the chain should be modified accordingly, with the aim to obtain a correct balance of the work between master and workers in all intermediate fronts (except, possibly, for the last one). Simple models of such chains were discussed in [16] and have been revisited in [17]. In Fig. 5, we report the simulated speed-ups with varying *npiv* when this generalized approach is applied, with *eqRows* for RL and *eqFlops* for LL. For both RL and LL, the speed-ups are much less sensitive to

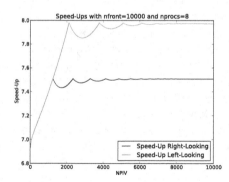

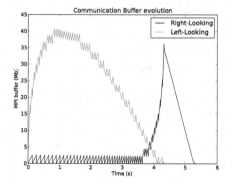

Fig. 5. Simulated generalized 1D factorization (*nfront* =10000, *nproc* =8) with varying *npiv*. LL (resp. RL) uses *eqFlops* (resp. *eqRows*).

Fig. 6. Amount of data sent but not ready to be received using RL an LL algorithms with *eqFlops* (*nproc* =8, *nfront* =10000, *npiv* =2155).

npiv (compared to Fig. 4) because each intermediate 1D factorization is now well-balanced. RL speed-ups are not as good as LL ones because of idle times on the master. When targeting an entire sparse matrix factorization rather than focusing on a single front or a chain of fronts, new kinds of load balancing issues arise, which are handled in our context study by a dynamic and asynchronous scheduling approach, which adapts to the load of the processes.

3   Modeling Communications

Memory for Communication. Assuming that sends are performed as soon as possible, Fig. 6 represents the evolution of the memory utilization in the send buffer for LL and RL factorizations, both with *eqFlops*. This send buffer is the place in memory where panels computed by the master are temporarily stored (contiguously) and sent using non-blocking primitives; when the workers start receiving, send buffer can often be freed. This allows for an overlap of computations and communications, and allows the main process to manage its memory independently of the advancement of communications. The memory utilization in the send buffer then represents the volume of data that has been sent and that is not received yet. Its size needs to be controlled and limited: a full send buffer implies in practice that the sender will wait for receptions to occur before being able to perform a new send. Most of the time, the buffer in the RL variant only contains one panel, immediately consumed by the workers; When master computations shrink (for the last panels), the master rapidly produces many panels that cannot be consumed immediately. In contrast, the LL variant always has enough panels ready to be sent. This is because RL with *eqFlops* is not able to correctly feed the workers, whereas the LL does. Second, the peak of buffer memory used for RL is 36 MB while it is 41 MB for LL. The scheduling advantage of LL thus comes at the price of a higher buffer memory usage. However, this additional memory becomes significant in comparison to the total memory used by the master process for the factorization (*nfront* * *npiv* * *sizeof*(*double*) = 172 Mb). Send buffers may have a given limited size in practice, smaller than the peaks from Fig. 6 (36 MB and 41 MB for RL and LL variants, respectively). If only a few panels can fit in buffer memory, the master must wait when the send buffer is full, leading to some performance loss. Instead, we prefer to copy new panels to the send buffer only when space is available in the buffer, independently of the fact that many more panels may have been computed. This study also shows that, in order to control buffer memory, messages should *not* be sent as soon as possible (but should still be sent early enough so that receivers do not have to wait).

Limited Bandwidth and Asynchronous Collective Communications. We observed experimental results to be very similar to those of the model, as long as the ratio between computations and communications remains large enough (*nfront* relatively large compared to *nproc*). Strong scaling, i.e., increasing *nproc* for a given *nfront*, globally increases the amount of communications while keeping the amount of computations identical. The master process sends a copy of each panel to more workers, decreasing the bandwidth dedicated to the transmission of a panel to each worker: the maximal master bandwidth is divided by *nworkers* in this one-to-many communication pattern, making the communication of the panels from master to workers a possible bottleneck.

Many efficient broadcast implementations exist for MPI [21], and asynchronous collective communications are part of the MPI-3 standard. However the semantic of these operations requires that all the processes involved in the collective operation call the same function (MPI_IBCAST). This is constraining for our asynchronous approach which is such that any process, at any time, receives and treats any kind of message and task: we want to keep a generic approach where processes do not know in advance if the next message to receive in the main reception buffer is a factored panel or some other message. Furthermore, we need an asynchronous, pipelined broadcast algorithm which means that a binomial broadcast tree would not be appropriate since once a process has received a panel and forwarded it, its bandwidth will be needed to process next panel. For these reasons, we have designed our own asynchronous pipelined broadcast algorithm based on MPI_ISEND calls using a classical w-ary broadcast tree, as illustrated in Fig. 7(b). The Gantt charts of Fig. 7 show the impact of the communication patterns with limited bandwidth per process, using our Python simulator. With the baseline communication algorithm, the workers are most often idle, spending their time waiting for the communications to finish, before doing the corresponding computations, whereas the tree-based (here using a binary tree) has a perfect behaviour: the Gantt chart of the worker is only slightly translated

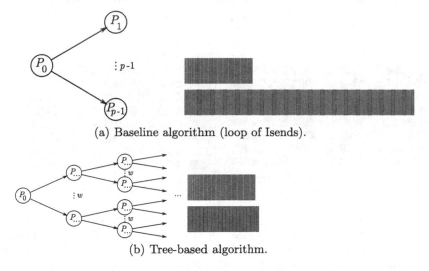

(a) Baseline algorithm (loop of Isends).

(b) Tree-based algorithm.

Fig. 7. Influence of the IBcast communication pattern with a limited bandwidth per proc (γ=1.2 Gb/s, α=10 GFlops/s) on LL algorithm with *nfront* = 10000, *npan* = 32, *nproc* = 32 and *npiv* chosen to balance work (idle times in red) (Color figure online).

in time (due to the time it takes to receive the first panel) and the remaining communications overlap well with computations. When further increasing *nproc* or with more cores per process, we did not always observe such a perfect overlap of communications and computations, but the tree-based algorithm always led to an overall transmission time for each panel of $\frac{nfront \times npan \times w \times \log_w(nproc)}{\gamma}$, much smaller than that of the baseline algorithm $\frac{nfront \times npan \times (nproc-1)}{\gamma}$. An IBcast-like scheme is thus of great importance when the number of processes grows.

4   Preliminary Experimental Results

In order to study the left-looking and right-looking variants of the 1D pipelined factorization algorithm from Sect. 2 on arbitrary fronts, we generalized the asynchronous factorization algorithms available in the MUMPS solver [3] in order to implement left-looking and right-looking variants with several levels of blocking. We use a Sandy Bridge-based cluster with 4×8 core nodes (*ada*, from IDRIS) as well as a Xeon-based SGI Altix ICE 8200 with 2×4 core nodes (*hyperion*, from CALMIP) Intel BLAS (MKL) and MPI libraries are used and, because asynchronous communications only progressed inside MPI calls, we use a progress thread [14] to force `MPI_TEST` calls every milisecond.

Figures 8(a) and 8(b) show the Gantt-charts of executions of a dense partial right-looking LU factorization on a front of size *nfront* = 10000 with *nproc* = 8 MPI processes, with a number of pivots to be eliminated following *eqFlops* and *eqRows*, respectively. We can see that Fig. 8(a) is very similar to what our model predicted (See Fig. 3). Moreover, we can see on Fig. 8(b) that the fact of respecting *eqRows* in the RL variant makes the workers wait much less than in the *eqFlops* case, which confirms the observations made thanks to our models.

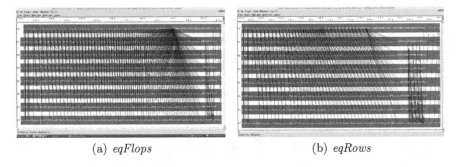

(a) *eqFlops*   (b) *eqRows*

Fig. 8. Gannt-chart of execution of a dense partial right-looking LU factorization on a front of size *nfront* = 10000 with *nproc* = 8 MPI processes, with a number of pivots to be eliminated either respecting *eqFlops* (on the left) or *eqRows* (on the right), on a shared-memory node (to validate the communication-less model).

Other experiments with real-life Gantt charts confirmed that *eqRows* is more adapted to RL and *eqFlops* is more adapted to LL. However, and as mentioned before, due to the fact that computations on the master (that partly uses BLAS2) are slower than on the workers, *eqFlops* (in case of LL) has to be slightly modified and was replaced by *eqTime*, such that $\frac{flops_{master}}{GFlopsrate_{master}} = \frac{flops_{worker}}{GFlopsrate_{worker}}$.

Table 1 confirms the interest of a tree-based pipelined *IBcast* algorithm. It also illustrates the interest of using two levels of panels. In all cases, we used a RL algorithm for internal panels, that was observed to be more efficient than LL on small blocks. Also, and as predicted in the models, *eqFlops* (and *eqTime*) led to bad results for RL; this is why we use *eqRows* in that table. Remark that, although *eqTime* would have been better suited to LL, we used *eqRows* even for LL in order to be able to compare the times of RL and LL on a front with the same characteristics.

Table 1. Influence of *IBcast* and of double-blocking on the factorization time (seconds) of a front, for RL and LL variants on the most external panels; "-" in column *npan2* indicates that a single level of panels is used.

Machine	*nfront*	*nproc*	(*ncores*)	*IBcast* tree	*npan1*	*npan2*	RL	LL
ada	100000	64	(512)	No *IBcast*	32	-	35.7	29.8
ada	100000	64	(512)	depth 2	32	-	22.8	26.2
ada	100000	64	(512)	binary	32	-	21.8	22.0
ada	100000	64	(512)	binary	64	-	21.2	21.1
ada	100000	64	(512)	binary	32	64	20.5	19.8
hyperion	64000	8	(64)	binary	32	-	203	204
hyperion	64000	8	(64)	binary	128	-	117	110
hyperion	64000	8	(64)	binary	64	128	97	93

Table 2 shows the impact of the asynchronous broadcast algorithm on the performance for a generalized frontal matrix with a binary *IBcast* tree when two levels of panels are used. It is interesting to note that *IBcast* gains are larger when more cores are used per process, showing that communications become more critical in that case. When considering the factorization of an entire sparse matrix in a limited-memory environment [16], more workers have to be mapped on each front of the assembly tree. On 128 MPI processes of *hyperion*, on the factorization of an entire sparse matrix arising from a 3D finite-difference Laplacian problem

Table 2. Influence of *IBcast* on *hyperion* with *nfront* =*npiv* =64000. Factorization times in seconds.

Cores	Cores/ MPI process	Without	With
64	1	1702	1341
512	8	1380	404

on a 128^3 grid, we observed a time reduction from 805 to 505 s thanks to *IBcast* (see [17] for further results).

5   Conclusion

We modeled a dense asynchronous kernel for multifrontal factorizations, targeting large matrices and large numbers of cores. We studied both communication and computation aspects. The approach allows for standard threshold numerical pivoting, and can be integrated in a fully asynchronous environment with dynamic, distributed schedulers. Such an environment is precisely the one of the MUMPS solver [3], on which this work was shown to have a strong performance impact.

In the future, we plan to further optimize multithreaded kernels (inside each MPI process), and optimize the communication volume when remapping needs to be done between two successive pipelined factorizations. Topology-aware broadcast algorithms [18] are also a promising approach to further improve the cost of broadcasting factorized panels. Moreover, comparisons between models and experiments of dense factorizations will allow us to improve the performance results on full sparse multifrontal factorizations. Comparison with techniques used in HPL[1] would also be interesting.

Acknowledgement. This work was granted access to the HPC resources of CALMIP under the allocation 2013-0989 and GENCI/IDRIS resources under allocation x2013065063.

References

1. Agullo, E., Demmel, J., Dongarra, J., Hadri, B., Kurzak, J., Langou, J., Ltaief, H., Luszczek, P., Tomov, S.: Numerical linear algebra on emerging architectures: the PLASMA and MAGMA projects. J. Phys. Conf. Ser. **180**(1), 012037 (2009)
2. Amestoy, P.R., Buttari, A., Duff, I.S., Guermouche, A., L'Excellent, J.-Y., Uçar, B.: The multifrontal method. In: Padua, D. (ed.) Encyclopedia of Parallel Computing, pp. 1209–1216. Springer, Heidelberg (2011)
3. Amestoy, P.R., Duff, I.S., Koster, J., L'Excellent, J.-Y.: A fully asynchronous multifrontal solver using distributed dynamic scheduling. SIAM J. Matrix Anal. Appl. **23**(1), 15–41 (2001)
4. Augonnet, C., Thibault, S., Namyst, R., Wacrenier, P.-A.: StarPU: a unified platform for task scheduling on heterogeneous multicore architectures. Concurrency Comput.: Pract. Experience **23**(2), 187–198 (2011). Special Issue: Euro-Par 2009
5. Bosilca, G., Bouteiller, A., Danalis, A., Faverge, M., Haidar, A., Herault, T., Kurzak, J., Langou, J., Lemarinier, P., Ltaief, H., Luszczek, P., Yarkhan, A., Dongarra, J.J.: distibuted dense numerical linear algebra algorithms on massively parallel architectures: DPLASMA. In: Proceedings of the 25th IEEE International Symposium on Parallel & Distributed Processing Workshops and Ph.D. Forum (IPDPSW'11). PDSEC 2011, pp. 1432–1441. Anchorage, USA (2011)

---

[1] http://www.netlib.org/hpl/.

6. Bosilca,G., Bouteiller, A., Danalis, A., Herault, T., Lemarinier, P., Dongarra, J.: DAGuE: A generic distributed DAG engine for high performance computing. In: 16th International Workshop on High-Level Parallel Programming Models and Supportive Environments (HIPS'11) (2011)
7. Buttari, A., Langou, J., Kurzak, J., Dongarra, J.: A class of parallel tiled linear algebra algorithms for multicore architectures. Parallel Comput. **35**(1), 38–53 (2009)
8. Choi, J., Dongarra, J.J., Ostrouchov, L.S., Petitet, A.P., Walker, D.W., Whaley, R.C.: Design and implementation of the ScaLAPACK LU, QR, and Cholesky factorization routines. Sci. Program. **5**(3), 173–184 (1996)
9. Desprez, F., Dongarra, J.J., Tourancheau, B.: Performance complexity of LU factorization with efficient pipelining and overlap on a multiprocessor. LAPACK working note 67, Computer Science Department, University of Tennessee, Knoxville, Tennessee (1994)
10. Duff, I.S., Erisman, A.M., Reid, J.K.: Direct Methods for Sparse Matrices. Oxford University Press, London (1986)
11. Duff, I.S., Reid, J.K.: The multifrontal solution of unsymmetric sets of linear systems. SIAM J. Sci. Stat. Comput. **5**, 633–641 (1984)
12. Golub, G.H., Van Loan, C.F.: Matrix Computations, 2nd edn. Johns Hopkins Press, Baltimore (1989)
13. Grigori, L., Demmel, J., Xiang, H.: CALU: a communication optimal LU factorization algorithm. SIAM J. Matrix Anal. Appl. **32**(4), 1317–1350 (2011)
14. Hoefler, T., Lumsdaine, A.: Message progression in parallel computing - to thread or not to thread? In: IEEE International Conference on Cluster Computing, pp. 213–222 (2008)
15. Liu, J.W.H.: The multifrontal method for sparse matrix solution: theory and practice. SIAM Rev. **34**, 82–109 (1992)
16. Rouet, F.-H.: Memory and performance issues in parallel multifrontal factorizations and triangular solutions with sparse right-hand sides. Ph.D. thesis, Institut National Polytechnique de Toulouse, October 2012
17. Sid-Lakhdar, W.M.: Scaling multifrontal methods for the solution of large sparse linear systems on hybrid shared-distributed memory architectures. Ph.D. dissertation, ENS Lyon (2014, In preparation)
18. Solomonik, E., Bhatele, A., Demmel, J.: Improving communication performance in dense linear algebra via topology aware collectives. In: Proceedings of 2011 International Conference for High Performance Computing, Networking, Storage and Analysis, SC 2011, pp. 77:1–77:11. ACM, New York (2011)
19. Solomonik, E., Demmel, J.: Communication-optimal parallel 2.5D matrix multiplication and LU factorization algorithms. In: Jeannot, E., Namyst, R., Roman, J. (eds.) Euro-Par 2011, Part II. LNCS, vol. 6853, pp. 90–109. Springer, Heidelberg (2011)
20. Toledo, S.: Locality of reference in lu decomposition with partial pivoting. SIAM J. Matrix Anal. Appl. **18**(4), 1065–1081 (1997)
21. Wadsworht, D.M., Chen, Z.: Performance of MPI broadcast algorithms. In: Proceedings of the 22nd International Parallel and Distributed Processing Symposium (IPDPS 2008), pp. 1–7 (2008)

Performance Tuning

Performance Characteristics of HYDRA – A Multi-physics Simulation Code from LLNL

Steven H. Langer[(✉)], Ian Karlin, and Michael M. Marinak

Lawrence Livermore National Laboratory, Livermore 94551, USA
{langer1,karlin1,marinak1}@llnl.gov

Abstract. HYDRA simulates a variety of experiments carried out at the National Ignition Facility and other high energy density physics facilities. It has packages to simulate radiation transfer, atomic physics, hydrodynamics, laser propagation, and a number of other physics effects. HYDRA has over one million lines of code, includes MPI and thread-level (OpenMP and pthreads) parallelism, has run on a variety of platforms for two decades, and is undergoing active development.

In this paper, we demonstrate that HYDRA's thread-based load balancing approach is very effective. Hardware counters from IBM Blue Gene/Q runs show that none of HYDRA's packages are memory bandwidth limited, a few come close to the maximum integer instruction issue rate, and all are well below the maximum floating point issue rate.

Keywords: Large-scale simulations in CS&E · Multiscale and multiphysics problems · Performance analysis

1 HYDRA - A Multi-physics Simulation Code

The goal of this paper is to introduce readers to a complex "multi-physics" code, discuss some of the techniques used to improve performance, and use data from hardware counters to provide insight into the bottlenecks controlling the performance. We chose HYDRA [5,6], which is used to simulate experiments conducted at the National Ignition Facility (NIF) [7] and other pulsed laser facilities, as our test code. The laser deposits a large amount of energy in a small volume, so HYDRA is focused on simulating the processes of high energy density physics.

HYDRA is a "multi-physics" simulation code. Figure 1 shows the many physics packages in HYDRA and their interconnections. HYDRA has characteristics similar to other multi-physics codes at LLNL. It consists of over a million lines of code, has run on a variety of platforms for two decades, and is still undergoing active development. HYDRA runs a wide range of simulations and only a subset of the physics packages are used in any given run.

The rights of this work are transferred to the extent transferable according to title 17 § 105 U.S.C.

© Springer International Publishing Switzerland 2015
M. Daydé et al. (Eds.): VECPAR 2014, LNCS 8969, pp. 173–181, 2015.
DOI: 10.1007/978-3-319-17353-5_15

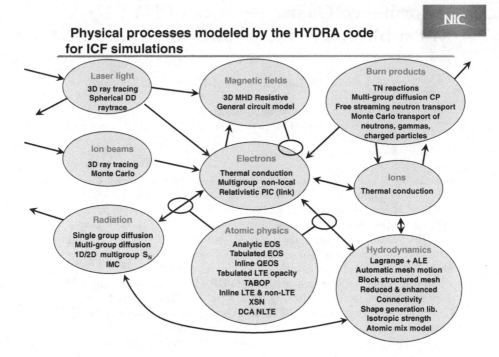

Fig. 1. HYDRA has many physics packages so that it can simulate a broad range of experiments, including those performed on the National Ignition Facility Laser.

This is the first paper to present a performance analysis of HYDRA. Single-physics codes may have a single loop which consumes over 90 % of the run time. HYDRA and other multi-physics codes have dozens of "hot loops", and the hot loops change from one run to the next. Multi-physics codes use a programming approach focused on portability and programmer productivity. Performance is very important, but optimizations need to work for a variety of systems. Modifying the code to increase the percentage of stride one accesses or enable generic SIMD utilization is worthwhile, but tuning for a particular SIMD unit is not.

HYDRA solves a set of coupled partial differential equations (PDEs) for time-dependent fields on a grid in three spatial dimensions. A large number of zones are often required to resolve small features. The temperature, density, and velocity depend only on the spatial coordinates, but the radiation field also depends on the photon energy. Simulations often use 100–200 energy bins so the radiation field and opacity arrays may dominate memory usage. The equations are solved using the method of operator splitting [2]. This essentially means having one function call (or one loop nest) for each term in the PDEs.

There is synchronization between all MPI processes at the end of each operator. This approach is referred to as "bulk synchronous" programming [9]. Bulk synchronous programs have loops which are much simpler than if all terms were evaluated in a single very large loop. This makes the code easier to write and maintain. Each team member specializes in a few areas of physics and rarely needs to look at code related to other physics packages.

The operators in HYDRA are applied to full domains and the arrays they operate on are large compared to cache. That means each operator pulls its input arrays from DRAM into the cache. It then performs calculations, and stores the updated arrays back to DRAM. Most fields are used by multiple operators, so they may make multiple round trips between DRAM and cache every time step. Operators using iterative methods may pull arrays into cache multiple times. A system needs to have enough memory bandwidth to fetch arrays in a time short compared to the compute time for a single operator, not the compute time for a whole time step.

2   HYDRA Characterization

HYDRA uses a block structured mesh. The mesh has one or more user blocks which correspond to major components of the object being simulated. For example, the capsule might be the first user block and the hohlraum wall the second user block in a NIF simulation. User blocks have curvilinear coordinates and a regular 3D grid topology. This is sometimes referred to as an "ijk grid". There is a one-to-one match of faces on adjacent user blocks. Most zones are surrounded by exactly 26 other zones. The number of neighboring zones may be more or less than 26 at the corner of a user block ("enhanced" and "reduced" connectivity). HYDRA uses domain decomposition of the spatial grid to implement MPI parallelism. User blocks are decomposed into multiple ijk MPI domains.

All major physics packages also have thread-level parallelism. In the case of hydrodynamics and some other packages, threading is over domains. If there are 4 hardware threads available per MPI process, the user requests 4 domains per process and one thread handles each domain. This threading is implemented via OpenMP directives and is done at a level high enough that OpenMP thread synchronization time is not an issue. There are a number of important physics packages where the computational cost of updating a zone varies by large amounts from domain to domain. The regions with the highest work load shift throughout the course of a run. Some HYDRA packages use a more complex threading approach to deal with this type of load imbalance.

The DCA package computes frequency-dependent opacities for all zones. Some zones require more work than others, particularly when the matter is not in local thermodynamic equilibrium. The DCA package varies the number of OpenMP threads per domain (based on timing from the last time step) so that the work per thread is roughly constant. As an example, HYDRA might have 8 MPI domains on an Intel Sandy Bridge node with 16 cores and 32 hardware threads. If one domain has much more DCA work than the other seven, it might be assigned 32 threads while the other domains have one thread each. This approach evens out the work per hardware thread on a single node, but it does not help when there is a large imbalance in the DCA work on different nodes.

Threads are statically bound to processes on a BGQ system, so the dynamic load balancing in DCA is turned off.

The threads in HYDRA's laser ray trace and IMC (Implicit Monte Carlo) packages cooperatively process a set of domains. On Linux clusters, two MPI processes per node are typically designated as "masters". The other processes on a node send their domains to the master processes and then become inactive. On Blue Gene/Q systems, each process makes several "replicates" of its domain. The IMC particles or laser rays for a domain are split among the replicates. In either case, the active processes have several domains. A genetic algorithm shuffles domains around until each process has a nearly constant amount of work (based on the last time step). A process assigned a "difficult" domain is also assigned several "easy" domains. Load balancing works well with 4 or more domains per active process.

These packages use pthreads with thread specialization. Each active process has a thread which handles all MPI message passing, so a thread safe MPI is not required. Another thread handles all updates of the energy deposition array (recording the net transfer of energy between the matter in a zone and the laser rays or IMC particles passing through it), so locks on the deposition array are not required. The remaining threads push IMC photons or trace laser rays.

3   Dynamic Load Balancing

The physics equations solved by HYDRA often require more work in some zones than in others. If the problem is divided up into equal sized domains and no provision is made for load imbalance, the run time will be much longer than for a well balanced job. The dynamic load balancing results presented below were obtained from a standard NIF hohlraum simulation on a cluster with dual socket Intel Sandy Bridge nodes and an Infiniband QDR interconnect.

Table 1. This table shows the time without and with dynamic load balancing for the laser, IMC, and DCA packages during a HYDRA NIF simulation. The speedup for the laser package was greater earlier in the run. The final line is the total physics time for the run. A 50 % speedup in the physics time is much appreciated by HYDRA's users.

Package	Time (sec)	Time (sec)	Speedup ratio
	No balance	Balanced	
Laser	162.2	130.6	1.24
IMC	392.6	234.4	1.68
DCA	6.8	3.6	1.90
Total	557.9	363.4	1.51

The results in Table 1 show that load balancing cuts the run time for some packages almost in half and reduces the overall run time of the NIF simulation to roughly two thirds of what it would be without load balancing. This is a large improvement compared to what is typically obtained by adjusting compiler optimization flags. The load balance varies from one simulation to the next. In one recent simulation, DCA load balancing reduced the run time to roughly 10 % of the time with no load balancing (i.e., a 10X speedup). Other multi-physics codes might benefit from adopting a load balancing approach like HYDRA's.

4   Studies on Blue Gene/Q

4.1   BGQ Overview

IBM's Blue Gene/Q was chosen as the system on which to gather performance data. The clock speed is 1.6 GHz and the chip uses the Power instruction set. A BGQ chip has 16 cores with 4 hardware threads each. Two threads per core are required to reach the maximum instruction issue rate of 16 integer instructions and 16 floating point instructions per node per cycle. The BGQ has a 4-wide SIMD floating point unit and has a fused multiply-add (FMA) instruction. Floating point instructions may perform from 1 to 8 floating point operations. The L1 cache is 16 kB per core. Each core has a 2 MB slice of L2 cache operated as part of a 32 MB shared cache. It has a high latency (roughly 50 ns) because it uses eDRAM and requires extra logic to bind the slices together. The Intel Sandy Bridge has a 11 ns latency to its L3 cache. The integer unit on the BGQ chip handles loads, stores, integer arithmetic, address computations, and a number of other instructions. Codes operating on arrays of floating point numbers issue many integer instructions as they load and store array elements and compute addresses. A BGQ system has a streams bandwidth [3] of 28 GB/s per node. The memory space is flat, so there is no need to worry about NUMA effects.

4.2   The BGQ Test Codes

HYDRA tests were run with 4 processes per node on 16 BGQ nodes. The 64 hardware threads on a node were equally divided amongst the 4 processes. Data from three other codes is provided to help assess whether HYDRA has unique performance bottlenecks.

pF3D [1,4,8] is a massively parallel code which simulates laser-plasma interactions in experiments using the National Ignition Facility laser and other high power lasers. pF3D has fewer packages than HYDRA, but still has 2 dozen performance critical loops. pF3D operates on 3D arrays which makes it easy to use stride one memory access in many (but not all) loops.

MCB is a Monte Carlo mini-app used in investigating new computer systems and new programming approaches. It is dominated by integer computation and has the erratic branching character of all Monte Carlo radiation transport codes.

microK is a set of simple vector loops used to measure the performance impact of falling out of cache, speedups due to using SIMD instructions, and other processor features. The loops in microK are simple so it is fairly easy to optimize them on a new system. The microK runs used one MPI process with 32 OpenMP threads on a single node.

Table 2. This table shows performance metrics for two HYDRA test problems. Metrics from three other codes are shown for reference. The polynomial kernel is the only one which issues more floating point than integer instructions. The polynomial and dot product kernels use nearly the full memory bandwidth (28 GB/s) for large vectors, but all other tests use no more than 11 %. For small vectors, microK runs completely out of the L2 cache so we do not report any DRAM related numbers.

Package	Time (sec)	Int Instr per cycle	FP Instr per cycle	FLOP per Instr	DRAM BW (GB/s)	DRAM xfer (GB)	L2 miss per line
Hydra hyd607							
Advect	0.66	5.74	1.90	1.59	3.21	2.13	1.31
EosOpac	0.32	1.70	0.48	1.32	0.46	0.146	2.17
Econd	0.90	10.03	0.39	1.53	0.94	0.85	0.86
Mtgrdif	17.20	8.74	0.25	1.58	0.85	14.65	0.73
Hydra nifburn							
Hydro	0.17	4.38	2.21	1.62	1.89	0.31	1.58
Advect	0.29	4.68	0.58	1.62	2.91	0.72	1.33
Econd	0.32	12.66	0.95	1.66	0.32	0.10	2.89
Laser	2.29	1.69	0.05	1.25	3.02	6.93	4.31
Imc	10.06	1.06	0.21	1.60	1.29	12.93	5.27
Burn	1.65	12.40	0.72	1.60	0.17	0.29	1.03
MCB							
Advance	19.52	4.52	0.18	1.31	0.25	4.92	1.01
Pf3d kernels							
Couple4	8.43	4.02	1.47	2.81	1.46	12.32	0.45
Absorbdt	1.02	4.61	1.13	1.77	1.21	1.23	0.43
Acadv	5.16	3.56	1.15	2.25	1.87	9.67	0.28
Advancefi	3.47	5.29	1.78	2.12	0.88	3.07	0.48
FFT	0.68	3.02	1.88	1.30	2.64	1.79	0.48
MicroK small							
Sdot	0.02	8.16	1.90	8.00			
Poly	0.02	4.04	9.28	8.00			
MicroK large							
Sdot	0.57	1.96	0.46	8.00	23.73	13.48	0.51
Poly	0.51	1.09	2.56	8.00	26.21	13.42	1.42

4.3   Performance Metrics

The HPM library written by Bob Walkup of IBM provides a simple way to gather the desired hardware counters. HPM start and stop calls were added around calls to physics packages in the time step loop. HPM reports how many times each event occurred in a package. An L2 miss is recorded for every 128 byte cache line loaded and a flush for every line stored. This makes it easy to calculate the DRAM read and write bandwidth.

Table 2 reports performance metrics from BGQ runs of the test applications. Two HYDRA test problems were run to show how the time spent in physics packages and the set of packages used varies from problem to problem.

The hyd607 test problem performs a capsule-only simulation of a NIF implosion experiment. Most time is spent in the multi-group diffusion package (mtgrdif), with roughly 10 % of the time spent on electron heat conduction, advection, equation of state, and opacities. The nifburn test problem performs an integrated simulation of the capsule and the surrounding hohlraum for a NIF experiment. Domain replication was employed to allow load balancing of the laser and IMC packages. Most time is spent in the laser and IMC packages. The hydrodynamics package, advection associated with ALE remaps, electron heat conduction, and fusion burn combine to consume about 15 % of the run time.

The BGQ compiler generates a fairly high fraction (30 % or more) of FMA instructions for all test codes. The BGQ compiler has difficulty generating SIMD instructions unless the code is annotated with BGQ-specific alignment directives. The simple loops in microK allowed us to add alignment directives and achieve nearly a 100 % SIMD fraction. It is impractical to add those directives to a large code, so the SIMD fraction is low for HYDRA, pF3D, and MCB. Some of the pF3D kernels deliver more than 2 FLOPs per instruction because they call IBM's "hand written" sin, exp, etc. special functions.

Floating point instruction issue rates are not a bottleneck for HYDRA, pF3D, or MCB. Only the polynomial kernel issues more floating point instructions than integer instructions (on the BGQ). HYDRA's ccond, burn and mtgrdiff packages and the dot product of short vectors all execute more than 8 integer instructions per cycle and their performance may be limited by integer issue rates.

MicroK results are reported for vectors which fit in the L2 cache (64 K elements per thread) and for vectors large enough (512 K elements per thread) that they must be fetched from DRAM. The polynomial kernel achieves over 50 % of the peak floating point performance for short vectors but only 16 % of peak for long vectors. The microK kernels are memory bandwidth limited for large vectors (the bandwidth is close to the 28 GB/s streams bandwidth). The highest memory bandwidth for HYDRA, pF3D, or MCB is 3.2 GB/s, so memory bandwidth is not a bottleneck for "production" codes.

The high cache and DRAM latency on a BGQ hurts the performance of the IMC, laser ray trace, and EOS and opacity lookup packages in HYDRA. The 50 ns latency to the L2 cache on a BGQ is much larger than the 11 ns latency to the L3 cache on a Sandy Bridge. The L1P unit on a BGQ prefetches from L2 to L1 to try and hide latency. The L1P hit rate for the IMC and laser packages is 1–2 %. HYDRA suffers a 50 ns delay on almost all L2 accesses, and the effective bandwidth of L2 will be low. The DRAM latency on a BGQ is 220 ns versus 70 ns for a Sandy Bridge. The high latency on the BGQ will prevent HYDRA from achieving full memory bandwidth unless prefetching from DRAM to L2 works well. We have not measured the DRAM prefetch efficiency. The laser rays and IMC particles move from zone-to-zone in a manner which is hard to predict, so we expect that prefetch efficiency will be low.

The pF3D kernels issue from 3 to 5.3 integer instructions per cycle. That is low enough that they probably are not bottlenecked on the integer issue rate. The pF3D kernels all issue at least one floating point instruction per cycle per node which is several times more than the laser and IMC packages in HYDRA, but far below the peak. The pF3D kernels have low L2 miss rates because many inner loops access arrays in stride one order. It is not clear what is the key performance bottleneck for the pF3D kernels.

The tables include ratios of cache misses to lines read. A cache line is 128 bytes on a BGQ system. HYDRA performs most computations using double precision operands, so a line holds 16 numbers. A stride one loop should have one cache miss per L2 cache line read. A package which accesses large arrays randomly might have up to 16 misses per line. HYDRA's IMC package has a higher miss fraction than any other package in the table. That is not surprising given that the particle list has photons scattered almost randomly through the grid at the time the performance counters were read.

4.4   Memory Usage

The hyd607 test problem uses 1.7 GB of heap memory per node. The radiation diffusion package transfers 14.7 GB between DRAM and the processor during a time step, which shows that some arrays are read multiple times. The diffusion package solves a large sparse matrix using Hypre's iterative CG solver with hybrid AMG/diagonal pre-conditioner. The iteration is the reason arrays are fetched multiple times.

The nifburn test problem uses 2.7 GB of heap memory per node. The IMC package transfers 12.97 GB between DRAM and the processor during a time step, so some arrays are read multiple times. As the Monte Carlo particles randomly wander through the grid, they will pull the opacity array in multiple times.

HYDRA either has enough memory bandwidth on the BGQ or (more likely) is bottlenecked by memory latency. Future systems will have a lower ratio of DRAM bandwidth to peak performance. To deal with this "memory wall", these systems may include some in-package memory (IPM). IPM bandwidth will be much higher than external DRAM bandwidth, but its latency will be similar. If memory latency is the key bottleneck for HYDRA, IPM may not help performance significantly.

5   Conclusion

Our goal in this work was to investigate the performance characteristics of HYDRA, a multi-physics simulation code. We expect that other multi-physics codes from LLNL will have similar characteristics. We demonstrated that HYDRA's thread based load balancing strategy is very effective. DRAM bandwidth is not a limiting factor for any HYDRA package. The latency of DRAM or the L2 cache may be a bottleneck. Floating point issue rates are never a limiting factor for HYDRA, but integer issue rates may be a limiting factor for a few packages.

We demonstrated that the total memory traffic between the processor chip and DRAM is significantly greater than the total amount of memory in use by HYDRA, indicating that using IPM as a cache may have benefits for HYDRA on future systems.

This work was performed under the auspices of the U.S. Department of Energy by Lawrence Livermore National Laboratory under contract DE-AC52-07NA27344. Lawrence Livermore National Security, LLC. Document release LLNL-PROC-665891.

References

1. Berger, R.L., Lasinski, B.F., Langdon, A.B., Kaiser, T.B., Afeyan, B.B., Cohen, B.I., Still, C.H., Williams, E.A.: Influence of spatial and temporal laser beam smoothing on stimulated brillouin scattering in filamentary laser light. Phys. Rev. Lett. **75**(6), 1078–1081 (1995)
2. Holden, H., Karlsen, K.H., Lie, K.-A., Risebro, H.: Splitting Methods for Partial Differential Equations with Rough Solutions. European Mathematical Society, Zurich (2010)
3. McCalpin, J.D.: The streams benchmark home page. http://www.cs.virginia.edu/stream/
4. Langer, S., Still, B., Bremer, T., Hinkel, D., Langdon, B., Williams, E. A.: Cielo full-system simulations of multi-beam laser-plasma interaction in nif experiments. In: CUG 2011 proceedings (2011)
5. Marinak, M., Kerbel, G., Koning, J., Patel, M., Sepke, S., McKinley, M., O'Brien, M., Procassini, R., Munro. D.: Advances in hydra and its applications to simulations of inertial confinement fusion targets. In: EPJ Web of Conferences on Proceedings of the 2011 International Fusion Sciences and Applications Conference (IFSA 2013), vol. 59, p. 3011 (2013)
6. Marinak, M.M., Kerbel, G.D., Gentile, N.A., Jones, O., Munro, D., Pollaine, S., Dittrich, T.R., Haan, S.W.: Three-dimensional hydra simulations of national ignition facility targets. Phys. Plasmas **8**(4), 22755 (2001)
7. Moses, E.I., Boyd, R.N., Remington, B.A., Keane, C.J., Al-Ayat, R.: The national ignition facility: ushering in a new age for high energy density science. Phys. Plasmas **16**(041006), 1–13 (2009)
8. Still, C.H., Berger, R.L., Langdon, A.B., Hinkel, D.E., Suter, L.J., Williams, E.A.: Filamentation and forward brillouin scatter of entire smoothed and aberrated laser beams. Phys. Plasmas **7**(5), 2023–2032 (2000)
9. Valiant, L.G.: A bridging model for parallel computation. Commun. ACM **33**(8), 103–111 (1990)

Accelerating Computation of Eigenvectors in the Dense Nonsymmetric Eigenvalue Problem

Mark Gates[1]([✉]), Azzam Haidar[1], and Jack Dongarra[1,2,3]

[1] University of Tennessee, Knoxville, TN, USA
{mgates3,haidar,dongarra}@eecs.utk.edu
[2] Oak Ridge National Laboratory, Oak Ridge, TN, USA
[3] University of Manchester, Manchester, UK

Abstract. In the dense nonsymmetric eigenvalue problem, work has focused on the Hessenberg reduction and QR iteration, using efficient algorithms and fast, Level 3 BLAS. Comparatively, computation of eigenvectors performs poorly, limited to slow, Level 2 BLAS performance with little speedup on multi-core systems. It has thus become a dominant cost in the solution of the eigenvalue problem. To address this, we present improvements for the eigenvector computation to use Level 3 BLAS and parallelize the triangular solves, achieving good parallel scaling and accelerating the overall eigenvalue problem more than three-fold.

1 Introduction

Eigenvalue problems are fundamental for many engineering and physics applications. For example, image processing, facial recognition, vibration analysis of mechanical structures, seismic reflection tomography, and computing electron energy levels can all be expressed as eigenvalue problems. The eigenvalue problem is to find an eigenvalue λ and eigenvector x that satisfy $Ax = \lambda x$, where A is an $n \times n$ matrix. When the entire eigenvalue decomposition is computed we have $A = X\Lambda X^{-1}$, where Λ is a diagonal matrix of eigenvalues and X is a matrix of eigenvectors. In this paper we consider the case when A is dense and nonsymmetric. We concentrate on computing all the eigenvectors, and present optimizations that accelerate the overall eigenvalue problem more than three-fold.

The solution of the eigenvalue problem proceeds in three phases. First, the matrix is reduced to upper Hessenberg form by applying orthogonal Q matrices on the left and right, to form $H = Q_1^T A Q_1$. The second phase, QR iteration, is an iterative process that reduces the Hessenberg matrix to upper triangular Schur form, $T = Q_2^T H Q_2$. Being based on similarity transformations, the eigenvalues of A are the same as the eigenvalues of T, which are simply the diagonal entries of T. The third phase computes eigenvectors Z of the Schur form T via triangular solves, and then back-transforms them to eigenvectors X of the original matrix A. The eigenvectors of A are related to the eigenvectors of T by the orthogonal transformations used in the Hessenberg reduction and QR iteration as $X = Q_1 Q_2 Z$. This final phase computing eigenvectors will be the focus of this work.

© Springer International Publishing Switzerland 2015
M. Daydé et al. (Eds.): VECPAR 2014, LNCS 8969, pp. 182–191, 2015.
DOI: 10.1007/978-3-319-17353-5_16

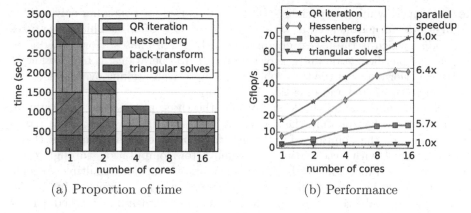

(a) Proportion of time   (b) Performance

Fig. 1. Parallel scaling of phases of eigenvector computation in existing LAPACK implementation, for $n = 14000$. Parallel speedup for 16 Intel cores compared to 1 core is annotated in right margin.

Figure 1 shows the time for each phase in the existing LAPACK implementation as the number of cores is increased. The performance of each phase depends on how many of its operations are done in efficient Level 3 BLAS versus in memory-bound Level 2 BLAS. The Hessenberg reduction is formulated [1,2] so that 80% of its floating point operations (flops) occur in Level 3 matrix-matrix multiply (gemm), with the remainder in memory-bound Level 2 matrix-vector multiply (gemv). This 20% of gemv operations bounds the Hessenberg performance to less than five times the gemv Gflop/s rate. A recent two-stage implementation reduces the amount of gemv operations [7]. A GPU accelerated version [10] is an additional 4 times faster than the 16-core performance. For large matrices, the Hessenberg reduction accounts for approximately 20% of the overall time using 16 cores, achieving close to 50 Gflop/s in Fig. 1b (cyan diamonds).

For QR iteration, we use the implicit multishift QR iteration [3,4] implemented in LAPACK as hseqr, which takes advantage of Level 3 BLAS for efficient computation. The QR iteration algorithm takes $O(n^3)$ flops, but being an iterative method, the exact count depends heavily on the convergence rate and techniques such as aggressive early deflation [4], which is also included in LAPACK's implementation. For small matrices, it can take $25n^3$ flops [5], but we found it to be $\frac{10}{3}n^3$ flops for large matrices. Distributed parallel versions of QR iteration also exist [6]. QR iteration can take over 50% of the time for small matrices, but for large matrices this reduces to about 15% of the time on 16 cores, and achieves 70 Gflop/s as shown in Fig. 1b (green stars).

In contrast to the first two phases, the final phase computing eigenvectors has previously been implemented only with Level 2 BLAS. In the LAPACK implementation (trevc), each eigenvector is computed with a triangular solve (latrs), then back-transformed with a matrix-vector product (gemv). In Fig. 1b, we see that the performance of the back-transformation (red boxes) is limited to 14 Gflop/s, while the specialized latrs routine used for triangular solves (blue triangles) is not parallelized at all and is limited to 2 Gflop/s. This phase takes $\frac{4}{3}n^3$ flops, which is the fewest operations of the three phases. However, due to the lack of parallelization and use of Level 2 BLAS operations, it has the lowest Gflop/s performance of the three phases, asymptotically taking over 60 % of the time on 16 cores (red and blue tiers in Fig. 1a).

Thus, despite having the least flops of the three phases, the computation of eigenvectors has become the dominant cost and limited the overall parallel speedup of the solution of the eigenvalue problem. This paper is therefore concerned with accelerating the eigenvector computation through three improvements. First, for the back-transformation, we block multiple Level 2 gemv products into an efficient Level 3 gemm product, discussed in Sect. 3. Second, we parallelize the triangular solves using a task-based scheduler, as described in Sect. 4. Finally, using a GPU, the back-transformation is further accelerated and done in parallel with the triangular solves, in Sect. 5. Combined, these improvements significantly increase the performance and scalability of the overall eigenvalue problem.

2   Eigenvector Computation

When eigenvectors are desired, the third phase computes eigenvectors of the triangular Schur form T, then back-transforms them to eigenvectors of the original matrix A. In LAPACK, this phase is implemented in the trevc (triangular eigenvector computation) routine. We will assume only right eigenvectors are desired; the computation of left eigenvectors is similar and amenable to the same techniques described here. After the Hessenberg and QR iteration phases, the diagonal entries of T are the eigenvalues λ_k of A. To determine the corresponding eigenvectors, we solve $Tz_k = \lambda_k z_k$ by considering the decomposition [5]

$$\begin{bmatrix} T_{11} & u & T_{13} \\ 0 & \lambda_k & v^T \\ 0 & 0 & T_{33} \end{bmatrix} \begin{bmatrix} \hat{z} \\ 1 \\ 0 \end{bmatrix} = \lambda_k \begin{bmatrix} \hat{z} \\ 1 \\ 0 \end{bmatrix}, \tag{1}$$

which yields $(T_{11} - \lambda_k I)\hat{z} = -u$. Thus computing each eigenvector z_k of T involves a $k-1 \times k-1$ triangular solve, for $k = 2, \ldots, n$. Each solve has a slightly different T matrix, with the diagonal modified by subtracting λ_k. The resulting eigenvector z_k of T must then be back-transformed by multiplying with the $Q = Q_1 Q_2$ formed in the Hessenberg and QR iteration phases to get the eigenvector $x_k = Qz_k$ of the original matrix A.

Note that if two eigenvalues, λ_k and λ_j $(k > j)$, are identical, then $T_{11} - \lambda_k I$ is singular. More generally, $T_{11} - \lambda_k I$ can be badly conditioned. Therefore, instead of using the standard BLAS triangular solver (trsv), it uses a specialized triangular solver (latrs) that scales columns to protect against overflow and can generate a consistent solution for a singular matrix.

This method works in complex arithmetic, however the case in real arithmetic is more complicated. For a real matrix A, the eigenvalues can still be complex, coming in conjugate pairs, λ_k and $\bar{\lambda}_k$. The eigenvectors are likewise conjugate pairs, z_k and $\bar{z}_k$. In real arithmetic, the closest that QR iteration can come to triangular Schur form is quasi-triangular real Schur form, which has 2×2 diagonal blocks corresponding to conjugate pairs of eigenvalues. A specialized quasi-triangular solver is required, which factors each 2×2 diagonal block, as well as protecting against overflow and dealing with singular matrices. In LAPACK this solver is implemented as part of the dtrevc routine.

3   Blocking Back-Transformation

It is well established for dense linear algebra that blocked, Level 3 BLAS implementations — which operate on multiple columns in each BLAS call — achieve much higher performance than non-blocked, Level 2 BLAS implementations. Therefore, our first step to improve the eigenvector computation is to block the n gemv operations for the back-transformation into n/n_b gemm operations, where n_b is the block size. This requires two $n \times n_b$ workspaces: the first, Z, for the vectors z_k, and the second, X, for the back-transformed vectors x_k before copying to the output V.

Pseudocode for the blocked back-transformation is shown in Algorithm 1, which includes the parallel solver described in Sect. 4. For each block, we loop over n_b columns, performing a triangular solve for each column and storing the resulting eigenvectors z_k in workspace Z. After filling up n_b columns of Z, a single gemm back-transforms all n_b vectors, storing the result in workspace X. The vectors are then normalized and copied to V. On input, the matrix $V = Q$. Recall from Eq. (1) that the bottom $n - k$ rows of eigenvector z_k are 0, so the last $n - k$ columns of Q are not needed for the gemm. Therefore, we start from $k = n$ and work down to $k = 1$, writing each block of eigenvectors to V over columns of Q after they are no longer needed.

The real case is similar, but has the minor complication that complex conjugate pairs of eigenvalues will generate conjugate pairs of eigenvectors, $z_k = a + bi$ and $\bar{z}_k = a - bi$, which are stored as two real columns, a and b, in Z. When the first eigenvalue of each pair is encountered, both columns are computed; then the next eigenvalue (its conjugate) is skipped. Once $n_b - 1$ columns are processed, if the next eigenvector is complex, there is not space to store the two resulting columns a and b, so it must be delayed until the next block.

4   Multi-threading Triangular Solver

After blocking the back-transform, the triangular solver remains a major bottleneck because it is not parallelized. Recall that the triangular matrix being solved is different for each eigenvector — the diagonal is modified by subtracting λ_k. This prevents blocking multiple eigenvectors together using a Level 3 BLAS trsm operation to solve multiple eigenvectors together.

In the complex case, LAPACK's ztrevc uses a safe triangular solver, zlatrs. Unlike the standard ztrsv BLAS routine, zlatrs uses column scaling to avoid numerical instability, and handles singular triangular matrices. Therefore, to not jeopardize the accuracy or stability of the eigensolver, we continue to rely on zlatrs instead of the optimized, multi-threaded ztrsv. However, processing T column-by-column prevents internal blocking and parallelizing individual calls to zlatrs, as is typically done for BLAS functions to achieve high performance. Instead, we observe that multiple triangular solves can occur in parallel. One obstacle to solving multiple eigenvectors in parallel is that a different λ_k is subtracted from the diagonal in each case, modifying T in memory. Our solution is to write a modified routine, zlatrsd (triangular solve with modified diagonal), which takes both the original unmodified T_{11} and the λ_k to subtract from the diagonal. The subtraction is done as the diagonal elements are used, without modifying T in memory. This allows us to pass the same T to each zlatrsd call and hence solve multiple eigenvectors in parallel, one in each thread.

As previously mentioned, the real case requires a special quasi-triangular solver to solve each 2×2 diagonal block. In the original LAPACK code, this quasi-triangular solver is embedded in the dtrevc routine. To support multi-threading, we refactor it into a new routine, dlaqtrsd, a quasi-triangular solver with modified diagonal. Unlike the complex case, instead of passing λ_k separately, dlaqtrsd computes it directly from the diagonal block of T. If λ_k is real, dlaqtrsd computes a single real eigenvector. If λ_k is one of a complex-conjugate pair, dlaqtrsd computes a complex eigenvector, as two real vectors. This implies

---

Algorithm 1. Multi-threaded eigenvector computation (complex-arithmetic).

1: **function** ZTREVC(n, T, V)
2:     // T is $n \times n$ upper triangular matrix.
3:     // V is $n \times n$ matrix; on input $V = Q$, on output V has eigenvectors.
4:     // Z and X are $n \times n_b$ workspaces, n_b is column block size.
5:     $k = n$
6:     **while** $k \geq 1$
7:         $j = n_b$
8:         **while** $j \geq 1$ and $k \geq 1$
9:             $\lambda_k = T_{k,\, k}$
10:             enqueue latrsd to solve $(T_{1:k-1,\, 1:k-1} - \lambda_k I)Z_{1:k-1,\, j} = -T_{1:k-1,\, k}$
11:             $Z_{k:n,\, j} = [\, 1,\, 0,\, \ldots,\, 0\,]^T$
12:             $j \mathrel{-}= 1;\quad k \mathrel{-}= 1$
13:         **end**
14:         sync queue
15:         $m = k + n_b - j$
16:         **for** $i = 1$ to n by $\lceil n/\mathrm{p} \rceil$
17:             $i_2 = \min(i + n_b - 1,\, n)$
18:             enqueue gemm to multiply $X_{i:i_2,\, j+1:n_b} = V_{i:i_2,\, 1:m} * Z_{1:m,\, j+1:n_b}$
19:         **end**
20:         sync queue
21:         normalize vectors in X and copy to $V_{1:n,\, k+1:k+n_b}$
22:     **end**
23: **end function**

---

an imbalance in time as some instances of dlaqtrsd solve a single real vector, while other instances solve a complex eigenvector. Also, each triangular solve is of size $k - 1 \times k - 1$, so they gradually decrease in size as k decrements.

To deal with multi-threading, we use a thread pool design pattern. As shown in Algorithm 1, the main thread inserts latrsd tasks into a queue. Worker threads pull tasks out of the queue and execute them. For this application, there are no dependencies to be tracked between the triangular solves. After a block of n_b vectors has been computed, we back-transform them with a gemm. We could call a multi-threaded gemm, as available in MKL, but to simplify thread management and avoid switching between our pthreads and MKL's threads, we found it more suitable to use the same thread pool for the gemm as for latrsd. For p threads, the gemm is split into p tasks, each task multiplying a single block row of Q with Z. After the gemm, the next block of n_b vectors is computed. Within each thread, the BLAS calls are single threaded.

5   GPU Acceleration

To further accelerate the eigenvector computation, we observe that the triangular solves and the back-transformation gemm can be done in parallel. In particular, the gemm can be done on a GPU while the CPU performs the triangular solves. Data transfers can also be done asynchronously and overlapped with the triangular solves. To facilitate this asynchronous computation, we double-buffer the CPU workspace Z, using it to store results from latrsd, then swapping with X, which is used to send data to the GPU while the next block of latrsd solves are performed with Z. The difference from Algorithm 1 is shown in Algorithm 2.

Algorithm 2. GPU accelerated back-transformation replaces lines 15–21 of Algorithm 1. Also, V is sent asynchronously to dV at start of ztrevc.

15:	// dV is $n \times n$ workspace on GPU.
16:	// dZ and dX are $n \times n_b$ workspaces on GPU.
17:	swap buffers Z and X
18:	async send X to dZ on GPU
19:	async gemm $dX_{1:n,\,j+1:n_b} = dV_{1:n,\,1:m} * dZ_{1:m,\,j+1:n_b}$ on GPU
20:	async receive dX to $V_{1:n,\,k+1:k+n_b}$ on CPU
21:	normalize vectors in V

---

6   Results

6.1   Intel Sandy Bridge and NVIDIA Kepler GPU

We performed tests on a machine with two 8-core, 2.6 GHz Intel Sandy Bridge Xeon E5-2670 CPUs and an 875 MHz NVIDIA Kepler K40 GPU. All tests used Intel MKL 11.0.5 for optimized, multi-threaded BLAS and numactl to interleave memory allocation across the NUMA nodes. Matrices were double precision with uniform random entries in $(0, 1)$. Inside our parallel trevc, we launch p pthreads and set MKL to be single-threaded; outside trevc, MKL uses the same number of threads, p.

Figure 2 shows the total eigenvalue problem (dgeev) time, broken down into four phases: QR iteration (bottom, green tier), Hessenberg reduction (2nd, cyan tier), triangular solves (3rd, blue tier), and back-transformation (top, red tier). The triangular solves and back-transformation together form the eigenvector computation. Columns are grouped by number of CPU threads. We compare results to two reference implementations: the LAPACK CPU version in the first column, and the MAGMA [8] GPU-accelerated version in the fourth column.

The improvement due to blocking the back-transformation is shown by the second column of each group in Fig. 2. The blocked back-transformation itself is up to 14 times faster than the non-blocked back-transformation using 16 threads. Further, the solid red line (squares) in Fig. 3 shows it has better parallel scaling, reaching a speedup of 12 times for 16 cores, compared to only 6 times for the LAPACK implementation. However, as the triangular solves are not yet parallelized, the overall improvement is limited, being at most 1.4 times faster, seen in the single threaded result in Fig. 2, and the red line (squares) in Fig. 4.

Parallelizing the triangular solves is the third column of each group in Fig. 2. For one thread, there is of course no parallel speedup, so the results are the same as the second column. With multiple threads, we see significant parallel speedup, up to 12.8 times for 16 threads, shown as the solid blue line (triangles) in Fig. 3. Combined with the blocked back-transformation, these two modifications significantly improve the solution of the overall eigenvalue problem by up to 2.5 times for 16 cores, shown by the blue line (triangles) in Fig. 4, and annotated

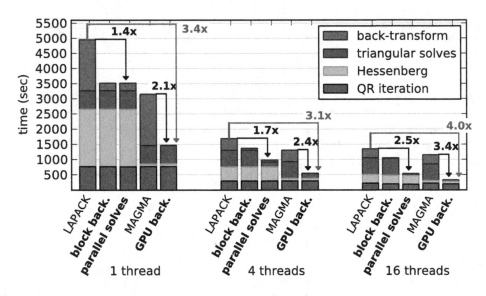

Fig. 2. Execution time of eigenvalue solver (dgeev) for matrix size $n = 16,000$.

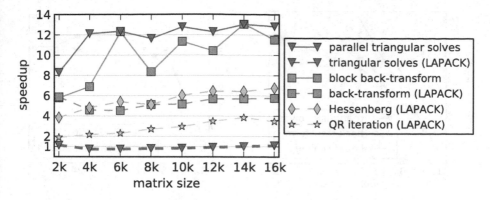

Fig. 3. Parallel speedup of each phase by itself, for $p = 16$, compared to $p = 1$.

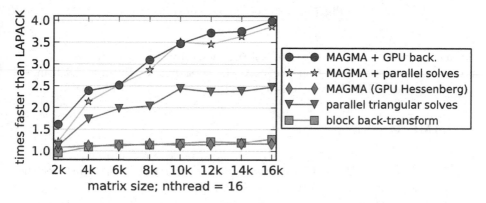

Fig. 4. Overall improvement of eigenvalue solver (dgeev) compared to LAPACK, after various improvements, using $p = 16$ threads.

with an arrow in Fig. 2. This is the total improvement available using only CPUs. Next we will look at the improvement also using GPUs.

The fourth column in Fig. 2 is the MAGMA reference time, which accelerates the Hessenberg reduction using the GPU. The MAGMA Hessenberg performance depends on the GPU, so is independent of the number of CPU threads. It is up to 3.4 times faster than the 16-core LAPACK Hessenberg. However, the LAPACK Hessenberg is only 20 % of the total time, so accelerating it reduces the total dgeev time by only 1.15 times, shown by the green line (diamonds) in Fig. 4. When combined with the new blocked back-transform and parallel triangular solves, the performance substantially increases, as shown by the orange line (stars) in Fig. 4, being up to 3.8 times faster than LAPACK.

Our final improvement is to move the back-transformation gemm to the GPU, shown as the fifth column of each group in Fig. 2. The gemm can be almost entirely overlapped with the triangular solves, practically eliminating time spent on the back-transformation. This is a minor additional improvement on top of the

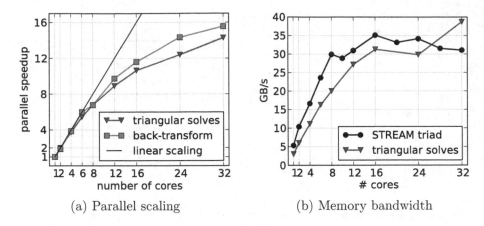

(a) Parallel scaling

(b) Memory bandwidth

Fig. 5. Results on AMD Piledriver.

blocked back-transform and parallel solves, shown by the magenta line (circles) in Fig. 4. The improvement from the MAGMA version using 16 cores is 3.4 times, while from the LAPACK version is 4.0 times, as annotated in Fig. 2.

6.2   AMD Piledriver

To study scalability to a larger number of cores, we also ran tests on a machine with four 8-core, 2.5 GHz AMD Piledriver Opteron 6380 CPUs. We compared the achieved memory bandwidth with the STREAM benchmark [9]. Due to cache effects, we cannot simply compute the achieved memory bandwidth based on each triangular solve reading its entire T matrix. Instead, we compute it based on cache misses, which we obtain from hardware counters using PAPI.

In Fig. 5, we observe near linear speedup on up to 6 cores. On 16 cores, we obtain similar results as on the 16-core Intel: the blocked back transformation achieved an 11.6 times speedup, while the parallel triangular solves achieved 10.6 times speedup. For a larger number of cores, we still see improvement, but at a reduced rate. As the number of cores increases, the memory bandwidth becomes saturated, causing the increase in speedup to drop off. Notice in Fig. 5b that the bandwidth for both the STREAM benchmark and the triangular solves flattens after 16 cores. The achieved bandwidth closely follows the STREAM benchmark bandwidth. Thus, the results scale up to the available memory bandwidth. To further improve the parallel scalability of the eigenvector computation, future work should focus on reusing data to reduce the required memory bandwidth.

7   Conclusion

It has been said that high performance computing is an exercise in chasing bottlenecks. Previously, the Hessenberg reduction and QR iteration have rightly been addressed as major bottlenecks in the solution of the nonsymmetric eigenvalue problem. Amdahl's Law requires that all phases of the algorithm receive attention. Indeed, while the Hessenberg was accelerated by 3.4 times with a GPU, the overall speedup was previously limited to 15 % (Fig. 4). We accelerated the remaining eigenvector computation phase by 12 times through introducing Level 3 BLAS and parallelizing the remaining Level 2 BLAS triangular solves. This improved the overall eigenvalue problem by 2.5 times for CPU-only code and 3.4 times for the GPU-accelerated version. The bottleneck is now moved back to QR iteration, which is natural as it has the most flops and its iterative nature makes it the most complicated and difficult phase to parallelize.

Acknowledgments. The results were obtained in part with the financial support of the Russian Scientific Fund, Agreement N14-11-00190; the National Science Foundation, U.S. Department of Energy, Intel, NVIDIA, and AMD.

References

1. Bischof, C.: A summary of block schemes for reducing a general matrix to Hessenberg form. Technical report, ANL/MCS-TM-175, Argonne National Lab (1993)
2. Bischof, C., Van Loan, C.: The WY representation for products of Householder matrices. SIAM J. Sci. Stat. Comput. **8**(1), s2–s13 (1987)
3. Braman, K., Byers, R., Mathias, R.: The multishift QR algorithm. part I: maintaining well-focused shifts and level 3 performance. SIAM J. Matrix Anal. Appl. **23**(4), 929–947 (2002)
4. Braman, K., Byers, R., Mathias, R.: The multishift QR algorithm. part II: aggressive early deflation. SIAM J. Matrix Anal. Appl. **23**(4), 948–973 (2002)
5. Golub, G., Van Loan, C.: Matrix Computations, 3rd edn. Johns Hopkins, Baltimore (1996)
6. Kågström, B., Kressner, D., Shao, M.: On aggressive early deflation in parallel variants of the QR algorithm. In: Jónasson, K. (ed.) PARA 2010, Part I. LNCS, vol. 7133, pp. 1–10. Springer, Heidelberg (2012)
7. Karlsson, L., Kågström, B.: Parallel two-stage reduction to Hessenberg form using dynamic scheduling on shared-memory architectures. Parallel Comput. **37**(12), 771–782 (2011)
8. MAGMA. http://icl.eecs.utk.edu/magma/
9. McCalpin, J.D.: Memory bandwidth and machine balance in current high performance computers. In: IEEE Computer Society Technical Committee on Computer Architecture (TCCA) Newsletter, December 1995
10. Tomov, S., Nath, R., Dongarra, J.: Accelerating the reduction to upper Hessenberg, tridiagonal, and bidiagonal forms through hybrid GPU-based computing. Parallel Comput. **36**(12), 645–654 (2010)

Low Byte/Flop Implementation of Iterative Solver for Sparse Matrices Derived from Stencil Computations

Kenji Ono[1,2,3]([⊠]), Shuichi Chiba[4], Shunsuke Inoue[4], and Kazuo Minami[1]

[1] RIKEN, Advanced Institute for Computational Science,
7-1-26, Minatojima-minami-machi, Chuo-ku, Kobe 650-0047, Japan
{keno,minami_kaz}@riken.jp
[2] Graduate School of System Informatics, Kobe University, Kobe, Japan
[3] Institute of Industrial Science, University of Tokyo, Bunkyō, Japan
[4] Fujitsu Limited, Tokyo, Japan
{shuc,inoue.shunsuke}@jp.fujitsu.com

Abstract. Practical simulators require high-performance iterative methods and efficient boundary conditions, especially in the field of computational fluid dynamics. In this paper, we propose a novel *bit-representation technique* to enhance the performance of such simulators. The technique is applied to an iterative kernel implementation that treats various boundary conditions in a stencil computation on a structured grid system. This approach reduces traffic from the main memory to CPU, and effectively utilizes Single Instruction–Multiple Data (SIMD) stream units with cache because of the bit-representation and compression of matrix elements. The proposed implementation also replaces if-branch statements with mask operations using the bit expression. This promotes the optimization of code during compilation and runtime. To evaluate the performance of the proposed implementation, we employ the Red–Black SOR and BiCGstab algorithms. Experimental results show that the proposed approach is up to 3.5 times faster than a naïve implementation on both the *Intel* and *Fujitsu Sparc* architectures.

Keywords: Sparse matrix · Boundary condition · SIMD · Bit operation · CFD

1 Introduction

Advances in computational capabilities have allowed us to increase the scale of many problems, and thus obtain more reliable solutions. Computational fluid dynamics, which is commonly used to design industrial products, involves large-scale linear systems with sparse matrices given by the pressure Poisson equation or implicit time integration. Iterative methods for such large-scale sparse matrices are a crucial building block for high-performance physical simulations. Recent computer architectures have been constructed to have a deep memory hierarchy,

© Springer International Publishing Switzerland 2015
M. Daydé et al. (Eds.): VECPAR 2014, LNCS 8969, pp. 192–205, 2015.
DOI: 10.1007/978-3-319-17353-5_17

which demands effective utilization of the small but high-speed cache for user programs. However, if a straightforward implementation of an iterative method is applied to such architectures, the code encounters memory-bandwidth limitations owing to the large number of load/store instructions relative to floating-point operations in each loop of the source code.

Using the *Roofline model* [1], which provides useful information for analyzing the performance of codes, it has been demonstrated that high operational intensity in an algorithm leads to high performance. Thus, one solution to the bandwidth limitations is to reduce memory traffic from the main memory to cache. To this end, many studies have investigated the Compressed Sparse Row (CSR) data format, run length encoding of CSR [2], and bit-representation schemes [3].

For engineers and researchers, it is essential to predict realistic phenomena and simultaneously reduce the time cost. Thus, simulations must consider various boundary conditions that reflect real-world effects as well as high-performance computation. However, the calculation of boundary conditions is generally complicated, and the presence of if-branch statements, table references, and indirect memory access makes it difficult to optimize the source code.

This paper presents a *bit-representation technique* that reduces memory traffic in an efficient implementation of iterative methods and boundary conditions. This technique expresses both the sparse matrix coefficients and a mask function that replaces if-branch statements using a bit sequence. The performance of the proposed approach is investigated on several architectures.

2   Basic Equations and Bit-Representation

We consider a pressure Poisson equation derived from the incompressible Navier–Stokes equation:

$$\nabla\,(\nabla p) \,=\, div\,(\frac{\partial \boldsymbol{u}}{\partial t}) \,\equiv\, \phi\,, \tag{1}$$

where p, $\boldsymbol{u}$, and ϕ represent the pressure, velocity vector, and a source term, respectively. This equation can be discretized using finite-difference or finite-volume methods on a Cartesian grid system. Hence, the Laplace operator of the equation is approximated by a 7-point stencil, and generates a linear system that has a large-scale sparse matrix.

Practical simulators must treat various types of boundary conditions and arbitrary positions in the computational domain to reproduce various real-world situations. This issue is another challenge to be addressed in computational fluid dynamics. In this context, we focus our attention on some basic boundary conditions. Almost all pressure boundary conditions can be reduced to Dirichlet or Neumann boundary conditions. For instance, we can substitute Neumann boundary conditions for ∇p in (1). Thus, introducing the Heaviside function H:

$$H \,=\, \begin{cases} 0 & (Boundary\,Condition) \\ 1 & (Fluid) \end{cases} \tag{2}$$

the pressure variable can be written as:

$$p = pH + (1 - H)p^{BC} \tag{3}$$

This expression is a mask function that replaces an if-branch statement in (2), and will promote optimization during compilation. A semi-discrete form of (1) can be combined with (2) and (3) to give

$$\sum_l \left(\nabla p\, H^N\right)_l \boldsymbol{n}_l = h\phi - \sum_l \left(1 - H_l^N\right) \nabla p_l^{BC} \boldsymbol{n}_l \tag{4}$$

where h, l, $\boldsymbol{n}$, and H^N denote the grid width, cell face location, outside normal vector at the cell face, and Heaviside function for the Neumann boundary condition, respectively. We can also introduce a Dirichlet boundary condition into (4) by replacing the pressure gradient term ∇p with a Heaviside function H^D. Equation (5) represents a Dirichlet condition on the face of cell i, as shown in Fig. 1.

$$\left(\nabla p\, H^N\right)_e = \frac{1}{h} \left\{ p_{i+1} H_e^D + \left(1 - H_e^D\right) p_{i+1}^{BC} - p_i \right\} H_e^N \tag{5}$$

Finally, we obtain a linear system in the form $A\boldsymbol{x} = \boldsymbol{b}$:

$$\begin{aligned}
&\sum_l \left(p\, H^D H^N\right)_l - p \sum_l H_l^N \\
&= h^2 \phi - h \sum_l \left(1 - H_l^N\right) \nabla p_l^{BC} \boldsymbol{n}_l - \sum_l \left(1 - H_l^D\right) p_l^{BC} H_l^N \boldsymbol{n}_l
\end{aligned} \tag{6}$$

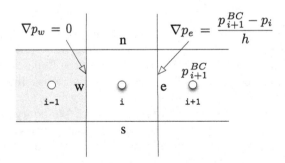

Fig. 1. Neumann and Dirichlet boundary conditions for two-dimensional cell i. A Neumann BC is applied on the west cell face, which is solidly shaded. A Dirichlet BC is employed on the east cell face, where the boundary value is given by the pressure p_{i+1}^{BC}.

The contribution of all boundary conditions is included in the RHS of (6), which is computed at the start of each iteration. Although the Heaviside function only

Fig. 2. Bit representation. Several bits required for the bit-representation are encoded into this array. This example includes diagonal (BC_Diag), non-diagonal (BC_Ndag_x), Neumann boundary (BC_N_x), Dirichlet boundary (BC_D_x), cell state (State), and activeness (Active) of a cell. Other bits are used for more complicated processes.

outputs a value of one or zero, the boundary values can be exactly reproduced in the RHS to distinguish between normal and boundary cell states. In (6), the second and third terms on the RHS can include any Neumann or Dirichlet conditions. It is obvious that the coefficients of the non-diagonal components $pH_l^D H_l^N$ are either 1 or 0, and the diagonal components $\sum H_l^N$ take values ranging from 0 to 6. Therefore, the coefficients of the non-diagonal components can be expressed in 1 bit, and those of the diagonal components can be expressed in 3 bits. That is, a total of 9 bits are sufficient to represent all coefficients in updating the pressure value of a cell. These coefficients are encoded in a 4-byte integer array, as shown in Fig. 2, with other functional flags. For instance, BC_Diag and BC_Ndag_x express the diagonal and non-diagonal elements of the coefficient in the sparse matrix A, where x denotes the six outward directions of the cell faces, denoted as W, E, S, N, B, and T. The two types of boundary conditions for each direction are encoded as, e.g., BC_N_x and BC_D_x. This bit sequence is prepared before the flow calculation. Bits can be encoded into an array b using a simple shift operation, e.g., b|=(0x1<<BC_Ndag_S) for the south direction of a cell. Unlike Tang's approach [3], the proposed method does not require any extra calculations, as we do not use compression.

This is the key idea for handling matrices with different coefficients in every row in the stencil computation.

3   Implementation of Iterative Algorithms with Bit-Compression

Based on the idea described in Sect. 2, we investigated an efficient implementation of (6) for the conventional Red–Black successive over-relaxation (SOR) method and the biconjugate gradient stabilized (BiCGstab) method [4].

3.1   Implementation of Red–Black SOR

We present two implementations of the Red–Black SOR method. The first is the proposed bit-representation (hereafter, *bit-reps*), and the second is a *naïve* code that stores all coefficients in floating-point arrays, where p, b, bp, pn denote pressure (solution vector x), the RHS vector b of a derived linear system

$A\boldsymbol{x} = \boldsymbol{b}$, a utility array of the mask function for a cell (indicates active or inactive), and the non-zero coefficients of matrix A for the *naïve* method, respectively. In the case of *bit-reps*, in particular, the array bp includes compressed coefficient information. The loop in both codes contains 31 floating-point arithmetic operations[1], whereas there are 5 loads/1 store[2] in the *bit-reps* code and 12 loads/1 store in the *naïve* system. In the *naïve* code, the variable array pn has consecutive memory access, so that the SIMD units work effectively.

Implemented Fortran code of the Red–Black SOR method

[Bit-reps code]

```
do color=0,1
  do k=1,kx
  do j=1,jx
  do i=1+mod(k+j+color,2), ix, 2
    idx = bp(i,j,k)
    c_e = real( ibits(idx, BC_Ndag_E, 1) )
    c_w = real( ibits(idx, BC_Ndag_W, 1) )
    c_n = real( ibits(idx, BC_Ndag_N, 1) )
    c_s = real( ibits(idx, BC_Ndag_S, 1) )
    c_t = real( ibits(idx, BC_Ndag_T, 1) )
    c_b = real( ibits(idx, BC_Ndag_B, 1) )
    d0  = real( ibits(idx, BC_Diag+0, 1) )
    d1  = real( ibits(idx, BC_Diag+1, 1) )
    d2  = real( ibits(idx, BC_Diag+2, 1) )
    dd  = d2*4.0 + d1*2.0 + d0
    pp = p(i,j,k)
    ss = c_e * p(i+1,j  ,k  ) + c_w * p(i-1,j  ,k  )
       + c_n * p(i  ,j+1,k  ) + c_s * p(i  ,j-1,k  )
       + c_t * p(i  ,j  ,k+1) + c_b * p(i  ,j  ,k-1)
    dp = ( (ss + b(i,j,k) ) / dd - pp ) * omg
    p(i,j,k) = pp + dp
    res = res + dble(dp*dp) * dble( ibits(idx, Active, 1) )
  end do
  end do
  end do
end do
```

[Naive code]

```
do color=0,1
```

---

[1] Here, we count 1 flop for multiplication, addition, and subtraction operators, and 8 flops for the division operator.

[2] The number of loads depends on the size of the cache line. In this case, there are 3 loads for array p owing to the L3 cache.

```
do k=1,kx
do j=1,jx
do i=1+mod(k+j+color,2), ix, 2
   c_w = pn(i,j,k,1)
   c_e = pn(i,j,k,2)
   c_s = pn(i,j,k,3)
   c_n = pn(i,j,k,4)
   c_b = pn(i,j,k,5)
   c_t = pn(i,j,k,6)
   dd  = pn(i,j,k,7)
   pp = p(i,j,k)
   ss = c_e * p(i+1,j  ,k  ) + c_w * p(i-1,j  ,k  )
      + c_n * p(i  ,j+1,k  ) + c_s * p(i  ,j-1,k  )
      + c_t * p(i  ,j  ,k+1) + c_b * p(i  ,j  ,k-1)
   dp = ( (ss + b(i,j,k) ) / dd - pp ) * omg
   p(i,j,k) = pp + dp
   res = res + dble(dp*dp) * dble( ibits(bp(i,j,k), Active, 1) )
   end do
   end do
   end do
end do
```

The above code clearly shows the number of arrays required in both implementations, i.e., ten scalar arrays for the *naïve* implementation, and just three for *bit-reps*. Table 1 shows the memory footprint of both implementations for the cases evaluated in Sect. 4. It is obvious that the *bit-reps* implementation demands one-third of the memory capacity needed by the *naïve* implementation. This has a significant impact on the performance of cache-based architectures.

Table 1. Comparison of the memory footprint (MB) required for the *naïve* and *bit-reps* implementations. These memory capacities include the size of the halo region.

Problem size (/w halo $= 2$)	16^3	32^3	64^3	128^3	256^3
Naïve	0.3	1.8	12.0	87.7	670.5
Bit-reps	0.1	0.5	3.6	26.3	201.1

3.2 Implementation of BiCGstab

Next, we examined the application of *bit-reps* to a non-preconditioned BiCGstab algorithm. The proposed *bit-reps* technique was applied to BiCGstab, which is a Krylov subspace method, because our implementation effectively accelerates the computation of the matrix–vector product Ax. The matrix–vector calculation is executed in lines 2, 7, and 10 of the following algorithm.

1: Start with an initial guess x_0

2: Compute $r_0 = b - Ax_0$
3: Choose an arbitrary vector r_0^* such that $\rho_0 = (r_0^* \cdot r_0) \neq 0$, e.g., $r_0^* = r_0$
4: $p_0 = r_0$
5: k=0
6: **repeat**
7:     $q = Ap_k$
8:     $\alpha = \rho_k/(r_0^* \cdot q)$
9:     $s = r_k - \alpha q$
10:     $t = As$
11:     $\omega = (t \cdot s)/(t \cdot t)$
12:     $x_{k+1} = x_k + \alpha p_k + \omega s$
13:     $r_{k+1} = s - \omega t$
14:     $\rho_{k+1} = (r_0^* \cdot r_{k+1})$
15:     $\beta = (\alpha/\omega)(\rho_{k+1}/\rho_k)$
16:     $p_{k+1} = r_{k+1} + \beta(p_k - \omega q)$
17:     $\rho_k \leftarrow \rho_{k+1}$
18:     k++
19: **until** $\|r_{k+1}\|_2/\|b\|_2 < \epsilon$

There are some limitations to our *bit-reps* implementation in terms of which preconditioners can be introduced. For instance, the incomplete Cholesky decomposition preconditioner cannot be employed, because this changes the matrix coefficients. In addition, straightforward scaling, which is one of the simplest preconditioning techniques, cannot be used for the same reason. Of course, we could implement scaling by exploiting an additional array to retain the scaling coefficients, but this will cause a deterioration in performance. Thus, only Gauss–Seidel smoothers can be applied for preconditioning, and, as described in Sect. 4.2, these will enhance performance.

4   Evaluation of Proposed Method and Discussion

4.1   Evaluation Problems

The performance of *bit-reps* is evaluated for two different linear systems, derived from a simple steady-state heat transfer equation and an unsteady incompressible Navier–Stokes equation.

Heat Transfer Problem. We consider the Laplace equation $\nabla^2 \varphi = 0$ under the conditions:

$$\varphi(x, y, z) = \begin{cases} \sin(\pi x)\sin(\pi y) & \text{on } (x, y, 0) \\ \sin(\pi x)\sin(\pi y) & \text{on } (x, y, 1) \\ 0 & \text{on other boundaries.} \end{cases} \tag{7}$$

on the Cartesian grid system, and investigate the convergence rate and performance of Red–Black SOR. The exact solution is given by:

$$\varphi(x, y, z) = \frac{1}{\sin(\sqrt{2}\pi)} \sin(\pi x)\sin(\pi y) \left\{ \sinh(\sqrt{2}\pi z) - \sinh(\sqrt{2}\pi(z-1)) \right\}. \quad (8)$$

A norm to determine the convergence is defined by:

$$\sqrt{\sum_{i,j,k} \left(|\varphi^m - \varphi_{exact}| \right)_{i,j,k}}, \quad (9)$$

where m indicates the number of iterations. The computational space is defined by the unit cube $\Omega = [0,1]^3$ divided into $NX \times NX \times NX$ cells, where NX is the dimension size. Details of the problem setting are given in [5].

Three-Dimensional Lid-Driven Cavity Flow Problem. A three-dimensional unsteady fluid simulator is employed to examine the performance of Red–Black SOR and BiCGstab under the *bit-reps* implementation. The source code is available from [7]. In this test, the number of iterations is fixed to 100 to compare the performance of the BiCGstab and Red–Black SOR subroutines.

4.2   Performance of Red–Black SOR

The performance of *bit-reps* and the *naïve* implementation was examined on *Intel* and *Sparc* architectures (see Table 2). Three-dimensional unsteady cavity flows were computed in single precision to evaluate the performance. The *Performance Monitor library (PMlib)* [6,8] was employed to measure the GFlops/s attained by each code. *PMlib* is designed to calculate performance statistics based on the user's declaration of the Flop count and the measured timing between marked sections in the source code. This approach is not perfect, but offers portability across many platforms, even with no means of accessing the hardware performance counter [6]. Figure 3 shows the serial measured performance for different problem sizes. In the *Intel* architectures, we can see that the performance of the *naïve* code worsens when the problem size exceeds the cache, i.e., beyond 64^3 in this case. On the other hand, the *bit-reps* code maintains the same level of performance. Because the *bit-reps* code has lower memory requirements, any degradation in performance seems to be delayed as the problem size increases.

Table 2. Specification of evaluation machines. TRIAD scores are measured by the STREAM benchmark [9].

Architecture	Clock (GHz)	Core	CPU	Peak (GFlops/s)	Cache (MB)	Memory (GB)	Theoretical BW (GB/s)	TRIAD (GB/s)
Xeon X5650	2.66	6	2	127.7	12	16	64	22
Xeon E5-2670	2.6	8	2	166.4	20	64	102	59
Xeon E5-1680	3.0	8	1	96.0	25	64	—	49
Sparc VIIIfx	2.0	8	1	128.0	6	16	64	36
Sparc IXfx	1.85	16	1	236.5	12	32	85	50

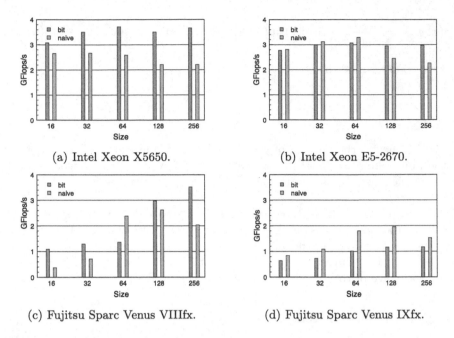

(a) Intel Xeon X5650.

(b) Intel Xeon E5-2670.

(c) Fujitsu Sparc Venus VIIIfx.

(d) Fujitsu Sparc Venus IXfx.

Fig. 3. Comparison of serial performance of each machine. The problem size varies from 16^3 to 256^3.

The *Sparc* architecture produces different behavior to the *Intel*, as the performance tends to improve as the problem size increases. We found that the *bit-reps* code produced superior performance to that of the *naïve* code on *Sparc VIIIfx*. In particular, the performance of *Sparc VIIIfx* is much better than that of *Sparc IXfx*. This phenomenon is a result of the assembler code. That is, the compiler on *Sparc VIIIfx* is optimized to decode shift operations in order to issue the SIMD instructions, but this is not the case on *Sparc IXfx*.

The measured thread performance is shown in Fig. 4. We chose a calculation size of 256^3 so that the data must be placed out of cache. Figure 4 indicates that memory-bound behavior occurs under the *naïve* code when there are more than four threads (*Intel*) and six threads (*Sparc VIIIfx*), but this is not observed on the *Sparc IXfx*. Although the performance of the *naïve* code on *Sparc IXfx* seems to have good scalability, this is because of the unoptimized code described above. In contrast, the *bit-reps* code exhibits a significant effect from suppressing the memory traffic, and achieves remarkable performance gain compared to the *naïve* implementation in all cases. In particular, the highest performance is mostly obtained with the maximum number of threads, a direct result of the designed bit-compression scheme.

Thus, the *bit-reps* code exhibits higher performance than the *naïve* code for all of the architectures evaluated here. We have found that the *bit-reps* code

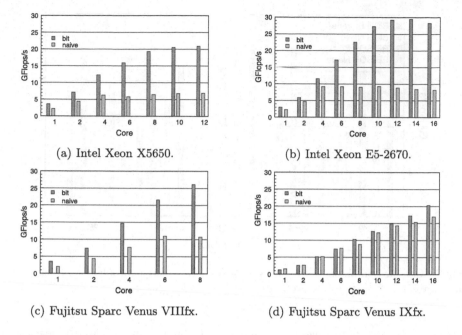

(a) Intel Xeon X5650.

(b) Intel Xeon E5-2670.

(c) Fujitsu Sparc Venus VIIIfx.

(d) Fujitsu Sparc Venus IXfx.

Fig. 4. Comparison of thread parallel performance of each machine with a problem size of 256^3.

significantly improves performance, and achieves over 17 % of the theoretical peak performance on both *Intel* and *Sparc* architectures. One reason for the higher performance attained by the *Intel* CPUs is that they have SIMD computing units for integer arithmetic in addition to floating-point arithmetic, and the optimized compiler enables both to work well.

4.3   Performance of BiCGstab Method

Three-dimensional unsteady cavity flows were computed in single precision to evaluate the performance of the *bit-reps* and *naïve* implementations of BiCGstab. The measured performance data were compared with those from Red–Black SOR, as shown in Fig. 5, for various problem sizes. It can be seen that Red–Black SOR with *bit-reps* gives the best performance, whereas Red–Black SOR with the *naïve* code produces the worst performance. The performance of BiCGstab falls between these two extremes, with the *bit-reps* version giving superior performance to the *naïve* version. The benefits of the *bit-reps* implementation appear for problem sizes of 64^3 and above. One possible reason for this is the memory requirement, because BiCGstab requires more than twice the memory of Red–Black SOR.

Because the performance difference is observed between problem sizes of 64^3 and 128^3, we investigate the thread performance for these two cases in Fig. 6. In the case of 64^3, the performance of BiCGstab approaches that of Red–Black SOR

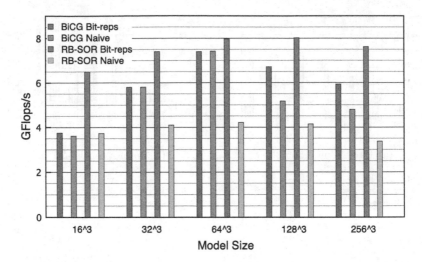

Fig. 5. Comparison of single-thread performance for different problem sizes. Sampled data are taken from the iterative part only.

with *bit-reps*. As the problem size increases, the performance rapidly degrades, with the exception of Red–Black SOR with *bit-reps*.

Next, the performance of BiCGstab sub-tasks such as inner products, matrix–vector multiplication, and TRIAD calculations was examined. Table 3 shows that solving the Poisson equation takes more than 90 % of the fluid simulator's total execution time. We can see that the sub-tasks involved in calculating $A\boldsymbol{x}$ and updating X require almost 60 %. The Byte/Flop (byte per flop) of these tasks is low, and therefore higher performance is expected. In fact, Fig. 7 reveals that sub-tasks involved in the $A\boldsymbol{x}$ computation operate at over 8 GFlops/s. In contrast, and as expected, high Byte/Flop tasks represent relatively low performance.

Because the BiCGstab algorithm is a combination of sub-tasks, it is not easy to predict the performance level, especially in multi-thread environments. On the contrary, the main task in Red–Black SOR is the computation of $A\boldsymbol{x}$. Therefore, it is easy to understand the behavior of the Red–Black SOR code.

Thus far, we have described how to implement the *bit-reps* code, and demonstrated its effectiveness in classical and modern iterative methods for Poisson's equation, as derived from the incompressible Navier–Stokes equation and the heat transfer equation. Usually, incompressible flow fields are solved by coupling Poisson's equation with other momentum equations. Our *bit-reps* implementation can be applied to a linear system derived from Poisson's equation or from implicit time integration. Of course, the same idea can be introduced for other equations, e.g., momentum equations, energy equations, and the equation of continuity, and a similar effect will be observed.

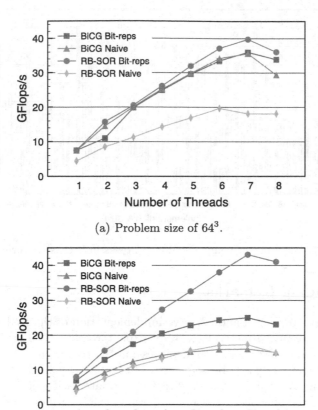

(a) Problem size of 64^3.

(b) Problem size of 128^3.

Fig. 6. Comparison of multi-thread performance.

Table 3. Ratio of execution time and Byte/Flop for each kernel task in BiCGstab *bit-reps* calculation. Data was measured on one Intel Xeon E5-1680 thread with Intel compiler v.15.0.0.077. Note that "Residual" includes the calculation of $A\boldsymbol{x}$.

Sub-task	Ratio (%)	B/F	Line number in Algorithm
$A\boldsymbol{x}$	45.9	1.1	7, 10
Dot	18.3	4.0	3, 8, 11, 14
Update X	14.9	1.7	12
Z = aX + Y	8.4	8.0	9, 13
Update P	5.1	4.0	16
Residual	0.3	1.3	2
Others	7.1	—	—

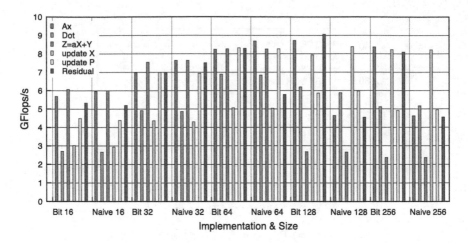

Fig. 7. Performance comparison of sub-tasks in BiCGstab kernels.

5 Concluding Remarks

We have described a novel and effective implementation of the Red–Black SOR and BiCGstab iterative methods to fully exploit recent cache-based architectures. The proposed *bit-reps* method represents the coefficients of a large-scale sparse matrix compactly as a bit sequence. This approach effectively reduces the volume of memory traffic, which is directly associated with performance. The performance of the *bit-reps* code was compared to that of a *naïve* code on several architectures. We found that *bit-reps* improved performance by up to 3.5 times, and enabled 17 % of the theoretical peak to be achieved on both *Intel* and *Fujitsu Sparc* architectures. In addition to this performance gain, the *bit-reps* approach allows us to handle various arbitrary boundary conditions and arbitrary positions in the computational domain without significant performance degradation. This feature is particularly favorable for practical simulations.

Acknowledgments. We thank the RIKEN Advanced Institute for Computational Science for allowing us to use the K computer to obtain our results. Part of this research was supported by a grant for the "Strategic Program on HPCI Field No. 4: Industrial Innovations" from the Ministry of Education, Culture, Sports, Science, and Technology (MEXT) "Development and Use of Advanced, High-Performance, General-Purpose Supercomputers Project," and was carried out in partnership with the University of Tokyo.

References

1. Williams, S., Waterman, A., Patterson, D.: Roofline: an insightful visual performance model for multicore architectures. Commun. ACM **52**(4), 65–76 (2009)
2. Willcock, J., Lumsdaine, A.: Accelerating sparse matrix computations via data compression. In: Proceedings of the 20th Annual ICS 2006, pp. 307–316 (2006)

3. Tang, W.T., et al.: Accelerating sparse matrix-vector multiplication on GPUs using bit-representation-optimized schemes. In: Proceedings of SC 2013, vol. 26, pp. 1–12 (2013)
4. Van der Vorst, H.A.: Bi-CGSTAB: a fast and smoothly converging variant of Bi-CG for the solution of nonsymmetric linear systems. SIAM J. Sci. Stat. Comput. **13**(2), 631–644 (1992)
5. Yokokawa, M.: Vector-parallel processing of the successive overrelaxation method. Japan Atomic Energy Research Institute JAERI-M Report No. 88–017 (1988) (in Japanese)
6. Ono, K., Kawashima, Y., Kawanabe, T.: Data centric framework for large-scale high-performance parallel computation. Procedia Comput. Sci. **29**, 2336–2350 (2014)
7. http://avr-aics-riken.github.io/ffvc_package/
8. http://avr-aics-riken.github.io/PMlib/
9. http://www.cs.virginia.edu/stream

The Ninth International Workshop on Automatic Performance Tuning

Environment-Sensitive Performance Tuning for Distributed Service Orchestration

Yu Lin[1]([⊠]), Franjo Ivančić[2], Pallavi Joshi[2], Gogul Balakrishnan[2],
Malay Ganai[2], and Aarti Gupta[2]

[1] University of Illinois at Urbana-Champaign, Champaign, IL, USA
`yulin2@illinois.edu`
[2] NEC Laboratories America, Princeton, NJ, USA

Abstract. Modern distributed systems are designed to tolerate unreliable environments, i.e., they aim to provide services even when some failures happen in the underlying hardware or network. However, the impact of unreliable environments can be significant on the performance of the distributed systems, which should be considered when deploying the services. In this paper, we present an approach to optimize performance of the distributed systems under unreliable deployed environments, through searching for optimal configuration parameters. To simulate an unreliable environment, we inject several failures in the environment of a service application, such as a node crash in the cluster, network failures between nodes, resource contention in nodes, etc. Then, we use a search algorithm to find the optimal parameters automatically in the user-selected parameter space, under the unreliable environment we created. We have implemented our approach in a testing-based framework and applied it to several well-known distributed service systems.

Keywords: Distributed application · Disturbance action · Performance optimization

1 Introduction

Cloud computing—the on-demand use of remote hardware or software computing resources over a network—has emerged as one of the primary ways of deploying new applications and services at scale. Effectively tuning and optimizing performance of distributed services is essential for controlling costs and improving customer satisfaction. To enable the administrators optimize the performance, distributed applications often provide configuration parameters that can be tuned for a specific system deployment.

However, finding the best configuration for distributed systems is challenging. First, the parameter space can be large, and it is hard to find the best configuration manually. For example, Hadoop contains more than 190 parameters that are specified to control the behavior of a MapReduce job. Second,

---

F. Ivančić and G. Balakrishnan—Current affiliation: Google, Inc.

the best configuration can be different under different deploying environments, or even for different workloads. Third, since distributed services are deployed on a distributed system of networked computer and storage elements, failures may occur in the deploying environment. The best configuration can also be affected by anomalies in the deployed environment, such as in the network (link failures, link congestions, packet drops), node (process crash, memory contention, deadlocks, CPU overload), or disk (slow response, failures, corruption). These anomalies can impact both the correctness and performance of distributed services.

Researchers have proposed several ways to mitigate the above challenges. Some approaches [15] are proposed to search the optimal configuration automatically for a given workload in large parameter spaces. However, these approaches only target some specific distributed applications/systems and do not consider the possible effects of anomalies in the environment (i.e., they assume an ideal environment without any failures). On the other hand, since modern distributed applications are designed to tolerate various environmental anomalies to some extent, some perturbation-based testing approaches [17] are also proposed to test the robustness or availability of the distributed services under environmental anomalies. However, such designs or testing approaches focus on robustness or correctness of the distributed system, but do not consider performance. There is relatively little work on optimizing the performance of distributed systems in the presence of various environmental anomalies.

This paper proposes an approach to optimize the performance of distributed services in the presence of environmental anomalies, such as node VM crashes, network failure, etc. Our approach injects *disturbance actions*, which simulate the environmental anomalies, into the environment. Examples of disturbance actions include shutting down a node in a cluster or in a cloud, disconnecting the link between nodes, limiting the resources a node can use, etc. Then, we use a search algorithm to find the optimal configuration for a distributed system that is deployed in the disturbed environment we created. Since various disturbed environments can be created by applying different combinations of disturbance actions, and it is impractical to optimize performance over all combinations, we also propose two strategies to select disturbed environments. One is to optimize the disturbed environment that leads to the worst performance (i.e., min-max game strategy), while the other is to optimize the performance based on the estimated probabilities of disturbance actions (i.e., weighted average strategy).

The intuition behind our approach is that the environmental anomalies can affect the performance of distributed systems. The configuration for best performance in an ideal environment (i.e., which has no anomalies) is not necessarily the same as in a real environment (i.e., in which anomalies may happen). Thus, system administrators should consider the environment anomalies when they are trying to tune performance.

This paper makes the following contributions:

- We formalize the problem of finding good configuration parameters for performance optimization of distributed services in the presence of environmental anomalies such as network failures or delays, node crashes, and multi-tenant resource contentions.

– We have implemented our approach for arbitrary distributed systems in a framework called RIOT, which is based on an automatic service orchestration framework Juju [2]. We have extended Juju to allow deployment-time queries on performance and health of the services, and combine automatic service deployment with anomaly injections.
– We present promising experiments on some popular distributed applications, including Apache Hadoop, HBase and ZooKeeper.

2   Problem Definition

In this section, we formalize the problem of optimizing configuration for performance of distributed applications in the presence of environmental anomalies.

We consider software performance to include the response time and system resource (e.g., CPU/memory) consumption [21]. We assume that the performance of a distributed application depends on the deployed environment (e.g., virtualized or cloud), the workload it runs, and the configuration. Thus, the performance can be denoted by formula (1), in which P is the performance, E is the environment, Ψ is the workload, and c is the configuration. P is denoted as a function F of E, Ψ and c, and larger P means worse performance.

$$P = F(E, \Psi, c) \tag{1}$$

In our approach, we inject a set of disturbance actions in the environment to simulate the environmental anomalies. Thus, the environment E can be denoted by a function Γ over a set a that contains disturbance actions (as shown in formula (2)). Set a_{all} denotes all the disturbance actions that may occur in the environment and a can be a subset of a_{all}. Note that the ideal (or undisturbed) environment is when $a = \emptyset$.

$$E = \Gamma(a), \text{where } a \subseteq a_{all} \text{ and } a_{all} = \{a_1, a_2, ..., a_n\} \tag{2}$$

For a given workload Ψ, our objective is to find the optimal configuration c_{opt} which leads to the best performance in the parameter space S, under an environment E (as shown in formula (3)).

$$c_{opt} = \underset{c \in S}{\arg\min} F(E, \Psi, c), \text{where } c = \{c_1, c_2, ...\} \tag{3}$$

However, for a disturbance action set of size n, we may need to consider 2^n environments, taking into account all combinations. If we further allow multiple occurrences of each disturbance action, the combinatorial problem becomes even worse. It is not practical to find the best configuration for each possible environment. Thus, we propose two strategies to select the environments to consider.

For a given action set a_{all}, let A denote a set of disturbance actions whose elements are subsets of a_{all} (formula 4). A can be selected by a user, based on her knowledge of the actual deployed environment.

$$A = \{a_1, a_2, ..., a_m\}, \text{where } a_i \subseteq a_{all} \tag{4}$$

$$a_{worst} = \operatorname*{argmax}_{a \in A} F(\Gamma(a), \Psi, c_{opt}^{ideal}), \text{where } c_{opt}^{ideal} = \operatorname*{argmin}_{c \in S} F(E_i, \Psi, c) \tag{5}$$

Based on the selected A, we can apply the following strategies:

(1) Min-max game strategy: Optimize configuration for the disturbance set a_{worst} which leads to the worst performance ($a_{worst} \in A$). a_{worst} can be denoted by formula 5: given the optimal configuration c_{opt}^{ideal} for the ideal environment E_i, a_{worst} is the action set that degrades the performance most under configuration c_{opt}^{ideal}. Thus, in this strategy, the configuration to find can be denoted by $c_{opt} = \operatorname*{argmin}_{c \in S} F(\Gamma(\operatorname*{argmax}_{a \in A} F(\Gamma(a), \Psi, c_{opt}^{ideal})), \Psi, c)$. We use this strategy in our experiment, where we set A to $\{\{a_1\}, \{a_2\}, ..., \{a_n\}\}$ (i.e., each action set in A contains only one action).

(2) Weighted average strategy: Optimize configuration for expected performance using probabilistic estimations about the frequency of disturbances. In this strategy, the performance can be denoted as $P = \sum_{i=1}^{m} w_i * F(\Gamma(a_i), \Psi, c)$. The weight w_i can be viewed as the estimated probability that a disturbance set may occur in the deployed environment.

3   Environment-Sensitive Performance Tuning Framework

In this section, we present our implementation of the technique for environment-sensitive performance tuning. We first give an overview of the infrastructure. Then, we introduce the disturbance actions we implemented. Finally, we discuss the search algorithm and the way to measure performance.

3.1   Infrastructure Overview

Our approach relies on service orchestration frameworks, which automate the process of deploying, configuring and maintaining distributed services efficiently. We propose new service orchestration commands that enable us to query performance or correctness of the underlying deployed services. We also integrate the orchestration framework with disturbance action injection.

In this paper, we use Juju [2] as the service orchestration framework. Juju uses Charms [3] to deploy the application and supports many popular cloud environments, such as EC2 and OpenStack. Charms essentially comprise scripts which define the ways of deploying and managing the distributed services. Currently, Juju provides a large set of Charms which support hundreds of different distributed applications. For example, to deploy Hadoop, we have to deploy four Hadoop services:

hdfs-datanode, *hdfs-namenode*, *mapred-jobtracker* and *mapred-tasktracker*. There are relations between different services because some services depend on others (e.g., *mapred-jobtracker* and *mapred-tasktracker* depend on *hdfs-namenode*, while *hdfs-namenode* depends on *hdfs-datanode*). Note that each service can be deployed on certain number of nodes.

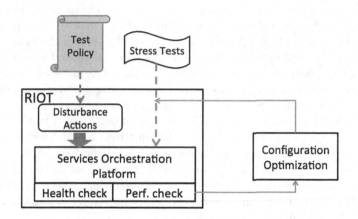

Fig. 1. An overview of RIOT.

Figure 1 shows an overview of RIOT, our framework for testing and tuning performance of distributed services. In addition to deploying distributed services, we create a test policy which defines a set of disturbance actions and the frequency of applying them. After applying the test policy, RIOT runs workloads (i.e., stress tests) under the disturbed environment, and queries the performance through the service orchestration commands (*performance check* in Fig. 1). The information to query is defined in Juju Charms. For example, in our experiments, we execute a script to query performance counters from Ganglia [1] in the Charms. Note that we also propose *health checks* for checking correctness, availability, or other requirements of interest. However, we target performance in this paper. After obtaining the performance in one execution, RIOT reconfigures the services dynamically through Juju to search for a better configuration, and the above process is repeated until some time limit.

3.2   Disturbance Actions

As a proof-of-concept, we implement the following disturbance actions in RIOT:

- *Shutting down a node*, which simulates a node crash in a VM/cluster/cloud.
- *Disconnecting two nodes*, which simulates a link failure.
- *Limiting the available memory in a node*, which simulates resource contention in a node.

Juju creates Linux LXC containers [5] and deploys the service units in the containers (i.e., the service processes are created by the containers). We shut down a node by stopping the corresponding LXC container. To disconnect two nodes, we use Linux traffic control to drop all packets sent between the two containers. We use LXC `cgroup` to limit the available memory, since this command allows specifying resource limits for a control group of processes. It is easy to define other disturbance actions, such as link congestions, packet delay using Linux traffic control tools, for example. Note that to determine which nodes an action should be injected to, one should consider the actual deployment scenario. For example, in a `Hadoop` deployment, suppose the administrator expects that the *mapred-tasktracker* service is unstable, she can shutdown one or some random task tracker nodes, since all task tracker nodes are functionally equivalent. However, for `ZooKeeper`, the administrator should distinguish leader node and follower nodes, since they are not functionally equivalent. All possible disturbance actions and their combinations constitute a_{all} defined in Sect. 2.

The RIOT framework allows the user to control the frequency of applying the disturbance actions. A user can inject and remove these disturbance actions multiple times (along with suitable sleeps in between) during their testing process to construct various disturbance scenarios. The disturbance actions to apply are specified in the test policy shown in Fig. 1.

3.3   Searching Configuration Parameters

As mentioned earlier, the search algorithm executes the stress tests repeatedly within a given time limit to search for the optimal configuration. For efficiency and effectiveness, the algorithm must minimize the number of stress test executions while finding near-optimal configurations. For this purpose, RIOT uses the *Recursive Random Search (RRS)* algorithm [23], which has been applied to black-box optimization problems. RRS first samples the search space randomly for a certain number of times, to identify regions that contain optimal solutions with high probability. These regions are then sampled recursively, and are either moved or shrunk gradually to local optima based on the samples. Then, RRS restarts random sampling to find better regions based on the result of the last iteration, and repeats the recursive search.

3.4   Performance Measurement

To compare the performance of different stress test executions, we have to measure and quantify the performance of an execution. We use Ganglia [1] as our performance measurement tool. Ganglia is a scalable distributed monitoring system for high-performance computing systems, including clouds. It provides around 50 different performance counters. RIOT deploys Ganglia along with the distributed services. During the execution of stress tests, RIOT queries the performance counters from Ganglia (i.e., the "performance check" module in Fig. 1) every 30 s, and calculates *the average of all nodes* for each performance counter. Note that when comparing performance, instead of simply adding absolute values with different units (e.g., adding memory usage which is in *byte* with execution time which is in *second*), RIOT first calculates the relative values that are relative to the first execution for each performance counter, and adds these relative values by using the weighted average. This weighted average is then used to represent the performance of one execution. A larger value means worse performance. Table 1 shows the performance counters and weights used in our experiments. These performance counters cover the usage of network, cpu, and memory[1]. Note that the weights we used are heuristic, and users can select other weights based on their estimated importance for each performance counter.

For example, suppose we have two executions. For the first execution, the averages of all nodes for each performance counter are 20 MB (bytes in), 10 MB (bytes out), 10 % (CPU user), 2 GB (Used memory), 1000 s (execution time); while for the second execution, the averages are 30 MB, 10 MB, 8 %, 1GB, 2000 s. Then the relative values of the first execution for each performance counter (relative to the first execution) are always 1; while for the second execution, the relative values are 1.5 (byte in), 1 (byte out), 0.8 (CPU), 0.5 (memory), 2 (execution time). Thus, the weighted average for these relative values is 1 for the first execution and 1.46 (i.e., $0.1 \times 1.5 + 0.1 \times 1 + 0.2 \times 0.8 + 0.1 \times 0.5 + 0.5 \times 2$) for the second execution, and the performance of the second execution is worse.

Table 1. Ganglia performance counters and weights used in the experiment.

Name	Description	Weight
Bytes in	Number of bytes in per second	0.1
Bytes out	Number of bytes out per second	0.1
CPU user	Percent of user CPU	0.2
Used memory	Amount of used memory	0.1
Execution time	Execution time of the stress tests	0.5

4   Experiments

We evaluate RIOT by answering the following research questions:
RQ1: Do the disturbance actions we inject affect the performance of the distributed systems?
RQ2: Can we optimize the performance for a distributed application deployed in disturbed environments?

4.1   Experimental Subjects

We applied RIOT on three well-known distributed applications in our experiments: Hadoop, HBase, and ZooKeeper. Hadoop is a framework for distributed

---

[1] Execution time is not obtained from Ganglia, but from Linux time command.

processing of large data sets using map-reduce programming models, while HBase is a distributed and scalable big data database for Hadoop. ZooKeeper is a centralized service for maintaining high-level configuration information, providing distributed synchronization and group services. We used versions 1.0.2, 0.92.1, and 3.3.5, respectively.

Table 2. Configuration parameters used in the experiments.

Application	#	Parameter	Values
Hadoop	c_1	number_of_datanode	3, 4, 5, 6, 7
	c_2	number_of_tasktracker	3, 4, 5, 6, 7
	c_3	io_sort_mb	80, 100, 120, 140, 160, 180, 200
	c_4	io_sort_factor	8, 9, 10, 11,12
Hadoop & HBase	c_5	dfs_block_size (MB)	32, 64, 128, 256
	c_6	datanode_max_xcievers	2048, 4096, 6144
	c_7	namenode_handler_count	8, 9, 10, 11,12
	c_8	heap (MB)	512, 1024, 1536, 2048
ZooKeeper	c_9	number_of_zookeeper_node	3, 4, 5, 6, 7
	c_{10}	default_group	0, 1, 2, 3, 4
	c_{11}	default_weight	1, 2, 3, 4, 5

As mentioned in Sect. 3.1, RIOT runs workloads (i.e., stress tests) on the applications to tune the performance. The workloads we used in experiments are shown in the first column of Table 3. The first four stress tests are used for Hadoop. These tests cover different layers of Hadoop: TestDFSIO is a read and write test for HDFS, TeraSort is a large-scale test that covers both map-reduce and HDFS layers, NNBench tests the NameNode hardware and configuration, MRBench focuses on the map-reduce layer by running many small map-reduce jobs. For HBase, we randomly generated HBase operations as workload, including adding/deleting tables, adding/deleting tuples, and queries. Each HBase node runs one randomly generated operation sequence. Similar to HBase, we randomly generated workloads for ZooKeeper, including stopping and re-starting the ZooKeeper service, creating new elements in the shared configuration state, re-setting values, querying some states, or deleting some states.

4.2   Experimental Setup

We did two sets of experiments. In experiment 1, we applied RIOT to the four stress tests of Hadoop. We ran stress tests with different disturbance actions, and different number of Hadoop task tracker nodes. As mentioned in Sect. 2, we injected only one action per execution and did not consider the combination of actions. Note that the number of nodes can be viewed as configuration parameter. Through this experiment, we can find out how the actions affect the performance and whether the performance for disturbed environments can be improved by changing configuration (i.e., number of task tracker nodes). Since this experiment does not aim to find an optimal configuration, we did not apply RRS. Instead, we added task tracker nodes manually for each execution.

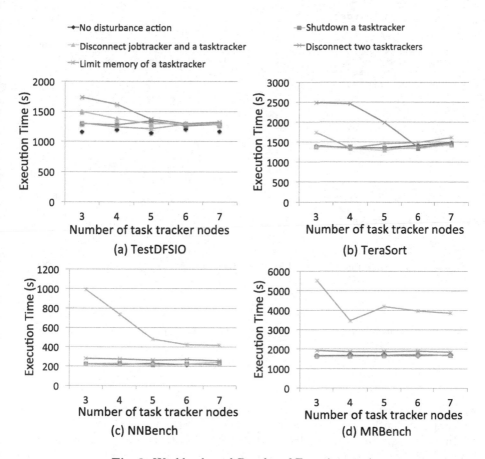

Fig. 2. Workloads and Results of Experiments 1.

In experiment 2, we applied RIOT to all the three distributed applications to automatically find optimal configurations for both ideal and disturbed environments. For each application, we select a set of popular configuration parameters as parameter space. These configuration parameters are shown in Table 2. Column 1 shows the application, and Column 2 shows the parameter number (for ease of reference). Columns 3 and 4 show parameter names and values (i.e., parameter space), respectively. Four parameters are used both by Hadoop and HBase. Note that we consider the number of nodes also as configuration parameters (e.g., number_of_datanode and number_of_tasktracker), and for different workloads we tune different configuration parameters. Similar to the first experiment,

Table 3. Workloads and Results of Experiments 2.

Stress test	a_{worst}	Performance value				Optimal configuration	
		E_i^{ic}	E_d^{ic}	E_d^{dc}	E_i^{dc}	E_i	E_d
TestDFSIO	disconnecting two **datanodes**	1	1.22	1.15	1.05	$c_1=5$, $c_3=100$,	$c_1=6$, $c_3=80$,
						$c_4=10$, $c_5=64$,	$c_4=12$, $c_5=32$,
						$c_6=4096$, $c_8=2048$	$c_6=4096$, $c_8=2048$
TeraSort	disconnecting two **task-trackers**	1	1.20	1.06	1.03	$c_2=3$, $c_3=100$,	$c_2=6$, $c_3=80$,
						$c_4=10$, $c_5=32$,	$c_4=10$, $c_5=32$,
						$c_8=2048$	$c_8=2048$
NNBench	limiting memory of a **task-tracker**	1	1.80	1.05	1.01	$c_2=3$, $c_3=120$,	$c_2=6$, $c_3=160$,
						$c_4=10$, $c_5=256$,	$c_4=12$, $c_5=256$,
						$c_8=1536$	$c_8=2048$
MRBench	limiting memory of a **task-tracker**	1	1.30	1.12	1.03	$c_2=3$, $c_3=100$,	$c_2=4$, $c_3=160$,
						$c_4=10$, $c_5=64$,	$c_4=10$, $c_5=64$,
						$c_8=1024$	$c_8=1536$
HBase	disconnecting a **Region-Server** with **Master**	1	1.53	1.06	1.10	$c_5=64$, $c_6=4096$,	$c_5=32$, $c_6=4096$,
						$c_7=10$, $c_8=1024$	$c_7=8$, $c_8=1536$
ZooKeeper	shutting down a node	1	1.07	1.07	1.05	$c_9=5$, $c_{10}=0$,	$c_9=7$, $c_{10}=4$,
						$c_{11}=2$	$c_{11}=1$

we injected three kinds of disturbance actions (described in Sect. 3.1), with one action per execution (i.e., we did not combine disturbance actions).

In both experiments, we deployed Hadoop with three data nodes and task trackers, and one name node and job tracker. For TestDFSIO, we read and wrote 10 files of size 1 GB; for TeraSort, we sorted 5 GB data; for NNBench, we created 1000 files of size 1 KB; for MRBench, we ran 50 small map-reduce jobs. For HBase, in order to focus on the performance of HBase itself, we used a stand-alone HBase installation with one master server and three region servers, without integrating it with Hadoop. We deployed ZooKeeper with five nodes. For HBase and ZooKeeper, we generated workloads with 1000 operations.

All the experiments are run in a cloud where each node has a 2.33 GHz 4-cores Intel Xeon processor, 2 GB memory and runs Ubuntu 12.04.1. The time limit for RRS is set to two hours.

4.3   Experimental Results

Experiment 1. Figure 2 shows the result of the first experiment. The four charts represent the results of executing the four Hadoop stress tests. The x-axis is the the number of task tracker nodes while the y-axis is the execution time of the stress tests. Note that to make the charts more illustrative, we only show the execution time instead of considering all the performance counters in Table 1. Each line in the charts represents the executions with one disturbance action (or no action) but different number of task tracker nodes.

Without injecting any disturbance actions, the performance for each stress test is relatively the same even if we increased the number of task tracker nodes. The reason could be the number of task trackers set for the initial ideal environment is already saturated for the workloads, so adding more task trackers do not affect the performance much. After we injected disturbance actions, some executions get an obvious performance degradation. For example, TestDFSIO and Tera-Sort are affected significantly by action *"disconnected two task trackers"*, while NNBench and MRBench are affected most by action *"memlimit a task trackers"*. However, for a given benchmark, there are also some disturbance actions which do not affect performance. Interestingly, for those actions that lead to worse performance, after adding more task tracker nodes, the performance becomes better for all the four stress tests.

From this experiment, we make the several observation. First, some disturbance actions affect the performance of the distributed application significantly while others do not. Second, different workloads are affect by different disturbance actions. Thus, performance tuning should target workloads rather than applications. Third, the performance for a disturbed environment can be improved by using better configurations.

Experiment 2. The result of the second experiment is shown in Table 3. The first column shows the workloads, in which the first four are Hadoop stress tests. We use the *Min-max game* strategy and set A to $\{\{a_1\}, \{a_2\}, ..., \{a_n\}\}$ in the experiments. The second column, a_{worst}, is the disturbance action set that leads to the worst performance. Columns 3–6 show four performance values (i.e., combining the ideal and disturbed environments E_i, E_d with two optimal configurations c_{opt}^{ideal} and $c_{opt}^{disturbed}$, where c_{opt}^{ideal} and $c_{opt}^{disturbed}$ represents the optimal configurations for E_i and E_d, respectively). E_i^{ic} is the *ideal* environment with its optimal configuration c_{opt}^{ideal}. E_d^{dc} is the *disturbed* environment injected by a_{worst}, and uses its new optimal configuration $c_{opt}^{disturbed}$. E_d^{ic} is the *disturbed* environment but uses c_{opt}^{ideal}, while E_i^{dc} is the *ideal* environment but uses $c_{opt}^{disturbed}$. Note that the performance value is a relative value based on the value of E_i^{ic}, so the performance value of E_i^{ic} is always 1. Columns 7 and 8 show the optimal configurations for E_i and E_d, respectively. We use the serial numbers in Table 2 to represent the parameter names (e.g., c_1 refers to number_of_datanode). Notice that for each workload, we only search for the configuration parameters related to it.

From these results, we make the following observations. First, as expected, the disturbance actions affect the performance even if we consider multiple performance counters, as seen by comparing E_i^{ic} and E_d^{ic}. This observation conforms to our first observation from experiment 1. An exception is ZooKeeper, for which the performance value is almost the same. This is likely because ZooKeeper is not compute-intensive like Hadoop, and the effect of environment anomalies is less on its main function of maintaining a quorum. Second, a comparison of Columns 6 and 7 shows that the optimal configurations for ideal and disturbed environments are different. Also, different workloads have different optimal configurations. Third, one should consider when to use an optimal configuration for a disturbed environment. By comparing the performance values, we can see that although the performance of E_d^{dc} is a bit worse than that of E_i^{ic}, there are obvious improvements when comparing with E_d^{ic} in almost all cases. Also, from E_i^{dc} we can observe that using $c_{opt}^{disturbed}$ in ideal environment only results in a subtle distinction on performance. This means that using the optimal configuration for a disturbed environment, rather than an ideal environment, can be a better choice in practice. The RIOT framework provides a flexible platform to find these configurations.

4.4   Threats to Validity and Limitations

Construct threats: First, different performance counters can be combined to represent the performance of a distributed application. The weights of the performance counters can differ on the basis of real deployment scenario. RIOT mitigates this threats by considering five commonly used performance counters [21] and using heuristic weights. Second, since the time limit for RRS is set to two hours, the reported configuration by RIOT is the optimal configuration found in two hours. However, a longer time limit may result in a better configuration. Third, we injected one disturbance action once in each execution. However, disturbed environments can be categorized on both the action type and the frequency we injected them. Injecting multiple actions with different frequencies per execution may give different results. Thus, in practice, we should construct as many disturbed environments as possible by combining disturbance actions and injecting them with different frequencies.

Internal threats: RIOT uses RRS to search optimal configurations and Ganglia to collect performance counters. The correctness of our implementation depends on the correctness of RRS and Ganglia. The authors are not aware of any bugs in RRS, Ganglia or the implementation of the RIOT techniques that may have affected the results of the experiments.

External threats: As it holds for any experimental study, the selected set of applications may not be fully representative. Disturbance actions' affect on performance may differ for different applications. To mitigate this limitation, we selected three applications with varying domains. Moreover, for each application, we used stress tests that test different components of the applications.

Other limitations: First, RIOT uses min-max game strategy to optimize the configuration. However, using other strategies, such as weighted average strategy, may lead to better tuning result. Second, since RIOT uses a black-box approach to tune the performance, it does not analyze why the performance is affected by disturbance actions. The reason could be related to an application's recovery algorithm, load balancing or error handling mechanism, etc. Analyzing the internal states of an application may also lead to better tuning result. Third, current RIOT implementation requires users to define test policies manually. However, we may also automatically generated test policy templates by mining commonly occurred disturbance actions in cloud.

5   Related Work

The closest related work in performance tuning of distributed systems is by Babu et al. on StarFish [8,14,15]. StarFish is a self-tuning analytics system for Hadoop. It uses model-based estimation to predict Hadoop job performance, and then uses the prediction to find good settings for configuration parameters. However, StarFish only targets Hadoop and assumes an ideal environment. Our approach considers the effect of environment anomalies and can be applied to arbitrary distributed systems.

Various stress testing [4,6,19] or fault injection based methods [7,9,10,12,13, 16–18,20,22] are used to measure performance or test robustness in distributed applications.

Stress testing frameworks, such as Selenium [6] and LoadRunner [4], test distributed applications under heavy load conditions to check robustness, availability, tolerance, error handling, etc. These frameworks are used to check if the system has noticeable defects under large and unpredictable network delays and heavy usage. Lubke et al. [19] provide an architecture that allows network emulation of standard client/server-based architectures. They provide an architecture, based on Dummynet [11], that allows network emulation of standard client/server-based architectures. The main goal, as is common for network emulators, is to get precise and accurate performance measurements. Our goal is to tune performance for arbitrary distributed service architectures.

There have been recent efforts for cloud testing that focus on testing of recovery functionality by either injecting disk/node/link failures [13] or specifying failure scenarios via testing policies [18]. To allow testers more control where faults should be injected, various test description languages have been proposed [9,10,12,16,20]. One such example is the tool LFI [20] for fault-injection based testing of recovery code when library calls fail. Our implementation also provides a way to define testing policies and inject failures.

Random fault-injection based methods are also frequently used for distributed application resiliency testing, such as chaos monkey testing [22] or Game-Day exercises [7]. One drawback of random application-level fault injection is that the injected faults may not be justifiable in practice. In our approach, the disturbance actions we injected guarantee that every observed performance degradation is justifiable and replicable.

Our recent work SETSUDO [17] targets robustness testing of cloud applications. It allows a tester to specify testing policies using application-dependent abstraction labels that expose internal states of the application. This requires instrumentation support and a scheduler that controls the ordering of certain execution events. Here, we treat the applications as black boxes without knowing the internal states of the application, and we focus on performance rather than robustness.

6   Conclusions

In this paper, we proposed a black-box approach to tune the performance of distributed services along with consideration of environmental anomalies. We implemented this approach in a framework called RIOT for arbitrary distributed systems. It extends the automatic service orchestration framework Juju to simulate and inject disturbance actions, and performs measurements and other health checks. We performed experiments on some popular distributed service applications, which show that environmental anomalies can significantly affect their performance, and the optimal configurations for ideal environment and disturbed environment are different. We also reported on how our platform is used to compare the optimal configuration for the ideal environment and an unreliable environment, and that choosing the latter provides less performance degradation in many cases. We believe practical deployment of distributed services can benefit from considering anomaly-aware configuration search.

References

1. Ganglia. http://ganglia.sourceforge.net/
2. Juju. https://juju.ubuntu.com/
3. Juju Charms. https://jujucharms.com/
4. LoadRunner. http://www.hp.com/go/LoadRunner
5. LXC. http://linuxcontainers.org/
6. Selenium. http://seleniumhq.org/
7. Allspaw, J.: Fault injection in production. Commun. ACM **55**(10), 48–52 (2012)
8. Babu, S.: Towards automatic optimization of mapreduce programs. In: SOCC 2010, pp. 137–142 (2010)
9. Banabic, R., Candea, G.: Fast black-box testing of system recovery code. In: Proceedings of the 7th ACM European Conference on Computer Systems, EuroSys 2012, pp. 281–294 (2012)
10. Broadwell, P., Sastry, N., Traupman, J.: FIG: A prototype tool for online verification of recovery. In: Workshop on Self-Healing, Adaptive and Self-Managed Systems (2002)
11. Carbone, M., Rizzo, L.: Dummynet revisited. SIGCOMM Comput. Commun. Rev. **40**(2), 12–20 (2010)
12. Dawson, S., Jahanian, F., Mitton, T.: Experiments on six commercial TCP implementations using a software fault injection tool. Softw. Pract. Exper. **27**(12), 1385–1410 (1997)

13. Gunawi, H., Do, T., Joshi, P., Alvaro, P., Hellerstein, J., Arpaci-Dusseau, A., Arpaci-Dusseau, R., Sen, K., Borthakur, D.: FATE and DESTINI: A framework for cloud recovery testing. In: Proceedings of the 8th USENIX Conference on Networked Systems Design and Implementation, NSDI 2011, pp. 18–18 (2011)
14. Herodotos, H., Babu, S.: Profiling, What-if analysis, and cost-based optimization of MapReduce programs. In: VLDB 2011, pp. 1111–1122 (2011)
15. Herodotou, H., Lim, H., Luo, G., Borisov, N., Dong, L., Cetin, F.B., Babu, S.: Starfish: A self-tuning system for big data analytics. In: CIDR 2011, pp. 261–272 (2011)
16. Hoarau, W., Tixeuil, S., Vauchelles, F.: FAIL-FCI: Versatile fault injection. Future Gener. Comput. Syst. **23**(7), 913–919 (2007)
17. Joshi, P., Ganai, M., Balakrishnan, B., Gupta, A., Papakonstantinou, N.: SET-SUDO: Perturbation-based testing framework for scalable distributed systems. In: Proceeding of the Conference on Timely Results in Operating Systems (2013)
18. Joshi, P., Gunawi, H., Sen, K.: PREFAIL: A programmable tool for multiple-failure injection. In: Proceedings of the 2011 ACM International Conference on Object Oriented Programming Systems Languages and Applications, OOPSLA 2011, pp. 171–188 (2011)
19. Lubke, R., Lungwitz, R., Schuster, D., Schill, A.: Large-scale tests of distributed systems with integrated emulation of advanced network behavior. WWW/Internet **10**(2), 138–151 (2013)
20. Marinescu, P., Candea, G.: Efficient testing of recovery code using fault injection. ACM Trans. Comput. Syst. **29**(4), 11:1–11:38 (2011)
21. Molyneaux, I.: The Art of Application Performance Testing: Help for Programmers and Quality Assurance. O'Reilly Media (2009)
22. Tseitlin, A.: The antifragile organization. CACM **56**(8), 40–44 (2013)
23. Ye, T., Kalyanaraman, S.: A recursive random search algorithm for large-scale network parameter configuration. In: SIGMETRICS 2003, pp. 196–205 (2003)

Historic Learning Approach for Auto-tuning OpenACC Accelerated Scientific Applications

Shahzeb Siddiqui[1], Fatemah AlZayer[1], and Saber Feki[2]([✉])

[1] Computer, Electrical and Mathematical Sciences and Engineering Division,
Extreme Computing Research Center, King Abdullah University of Science
and Technology, Thuwal, Saudi Arabia
[2] KAUST Supercomputing Laboratory, King Abdullah University of Science
and Technology, Thuwal, Saudi Arabia
saber.feki@kaust.edu.sa

Abstract. The performance optimization of scientific applications usually requires an in-depth knowledge of the hardware and software. A performance tuning mechanism is suggested to automatically tune OpenACC parameters to adapt to the execution environment on a given system. A historic learning based methodology is suggested to prune the parameter search space for a more efficient auto-tuning process. This approach is applied to tune the OpenACC gang and vector clauses for a better mapping of the compute kernels onto the underlying architecture. Our experiments show a significant performance improvement against the default compiler parameters and drastic reduction in tuning time compared to a brute force search-based approach.

1 Introduction

Accelerators are gradually becoming mainstream in supercomputing as their capability to significantly accelerate a large spectrum of scientific applications at a higher power efficiency has been clearly identified and proven. Moreover, with the introduction of high level programming models such as OpenACC [1] and OpenMP 4.0 [2], these devices are becoming more accessible and practical to use by a larger scientific community. OpenACC was announced in the ACM/IEEE Supercomputing Conference 2011 as a new standard for parallel programming targeting hardware accelerators. Although the goal of the standard is to increase programmer productivity by using compiler directives, getting the best performance of the target device is still tedious and requires a significant effort by the developer to tune some of the OpenACC annotations and the corresponding parameters. As a matter of fact, without tuning, a suboptimal performance is recorded on different applications while using the latest implementations of OpenACC compilers. In this work, we propose a new methodology for empirical tuning of OpenACC accelerated scientific applications to relieve the end user from this burden. The OpenACC gang and vector clauses are used to map nested loops to the underlying hardware architecture. The auto-tuning engine performs a search on the space of possible uses of these clauses and their corresponding attributes, for a better mapping and thus

© Springer International Publishing Switzerland 2015
M. Daydé et al. (Eds.): VECPAR 2014, LNCS 8969, pp. 224–235, 2015.
DOI: 10.1007/978-3-319-17353-5_19

to improve the performance of the offloaded kernel. However, when the number of parameters is quite big, the search space becomes significantly large, and the tuning procedure becomes very expensive and unpractical. We suggest a performance tuning strategy that reduces the cost of tuning by pruning the search space using the historic knowledge of previous tuning operations performed on similar problem sizes. This approach was presented and its effectiveness was demonstrated on auto-tuning MPI communication operations in the Abstract Data and Communication Library (ADCL) [3,4].

The remainder of the paper is organized as follows. Section 2 introduces our methodology to efficiently tune OpenACC clauses using a historic learning approach. In Sect. 3, the experimental setup is described and the performance results of the proposed tuning strategy are reported. Section 4 presents the related work in this research area. Finally, Sect. 5 summarizes our findings and ongoing work.

2   Performance Tuning Methodology

The OpenACC standard offers flexibility to the developer to further tune the *loop pragma* with the *gang* and *vector* clauses and therefore control the mapping of the nested loops to the underlying hardware compute units; that is the threads partitioning in the accelerator. The gang clause specifies in how many groups to aggregate the parallel threads generated by the parallelization of the given loop, which corresponds to the shape and size of the grid of blocks in the GPU environment. The vector clause specifies the granularity of the parallel threads per gang, which translates to the dimensionality and the size per dimension of each of the thread blocks in NVIDIA hardware. The latest compiler technology relies on heuristics to specify these parameters for a given application. Our analysis showed that a significant improvement could be further obtained by spending more effort in tuning these parameters. Our methodology for tackling this particular aspect is described next.

2.1   Tuning Methodology of OpenACC Loop Clauses

The strategy for auto-tuning proposed here is based on empirical evaluation of different uses of the OpenACC clauses gang and vector in nested OpenACC loops as depicted in Fig. 1 for the case of an OpenACC kernels pragma annotation. If the OpenACC parallel construct is to be used, the clauses *num_gang* and *vector_length* are the corresponding clauses for tuning the performance of the offloaded nested loops. For the first case, the performance tuning procedure is in two steps. In the first phase, the performance of different placements of the *gang* and *vector* clauses within the nested loops is evaluated and the best performing one is selected. At this initial phase, we keep the compiler choice of the numbers of gangs and vectors by omitting any specific values. Although, this placement of the gang and vector clauses is not part of the OpenACC standard, we found that making use of this feature in the PGI compiler resulted in an enhanced performance. Once the optimal placement of these clauses is determined, we explore in the second phase

`#pragma acc kernels`
`#pragma acc loop independent`
`gang(a),vector(b)`
`  for (x = 4 ; x < nx-4; x++) {`
`#pragma acc loop independent vector(c)`
`  for (y = 4; y < ny-4; y++) {`
`#pragma acc loop independent`
`gang(d),vector(e)`
`    for (z = 4; k < nz-4; z++) {`
`    U[x][y][z] = c1*V[x]][y][z] + .... } } }`
	`#pragma acc kernels`
	`#pragma acc loop independent`
	`gang(a),vector(b)`
	`  for (x = 4 ; x < nx-4; x++) {`
	`#pragma acc loop independent gang(c)`
	`  for (y = 4; y < ny-4; y++) {`
	`#pragma acc loop independent vector(d)`
	`    for (z = 4; k < nz-4; z++) {`
	`    U[x][y][z] = c1*V[x]][y][z] + .... } } }`

`#pragma acc kernels`
`#pragma acc loop independent`
`for (x = 4 ; x < nx-4; x++) {`
`#pragma acc loop independent`
`gang(a),vector(b)`
`  for (y = 4; y < ny-4; y++) {`
`#pragma acc loop independent`
`gang(c),vector(d)`
`    for (z = 4; k < nz-4; z++) {`
`    U[x][y][z] = c1*V[x]][y][z] + ....} } }`
	`#pragma acc kernels`
	`#pragma acc loop independent gang(a)`
	`for (x = 4 ; x < nx-4; x++) {`
	`#pragma acc loop independent`
	`gang(b),vector(c)`
	`  for (y = 4; y < ny-4; y++) {`
	`#pragma acc loop independent vector(d)`
	`    for (z = 4; k < nz-4; z++) {`
	`    U[x][y][z] = c1*V[x]][y][z] + ....} } }`

Fig. 1. Different placements of the gang and vector clauses in three nested loops.

different numbers of gangs and vectors other than the ones chosen by the compiler. This second phase will be sufficient by itself if the OpenACC parallel construct is used since only the number of gang and the vector length sizes are to be tuned. This tuning methodology is referred as the brute force tuning in the experimental results section. It is worth noting that the set of meaningful configurations is constrained by the specification of the accelerator. Despite this restriction, this parameter space can be very huge, and using an exhaustive search on all possible combinations could be considerably time-consuming and unreasonable. A historic learning based approach able to shrink the search space and accelerate the tuning process is detailed next. Without specifying width for last column:

2.2   Historic Learning Approach

Tuning OpenACC gang and vector clauses with an exhaustive search of the full parameter space is time consuming. In the case of an NVIDIA Kepler GPU, the total number of combinations of gang and vector values if applied to one level of the nested loops can easily reach up to 16,384 combinations. We suggest here a tuning methodology based on the previous tuning results of the same application, for different problem sizes, on the same hardware. Practically, a learning phase is needed for building a knowledge database, which consists in the best tuning parameters for various input sizes. Once such database is constructed, and given a new problem size PS_{new} for the same application, the nearest neighbor problem size in the knowledge base is identified using the Euclidian distance for example. This problem size is used as a reference for the suggested tuning approach and is referred as PS_{ref}. The optimal tuning parameters of PS_{ref} extracted from the knowledge base are used to define a subset of the parameters space to be explored by the tuning engine for PS_{new}. This search subspace consists in a smaller range of gang and vector values. The range of these values is centered on the optimal tuning parameters for PS_{ref} and we used five values in each direction for each tunable parameter in our first experiments. The search space of possible parameters combinations is then drastically reduced and the tuning procedure becomes much faster and thus more attractive. The tuning results of PS_{new} are then included to enrich the tuning knowledge database for future reuse. This tuning methodology is referred as the historic learning tuning in the experimental results section.

2.3   Tuning the Search Subspace Size

The initial results, as it will be detailed in the subsequent section, show that for few problem sizes, for which the reference problem size PS_{new} is not close enough, the performance results of the historic learning approach using the suggested search subspace are not very encouraging. In such cases, we propose to be less aggressive in the pruning process by using a larger number of values of each parameter (gang and vector) to define the size of the search subspace. We empirically identified a threshold of 250 in the Euclidian distance between PS_{ref} and PS_{new} above which, 10 values of each parameter are used to define the search subspace instead of the default value of 5 as described previously. Our experiments show that this approach helps to fill the gap between the speedups obtained by the brute force and the historic learning tuning approaches, and better performance is obtained without increasing the tuning time dramatically.

3   Experimental Results

The test bed hardware and software specifications and the test application used for this performance analysis are first described. Following that, we showcase the importance of the gang and vector clauses placement within nested loops to the performance tuning of OpenACC applications. The performance gain of applying the suggested tuning methodology on the test application as well as the tuning time reduction by using the historic learning approach is reported.

3.1   Test Bed Specifications and Test Application

The test bed used for our experiments consists in a dual socket CPU system hosting four NVIDIA Kepler K20c GPU cards. Each socket is an eight-core Sandy Bridge Intel(R) Xeon(R) CPU E5-2650, running at a clock speed of 2.00 GHz. The software stack consists in the PGI compiler version 12.9, and the NVIDIA CUDA driver 5.0. In our experiments, we used the isotropic finite difference kernel, which constitutes the building block for the Reverse Time Migration (RTM) and the Full Waveform Inversion (FWI) applications, extensively used by the oil and gas exploration industry for the velocity model building and seismic imaging of the sub-surface. The Reverse Time Migration application consists in a forward modeling and backward migration using a finite difference kernel that solves the acoustic wave equation.

$$\frac{1}{c^2}\frac{d^2 P}{dt^2} = \frac{d^2 P}{dx^2} + \frac{d^2 P}{dy^2} + \frac{d^2 P}{dz^2}$$

Where c is the velocity of the propagated wave and P is the wavefield pressure. The 3D finite difference stencil scheme is 8th order in space and 2nd order in time.

3.2   Importance of Gang and Vector Placement

The application of the brute force tuning methodology to a set of ten different problem sizes resulted in a performance improvement of up to 30 % while compared to the base code version tuned by the compiler. It is emphasized here that the optimal gang and vector clauses placement varies from one problem size to another. Table 1 shows for each 3D problem size, the chosen grid and block sizes by the compiler and the corresponding tuned parameters. The color code of each row corresponds to a different placement of the gang and vector clauses in the three nested loops as depicted in Fig. 1. We can conclude here that the first phase in the proposed tuning methodology is crucial as the clauses placement in the nested loops has a significant importance in the performance tuning of the OpenACC code.

3.3   Performance Tuning Results

In the brute force search-based tuning methodology, the performance of the compute kernel is evaluated with all possible gang and vector values allowed by the

Table 1. Compiler choices versus tuning results for gang and vector placement and values for different problem sizes.

3D Domain Size	Grid/block sizes chosen by PGI	Tuned grid/block size
128x128x128	grid:[2x30] block:[64x4]	grid:[30x120] block:[64x4]
256x256x256	grid:[4x62] block:[64x4]	grid:[248x6] block:[64x6]
512x512x512	grid:[8x126] block:[64x4]	grid:[504x63] block:[32x8]
640x640x640	grid:[10x158] block:[64x4]	grid:[10x316] block:[64x4x2]
128x128x640	grid:[10x30] block:[64x4]	grid:[10x64] block:[64x4]
128x640x128	grid:[2x158] block:[64x4]	grid:[4x256] block:[64x4]
640x128x128	grid:[2x30] block:[64x4]	grid:[2x316] block:[64x4x2]
640x640x128	grid:[2x158] block:[64x4]	grid:[2x316] block:[64x4x2]
640x128x640	grid:[10x30] block:[64x4]	grid:[10x316] block:[64x4x2]
28x640x640	grid:[10x158] block:[64x4]	grid:[10x316] block:[128x4]

hardware specification of the K20c GPU. In our experiment, the gang values were chosen at increments of 2, starting from 2 to 1,024. The vector values were chosen in multiples of 32 (warp size, as recommended by NVIDIA), starting from 32 up to 1,024. The total search space consists of 16,384 combinations. The brute force tuning method is a very time consuming process yet very simple in finding the best possible gang and vector tuple. The historic learning algorithm is used to identify a reference problem size PSref with a known solution (i.e. gang and vector tuple). The search space is then reduced by selecting a subset range of parameters keeping only the closest ten values of gang and vector with regard to the reference problem size solution. Therefore, the parameters space to be explored and evaluated is significantly reduced to only a 100 combinations. The brute force search tuning method was first applied to 25 problem sizes and the best tuning parameters are stored in a knowledge base. Then, another set of eight different problem sizes are tuned using both the brute force search and the historic learning tuning approaches. The performance speedups in comparison to the compiler tuned code version as well as the tuning time are recorded while using either of the tuning approaches for each of the eight new investigated problem sizes. As shown in Fig. 2, the tuning procedure using either the brute force search or the historic learning tuning method resulted in a better performance than the compiler default tuning. Indeed, a performance increase of up to 80 % is recorded relative to the performance of the base code. Moreover,

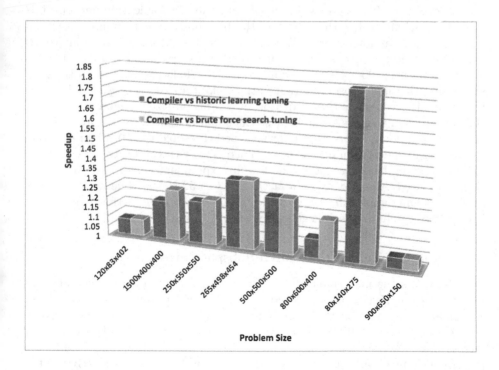

Fig. 2. Performance speedup analysis using the different tuning.

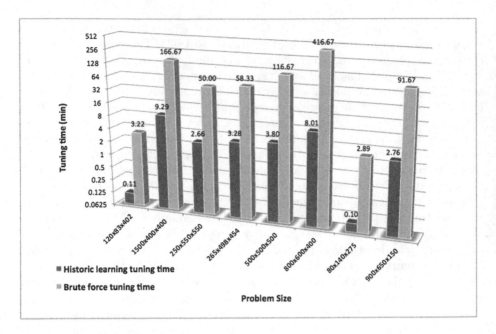

Fig. 3. Tuning time using the brute force and the historic learning tuning approach.

the performance of the code while tuned with the historic learning approach is in most cases within less than 1.5% of the best possible performance recorded while using a brute force search. The two cases where the historic learning tuned code was not performing as well as the brute force search tuned version are with the problem size $1500 \times 400 \times 400$ and $800 \times 600 \times 400$. The analysis of the data shows that the main reason for that is the lack of a close enough problem size in the database used for the prediction of a problem size of reference. Indeed, the two problem sizes have the highest Euclidian distances to the problem size of reference used and were above our specified threshold of 250. Figure 3 shows the required time for tuning a given problem size with the brute force and the historic learning based tuning methods. Our experiment shows that the tuning time is reduced dramatically by a factor of 18 to 52 times while using the new proposed tuning approach. At the same time, a comparable performance to the brute force search approach is generally achieved as detailed before.

3.4   Impact of Tuning the Search Subspace Size

Figure 4 shows the performance gain while using an extended search subspace for the cases of $1500 \times 400 \times 400$ and $800 \times 600 \times 400$ problem sizes, for which, the distance to the problem size of reference was 626.78 and 318.09 respectively. As these numbers are above the specified threshold, the search subspace is then increased as described previously. The results exhibit additional 15% and 14% performance improvements obtained for the two problem sizes respectively. It is also to be noted that the cost of this approach in terms of total tuning time is

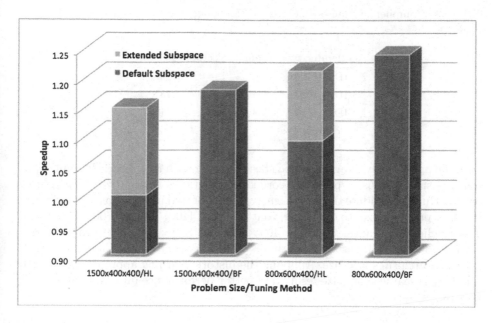

Fig. 4. Performance speedup using the different tuning methodologies using the default and extended subspace for $1500 \times 400 \times 400$ and $800 \times 600 \times 600$ problem sizes.

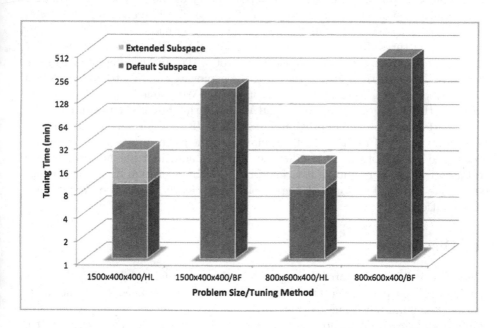

Fig. 5. Tuning time using the brute force and the historic learning tuning approach using the default and extended subspace for $1500 \times 400 \times 400$ and $800 \times 600 \times 600$ problem sizes.

Table 2. Compiler and auto-tuner choices for gang and vector values on different platforms.

Accelerator type	Gangh	Vector	Speedup
NVIDIA Fermi	256	144	1.29X
NVIDIA K20c	884	96	1.43X
NVIDIA K40	784	128	1.41X
PGI compiler	92	256	-

minimal and the advantage of a much faster tuning with the historic learning approach is still obvious as per the recorded tuning times summarized in Fig. 5. In order to further validate this approach, we incorporated in the second experiment, two new problem sizes to the historic data: $1400 \times 440 \times 350$ and $700 \times 550 \times 320$. The Euclidian distances to $1500 \times 400 \times 400$ and $800 \times 600 \times 400$ problem sizes are 118.74 and 137.48 respectively, that is below the specified threshold for extending the search subspace size. Applying the historic learning approach again with the default search space size resulted in better performance improvements of 7 % and 15 % respectively.

3.5   Adaptation to the Underlying Hardware

In this subsection we demonstrates the portability of the proposed auto-tuning methodology on different platforms. Table 2, shows the different gang and vector values chosen by the PGI compiler and our auto-tuner using the historic learning approach on three different generations of NVIDIA GPUs: Fermi, Kepler K20c and Kepler K40 for the problem size $100 \times 100 \times 100$. In all different platforms, the compiler heuristic is always selecting the same gang and vector values, 92 for the number of gangs and 256 as a vector length. We applied the proposed approach on the three generations of NVIDIA GPUs using historic data generated for each platform as detailed in previous subsections. The performance results shows a speedup ranging from 29 % to 43 % relative to the performance of the compiler-optimized version of the finite difference code. In addition, the gang and vector values selected by our auto-tuner vary from a GPU generation to another, which shows the ability of the proposed approach to efficiently adapt to different hardware platforms.

4   Discussion

The historic learning approach proposed in this paper for accelerating the auto-tuning of OpenACC accelerated kernels, has many advantages and some drawbacks. The main advantage is that it can cleverly prune the search space using hints from historic data and therefore reduce the tuning time considerably. It is to be noted here that the suggested methodology is independent of the used search algorithm. Another set of experiments is carried out using a random search algorithm applied to the entire search space and the pruned search space using historic learning. The results are very similar to what has been shown using the brute force search and further confirming the validity of the proposed approach.

In addition, the prediction accuracy of the machine learning algorithm used can improve over time as more data samples are created. However, we have noticed that with the lack of relevant historic data, the algorithm may result in a sub-optimal parameter values. As a remedy, we proposed a simple heuristic that can be applied in our case to tune the size of the pruned search sub-space depending on the quality of the historic data. Another drawback of this method is the non-portability of the historic data from a platform to another. The use of historic data generated on a given accelerator is unlikely to result in performance improvements on a different hardware.

5   Related Work

A significant research has been conducted in tuning applications written for GPUs. At the compiler level, another directive-based programming models called HMPP [6], is presented along with its tuning methodology. The CAPS OpenACC compiler also includes an auto-tuning driver that can explore the optimization space to tune kernel regions [15]. At the application level, researchers applied various tuning mechanisms to GPU codes such as sparse matrix-vector multiply [5], stencil computations [7,11], and computational electromagnetics [14]. Vuduc proposed in [12] a statistical approach for automatic performance tuning of matrix-matrix multiply operation. AtuneRT [13] is an application-independent auto-tuner, which optimizes GPU-specific parameters such as block size and loop-unrolling degree.

The historic learning approach was applied to the runtime tuning of MPI communications in the abstract data and communication library (ADCL) [8,9]. The notion of historic learning is also implemented to a limited extent in FFTW [10], namely with a feature called Wisdom. The user can export experiences gathered in previous runs into a file, and reload it at subsequent executions. However, the wisdom concept in FFTW lacks any notion of related problems, i.e. wisdom information can only be reused for exactly the same problem size that was used to generate it.

6   Conclusions and Ongoing Work

A historic learning-based approach for performance tuning of OpenACC accelerated applications is presented. The performance results obtained by the proposed tuning methodology on a finite difference kernel showed a significant performance gain relative to the compiler-tuned code and close to the optimal performance that can be obtained with an exhaustive search. Furthermore, the time needed for tuning is reduced drastically compared to the brute force tuning technique. Nevertheless, the main limitation of this approach is its dependency on the historic data that is crucial for a good prediction and therefore for achieving a performance close to the tuned code with a brute force search. To reduce this limitation, we suggested a heuristic for extending the search subspace depending on the Euclidian distance to the nearest neighbor identified in the historic data. This approach seems to enhance the performance for such cases with a minimal overhead on the tuning time.

Our ongoing work includes full automation of the tuning process including a code generator tool along with a parallel evaluation engine on a cluster with multiple GPUs. Further experiments will be carried out to validate the approach for a larger spectrum of scientific applications and on a variety of accelerator architectures including different NVIDIA GPU generations and Intels Xeon Phi coprocessors. Furthermore, machine learning algorithms such as K-nearest neighbors, Bayesian classifiers and support vector machines are being investigated to achieve a better prediction of the most similar problem sizes to use as a reference to shrink the search space.

Acknowledgments. The authors would like to thank NVIDIA for the hardware donation to KAUST as CUDA center of research and KAUST IT Research Computing for their support.

References

1. OpenACC Standard specification. www.openacc-standard.org
2. OpenMP 4.0 specification. www.openmp.org/mp-documents/OpenMP4.0.0.pdf
3. Gabriel, E., Feki, S., Benkert, K., Chaarawi, M.: The abstract data and communication library. J. Algorithms Comput. Technol. **2**(4), 581600 (2008)
4. Gabriel, E., Feki, S., Benkert, K., Resch, M.: Towards performance and portability through runtime adaption for high performance computing applications. In: International Supercomputing Conference, Dresden, Germany, June 2008
5. Choi, J.W., Singh, A., Vuduc, R.W.: Model-driven autotuning of sparse matrix-vector multiply on GPUs. In: Proceedings of the 15th Symposium on Principles and Practice of Parallel Programming
6. Dolbeau, R., Bihan, S., Bodin, F.: HMPP: a hybrid multi-core parallel programming environment. In: The 1st Workshop on General Purpose Processing on Graphics Processing Units, GPGPU (2007)
7. Siddiqui, S., Feki, S.: Predictive performance tuning of OpenACC accelerated applications, 29th International Conference, 22–26 June 2014, Leipzig, Germany. LNCS, vol. 8488, pp. 511–512 (2014)
8. Feki, S., Gabriel, E.: A historic knowledge based approach for dynamic optimization. In: Proceedings of the International Conference on Parallel Computing, pp. 389–396 (2009)
9. Feki, S., Gabriel, E.: Incorporating historic knowledge into a communication library for self-optimizing high performance computing applications. In: Second IEEE International Conference on Self-Adaptive and Self-Organizing Systems, Venice, Italy (2008)
10. Frigo, M., Johnson, S.: The design and implementation of FFTW3. Proceedings of IEEE **93**(2), 216–231 (2005)
11. Mametjanov, A., Lowell, M.C., Norris, B.: Autotuning stencil-based computations on GPUs, In: Cluster Conference, Beijing, China (2012)

12. Vuduc, R., Demmel, J.W., Bilmes, J.A.: Statistical models for empirical search-based performance tuning. Int. J. High Perform. Comput. Appl. **18**(1), 6594 (2004)
13. Tillmann, M., Karcher, T., Dachsbacher, C., Tichy, W.F.: Application-independent autotuning for GPUs. In: International Conference on Parallel Computing, Munich, Germany (2013)
14. Feki, S., Al-Jarro, A., Bagci, H.: Multi-GPU-based acceleration of the explicit time domain volume integral equation solver using MPI-OpenACC. In: IEEE International Symposium on Antennas and Propagation and USNC/URSI National Radio Science, Lake Buena Vista, Florida, USA (2013)
15. Bodin, F.: Using CAPS compiler on NVIDIA kepler and CARMA systems. In: Supercomputing, Salt Lake City, Utah, USA (2012)

Capturing the Expert: Generating Fast Matrix-Multiply Kernels with Spiral

Richard Veras$^{(\boxtimes)}$ and Franz Franchetti

Carnegie Mellon University, 5000 Forbes Avenue, Pittsburgh, PA, USA
rveras@andrew.cmu.edu

Abstract. Matrix-Matrix Multiplication (*MMM*) is a fundamental operation in scientific computing. Achieving the floating point peak with this operation requires expert knowledge of linear algebra and computer architecture to craft a tuned implementation, for a given microarchitecture. To do this an expert follows a mechanical process for implementing *MMM*, by deriving an algorithm from models found in the literature. Then, the expert applies optimizations which are well suited for the target architecture. Lastly, the expert expresses that implementation in assembly code. In this paper, we argue that this process is mechanical and can be captured in a rule based program generation system such as *Spiral*. We then show that given this machinery, *Spiral* can produce code for large size *MMM* implementations that are competitive with hand tuned code.

1 Introduction

Implementing a high performance implementation of *MMM* is difficult. A great body of work provides the mathematical machinery for determining the optimal blocking strategies analytically [1,2]. However, once this is determined, there is still the issue of mapping the implementation to the hardware. There are microarchitecture specific issues that must be addressed in order to reach high performance.

The Compiler: Compilers fall short of reaching peak performance for *MMM* on modern architectures. If a programmer provides a straightforward implementation of *MMM* with a triple nested loop (Fig. 1), the code generated by the compiler will not achieve the same high performance as the code an expert produces. One reason for this is that compilers are general purpose and trade off potential performance gains, in the code, for shorter compile times. As a result, the compiler contains optimizations which are well suited for the general case, but are not necessarily the optimizations needed for the specific case of *MMM*.

The Expert: The domain expert acts as a specialized compiler. She uses domain knowledge to optimize for both the operation and the hardware. How do we automate the expert? How do we capture this knowledge in an auto-generation framework? The goal of this paper is to formalize the necessary components that an expert would use, for generating high performance implementations *MMM*, in a rule based framework for automatic program generation.

© Springer International Publishing Switzerland 2015
M. Daydé et al. (Eds.): VECPAR 2014, LNCS 8969, pp. 236–244, 2015.
DOI: 10.1007/978-3-319-17353-5_20

In this paper, we exploit the fact that the expert uses a mechanical process for creating high performance *MMM* kernels and we formalize this process as a set of rules. We then feed these rules through the *Spiral* framework [7], which in turn, produces an autotuned high performance implementation of *MMM*. This implementation is comparable performance to an expert produced kernel. We base our work on the success of existing Automatic Program Generation Systems which replaced expert programmers in the fields of Digital Signal Processing (*DSP*) and Sparse Matrix-Vector Multiplication.

2 Related Work

The goal of our project is to capture the expert knowledge, behind implementing a high performance *MMM*, as rules in a rule in a system like *Spiral*, which in turn can generate and tune an implementation from those rules. The *Spiral* project [7] automates the expert programmer in the *DSP* field. They do this through a layering of Domain Specific Languages (*DSL*) and a database of rules for each *DSL* which captures the knowledge of the domain expert. In the case of *DSP*, they show that high performance code can be automatically generated for fixed and general sized operations. In [5], the authors extended *Spiral* for operations, like *MMM*, through a new language called OL. Their generated *MMM* kernels performed significantly faster than those produced by optimizing compilers, but fell short of the expert programmer because they did not capture the necessary rules that we will show in this paper. *LGen* [3] also generates high performance linear algebra kernels using SIMD instructions. They target small problems sizes for *BLAS* and *BLAS*-like operations where the matrices are L1 cache resident. We differ because we are targeting *MMM* kernels that are designed to scale to larger caches and act as primitives for memory resident *BLAS* operations. The *BTO* project [8] also generates linear algebra kernels, but they target sequences of level 2 and level 1 *BLAS* calls. By removing the interface into the *BLAS* and generating fused kernels they can achieve greater performance than the original sequence of *BLAS* calls. Our work is different because we are targeting *MMM* kernels whereas they focus on fused Matrix-Vector and Vector-Vector operations.

The *ATLAS* project [12], provides a framework for empirically determining the blocking dimensions for cache and registers for *MMM*. We differ from their project because we are generating code from rules given as inputs to a system, as opposed to using a system that is hard coded for a certain class of problems. The authors of [1] demonstrated that the algorithm implemented by *ATLAS* was not optimal for modern multi-level cache architectures, and presented a novel blocking strategy that achieves – both in theory and practice – near peak performance. This work was extended by the BLIS Framework [2] for all Level-3 *BLAS*, by providing a software architecture for minimizing the actual amount of code that an expert needs to produce for a given architecture. The AUGEMM project [4] took this last piece of the expert and automated the process of generating the kernels via templates in a *MMM* specific framework. We differ from their project because we generate our implementation from high level rules that are selected via search, instead of templates. The rules that we will describe in the next section capture the insight from these works. In the DxT project [9,10] the authors also use a rule based system and expert knowledge encoded as rules to generate optimized $BLAS$ routines, which are built on a set of compute and memory packing kernel primitives. Our work differs because they are generating $BLAS$ operations in terms of kernel primitives, whereas we are starting with and generating those same primitives in terms of SIMD vector instructions. Our purpose is to capture – in a rule based framework, $Spiral$– the expert knowledge that is needed to produce high performance MMM kernels for large size $BLAS$ operations, with the hopes of exploiting this knowledge in problems outside of the domain of linear algebra.

3   A New Spiral Operator Language

In this section, We illustrate our language for capturing the knowledge used by the expert in developing a high performance implementation of matrix-matrix multiplication (MMM). A matrix matrix multiplication can be mathematically described as $C = AB$ where A, B and C are general matrices of conformal sizes. In our operator language, the operation is $\mathrm{MMM}_{m,k,n}$ where the subscripts describe the sizes of the inputs (i.e., A is $m \times k$ and B is a $k \times n$ matrix).

3.1   A Naïve MMM Implementation

At the heart of our representation in Fig. 1 is the operator $\mathrm{Pt}(s, A, g)$ which is our non-overlapping loop operator (PtA is our overlapping loop, which accumulates the output). The result of each iteration of computation that is performed is written to a unique spot in memory. This operator is parameterized by three major components. The first is an *index mapping* function, g, for describing how elements are read from memory at each iteration. The second parameter is an operation, A, that is performed at each iteration. And the last parameter, is a second index mapping function, s, which describes how the result of A will be written to memory.

$$
\begin{array}{l|l}
\begin{array}{l}
\texttt{for}\,(\texttt{p=0; p < K; ++p)} \\
\quad \texttt{for}\,(\texttt{i=0; i < M; ++i)} \\
\qquad \texttt{for}\,(\texttt{j=0; j < N; ++j)} \\
\qquad\quad \texttt{C[j*M+i] +=} \\
\qquad\qquad \texttt{A[p*M+i]*} \\
\qquad\qquad \texttt{B[j*K+p];}
\end{array}
&
\begin{array}{l}
\mathrm{MMM}_{M,K,N} \rightarrow \\
\quad \mathrm{PtA}(h_{0,[1]}^{N\to N} \otimes h_{0,[1]}^{M\to M}, \\
\quad\ \mathrm{Pt}(h_{0,[1]}^{N\to N} \otimes h_{0,[1,1]}^{1\to M}, \\
\quad\quad\ \mathrm{Pt}(h_{0,[M,1]}^{1\to N} \otimes h_{0,[1]}^{1\to 1}, \\
\quad\quad\quad P_1 \\
\quad\quad\quad h_{0,[M]}^{1\to 1} \otimes h_{0,[1]}^{1\to 1} \times h_{0,[K,1]}^{1\to N} \otimes h_{0,[1]}^{1\to 1}), \\
\quad\quad\ h_{0,[1]}^{1\to 1} \otimes h_{0,[1,1]}^{1\to M} \times h_{0,[K]}^{N\to N} \otimes h_{0,[1]}^{1\to 1}), \\
\quad\ h_{0,[M,1]}^{1\to K} \otimes h_{0,[1]}^{M\to M} \times h_{0,[1]}^{N\to N} \otimes h_{0,[1,1]}^{1\to K})
\end{array}
\end{array}
$$

Fig. 1. On the left we have a naïve implementation of MMM using a triply nested loop. The inputs, A and B, are stored in column major ordering. On the right we have a representation of that implementation of MMM in our language.

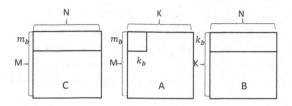

Fig. 2. The Goto/*BLIS* blocking structure.

For s and g, we use a *stride* operator [6] which captures the structure of the input or output. For example, $h_{0,[1,1]}^{n_b \to N} \otimes h_{0,[1]}^{M \to M}$ describes an $M \times N$ matrix which is linearized in memory in column major ordering (M is the leading dimension), and we are selecting out $m_b \times N$ adjacent sub-blocks from this matrix. The key point behind these stride functions is that they capture the structure of the dataset in memory.

3.2 What an Expert Really Does

The Goto/*BLIS* algorithm [1,2] which we will implement in our language, achieves high performance by exploiting several key factors: 1. blocking for last level cache reuse, 2. packing to minimize Translation Lookahead Buffer (TLB) misses, 3. data layout transformations to minimize Level 1 cache misses, 4. and efficient use of vector instructions [4]. We will take a step by step approach to show how we capture these insights in our language (Fig. 2).

Blocking. *Blocking*, or tiling, is a method for maximizing cache reuse by operating on blocks of an input rather than operating on scalar elements of the input. Because *MMM* is a computationally bounded problem, we want to insure that we maximize the cache reuse of our inputs so our performance is not restricted by the cost of moving data from memory into the processor [1]. We achieve blocking by adding an additional nesting of loop operators, Pt, around a smaller MMM and by using the stride operators to capture how the sub-blocks fit inside of the input matrices.

$$
\begin{aligned}
\text{MMM}_{M,K,N} \to \\
\text{PtA}(h_{0,[1]}^{N \to N} \otimes h_{0,[1]}^{M \to M}, \\
\text{Pt}(h_{0,[1]}^{N \to N} \otimes h_{0,[1,1]}^{m_b \to M}, \\
\text{Pt}(h_{0,[M/m_b,1]}^{n_r \to N} \otimes h_{0,[1]}^{m_b \to m_b}, \\
\text{Pt}(h_{0,[M/m_b]}^{n_r \to n_r} \otimes h_{0,[1,1]}^{m_r \to m_b}, \\
\text{MMM}_{m_r,k_b,n_r} \\
h_{0,[M/m_b]}^{k_b \to k_b} \otimes h_{0,[1,1]}^{m_r \to m_b} \times h_{0,[K/k_b]}^{n_r \to n_r} \otimes h_{0,[1]}^{k_b \to k_b}), \\
h_{0,[M/m_b]}^{k_b \to k_b} \otimes h_{0,[1]}^{m_b \to m_b} \times h_{0,[K/k_b,1]}^{n_r \to N} \otimes h_{0,[1]}^{k_b \to k_b}), \\
h_{0,[1]}^{k_b \to k_b} \otimes h_{0,[1,1]}^{m_b \to M} \times h_{0,[K/k_b]}^{N \to N} \otimes h_{0,[1]}^{k_b \to k_b}), \\
h_{0,[1,1]}^{k_b \to K} \otimes h_{0,[1]}^{M \to M} \times h_{0,[1]}^{N \to N} \otimes h_{0,[1,1]}^{k_b \to K})
\end{aligned}
$$

Using the formula above, we can express the blocking strategy used in the Goto/*BLIS* approach. For the sake of consistency we follow the same variable naming conventions for tile sizes (k_b, m_b, n_r, m_r) as those used in [1]. The essence of this formula, is that we can use the stride functions to express the data layout of the blocks in memory and use the loop operators to manipulate those blocks.

Packing Blocks into Contiguous Buffers. *Packing* takes blocking one step further by copying the sub-blocks into a contiguous buffer in memory. By placing the working set in a contiguous buffer, fewer TLB entries – a cache for address translation – are needed to address the working set and therefore fewer costly TLB misses occur which can severely hinder performance [1]. If we want to pack a $m_b \times k_b$ subblock of matrix A which is $M \times K$ we can use the following formula:

$$\text{PackA}_{M,m_b,k_b} \to$$
$$\text{Pt}(h_{0,[1,1]}^{1 \to k_b} \otimes h_{0,[1]}^{m_b \to m_b},$$
$$\text{Pt}(h_{0,[1]}^{1 \to 1} \otimes h_{0,[1,1]}^{1 \to m_b},$$
$$I_1,$$
$$h_{0,[M/m_b]}^{1 \to 1} \otimes h_{0,[1,1]}^{1 \to m_b}),$$
$$h_{0,[M/m_b,1]}^{1 \to k_b} \otimes h_{0,[1]}^{m_b \to m_b})$$

The formula above uses two nested loop operators to copy a sub-block of A into a contiguous buffer. Following this same construction we can pack an $k_b \times N$ subblock of matrix B which has the dimensions $K \times N$. We can then construct a new rule:

$$\text{MMM}_{m_b,k_b,N} \to \text{MMM}_{m_b,k_b,N} \circ (\text{PackA}_{M,m_b,k_b} \times I_{Nk_b})$$
$$\to \text{MMM}_{m_b,k_b,N} \circ (I_{m_b k_b} \times \text{PackB}_{K,k_b,N})$$

If you read these rules from right to left they express the behavior of packing a sub-block of one input (PackA and PackB) while leaving the other input unmodified $(I_{m_b k_b})$. After that they are passed onto the next step of computation, the MMM operator.

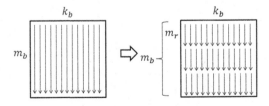

Fig. 3. The figuring on the left hand side shows how the elements of this sub-block of A will be stored if no data layout transformation is performed. Unfortunately, this is not the layout in which the elements will be traversed in the innermost loop of the *MMM* operation. We must rearrange the data such that each $m_r \times k_b$ block is stored contiguously and those blocks are themselves stored contiguously. This is explained in more detail in [2].

Data Layout Transformation of Packed Blocks. A *data layout transformation* is a reindexing of the inputs such that every memory access to that input is performed in unit stride. We can improves temporal locality in the cache by rearranging the elements of our sub-blocks of A and B as we pack them. The following formula captures the packing and data layout transformation where we read our input with one data layout and write our output in a different layout (Fig. 3):

$$
\begin{aligned}
\text{PackA}_{M,m_b,k_b} \to \\
\text{Pt}(h_{0,[1,1]}^{1\to m_b/m_r} \otimes h_{0,[1]}^{k_b\to k_b} \otimes h_{0,[1]}^{m_r\to m_r}, \\
\text{Pt}(h_{0,[1]}^{1\to 1} \otimes h_{0,[1,1]}^{1\to k_b} \otimes h_{0,[1]}^{m_r\to m_r}, \\
\text{Pt}(h_{0,[1]}^{1\to 1} \otimes h_{0,[1]}^{1\to 1} \otimes h_{0,[1]}^{m_r\to 1}, \\
I_1, \\
h_{0,[M/m_r]}^{1\to 1} \otimes h_{0,[1,1]}^{1\to m_r}), \\
h_{0,[M/m_r,1]}^{1\to k_b} \otimes h_{0,[1]}^{m_r\to m_r}), \\
h_{0,[M/m_b]}^{k_b\to k_b} \otimes h_{0,[1,1]}^{m_r\to m_b})
\end{aligned}
$$

We achieve our data layout transformation by modifying our method for packing through the introduction of a new index striding function for the sub-blocks. For example, our sub-block A originally has the initial layout described by this index mapping function, $h_{0,[1]}^{k_b\to k_b} \otimes h_{0,[1]}^{m_b\to m_b}$. Our goal is to reorder the elements into this index mapping function $h_{0,[1,1]}^{1\to m_b/m_r} \otimes h_{0,[1]}^{k_b\to k_b} \otimes h_{0,[1]}^{m_r\to m_r}$ which will allow us to access the elements in unit stride, in the innermost loop of our kernel Fig. 1.

Minimal Instruction SIMD Vector Broadcast. In order to achieve high performance on a processor that allows vector operations, we must use those vector operations. However, in the innermost loop of our *MMM* we need to multiply every element from the A matrix with every element from the B matrix. If we perform this task with vector operations, we would need a way to broadcast each element of one of the inputs into a vector. On some architectures this is sufficient, but on others this operation is expensive. An alternative to the broadcast instruction is to reorder the elements in the vector register for one input, in such a way that we are ultimately able to multiply each element from one input with the other. This is discussed in further detail in [4]. We capture this optimization through the addition of two special permutation operations: a permuted broadcast on a vector of length v, Y_v and a reordering operator that reverses the permutation, U_v.

$$
\begin{aligned}
\text{MMM}_{M,K,N} \to U_v \circ \\
\text{PtA}(h_{0,[1]}^{N\to N} \otimes h_{0,[1]}^{M\to M}, \\
\text{Pt}(h_{0,[1]}^{N\to N} \otimes h_{0,[1,1]}^{1\to M}, \\
\text{Pt}(h_{0,[M,1]}^{v\to N} \otimes h_{0,[1]}^{1\to 1}, \\
P_v \circ (Y_v \times I_v) \\
h_{0,[M]}^{1\to 1} \otimes h_{0,[1]}^{1\to 1} \times h_{0,[K,1]}^{v\to N} \otimes h_{0,[1]}^{1\to 1}), \\
h_{0,[1]}^{1\to 1} \otimes h_{0,[1,1]}^{1\to M} \times h_{0,[K]}^{N\to N} \otimes h_{0,[1]}^{1\to 1}), \\
h_{0,[M]}^{1\to K} \otimes h_{0,[1]}^{M\to M} \times h_{0,[1]}^{N\to N} \otimes h_{0,[1,1]}^{1\to K})
\end{aligned}
$$

Table 1. Here we have listed the translations between our scalar operators and C code. Given a formula B, we recursively apply $\mathrm{Code}\,(B, \mathbf{y}, \mathbf{x}, [j])$, until we have a final C coded implementation. The $\mathbf{y}$ and $\mathbf{x}$ refer to our output and input, and the variable j refers to loop variable in the current scope.

Operator	C Code Translation
$\mathrm{Code}\,(A, \mathbf{y}, \mathbf{x}, [i])$	A(y,x,[i]);
$\mathrm{Code}\,(A \circ B, \mathbf{z}, \mathbf{x}, [i])$	$\mathrm{Code}\,(A, \mathbf{z}, \mathbf{y}, [i])$; $\mathrm{Code}\,(B, \mathbf{y}, \mathbf{x}, [i])$;
$\mathrm{Code}\,(A \times B, \mathbf{y}, \mathbf{x}, [i])$	$\mathrm{Code}\,(A, \mathbf{y_0}, \mathbf{x_0}, [i])$; $\mathrm{Code}\,(B, \mathbf{y_1}, \mathbf{x_1}, [i])$;
$\mathrm{Code}\,(I_1, \mathbf{y}, \mathbf{x}, [i])$	*y = *x;
$\mathrm{Code}\,(P_1, \mathbf{y}, \mathbf{x}, [i])$	*y = *x0 * *x1;
$\mathrm{Code}\,(Pt(s, A, g), \mathbf{y}, \mathbf{x}, [i])$	for(j=1; j < rng(s)/dmn(s); ++j)
	$\mathrm{Code}\,(s, \hat{\mathbf{y}}, \mathbf{y}, [j])$
	$\mathrm{Code}\,(g, \hat{\mathbf{x}}, \mathbf{x}, [j])$
	$\mathrm{Code}\,(A, \hat{\mathbf{y}}, \hat{\mathbf{x}}, [j])$
$\mathrm{Code}\,(f, \mathbf{y}, \mathbf{x}, [j])$	y = &x[f(j)];

The permuted broadcast, Y_v, is applied at every iteration of the innermost loop, which would be the case with a normal vector broadcast, however the reordering operator, U_v, is applied outside of the inner loop. Depending on the architecture, the implementation of these operators may vary, however, they capture the essence that one of the inputs must be broadcasted and that the order in which it happens need not be preserved, because the results will be reordered before they are written to memory.

4   Code Generation

Spiral [7] is a rule based system for generating high performance code. Given a set of breakdown rules, which we described in the previous section, *Spiral* will recursively apply those rules until it has a family of fully expanded rule trees, which describe an implementation. Then, given a set of translation rules for converting those rule trees into code (Table 1). Lastly, via exhaustive empirical search, *Spiral* will find the best implementation in its search space.

5   Results

In this section we use *Spiral* to generate and optimize our implementation of *MMM* from the rules that we have defined. We then compare the performance achieved by the best generated code, found by *Spiral*'s exhaustive search, against Intel's Math Kernel Libraries (MKL) on several Intel architectures. In order to achieve high performance, there are several additional optimizations that must be performed when *Spiral* is generating the code. This includes: *memory prefetching* to minimize the number of last level cache misses, *software pipelining* [11]

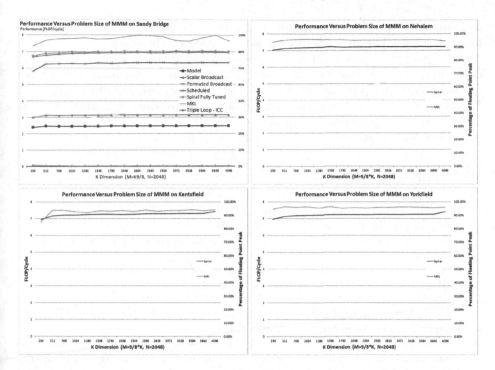

Fig. 4. Performance comparison of *Spiral* generated code versus expert written code on various architectures. In each of the charts, the top line represents that maximum floating point peak. **Top Left:** In this chart we show how the addition of a few low level transformations at code generation time can add the additional performance. Additionally, we show the performance of a triply nested implementation of *MMM*, described in Sect. 3, compiled with the Intel Compiler.

to hide the latency of bringing data elements from the cache, and *loop unrolling* which reduces the number of branch instructions and simplifies address computation. For our experiments, we compare the performance of our generated code against Intel MKL on the following microarchitectures: Intel Sandy Bridge, Nehalem, Kentsfield and Yorkfield. On the Yorkfield and Nehalem the performance of our code is within 5 % of the expert code and on the other architectures it is within 10 %. Further optimizations are necessary in order to match the expert produced code. Lastly, on the Sandy Bridge, we generated multiple implementations of *MMM* by incrementally adding additional optimizations. Starting from the best unoptimized implementation with a permuted broadcast, we achieve 80 % of the machine's peak. The addition or prefetching and software pipelining brings us closer to the expert (Fig. 4).

6   Conclusion

In this paper we suggest that an expert produces an efficient implementation of *MMM* via a mechanical process. We demonstrate that given a formal language and rule based program generation system one can capture this process in the form of rules. These rules help express the space of algorithms, which *Spiral* uses to generate *MMM* code. On a representative sample of modern systems, the code generated is competitive with expert programmers. We have shown that our language can capture high level optimization, which are described in Sect. 3. The next step is to capture these transformations as rewrite rules, which are not tied to a specific operation. In this way, we can extend these techniques on other operations such as stencil codes or signal processing codes.

References

1. Goto, K., van de Geijn, R.: Anatomy of high-performance matrix multiplication. ACM Trans. Math. Softw. **34**, 12:1–12:25 (2008)
2. Van Zee, F., van de Geijn, R.: BLIS: a framework for rapidly instantiating BLAS functionality. ACM Trans. Math. Softw. (2013)
3. Spampinato, D., Püschel, M.: A Basic Linear Algebra Compiler. ACM CG **23** (2014)
4. Qian, W., Xianyi, Z., Yunquan, Z., Yi, Q.: AUGEM: automatically generate high performance dense linear algebra kernels on x86 CPUs. In: International Conference on High Performance Computing (2013)
5. Franchetti, F., de Mesmay, F., McFarlin, D., Püschel, M.: Operator language: a program generation framework for fast kernels. In: Taha, W.M. (ed.) DSL 2009. LNCS, vol. 5658, pp. 385–409. Springer, Heidelberg (2009)
6. Franchetti, F., Püschel, M.: Formal loop merging for signal transforms. In: PLDI, pp. 315–326 (2005)
7. Püschel, M., Moura, J., Johnson, J., Padua, D., Veloso, M., Singer, B., Xiong, J., Franchetti, F., Gacic, A., Voronenko, Y., Chen, K., Johnson, R., Rizzolo, N.: SPIRAL: code generation for DSP transforms. In: Proceedings of IEEE on "Program Generation, Optimization and Adaptation", vol.93, pp. 232–275 (2005)
8. Siek, J., Karlin, I., Jessup, E.: Build to order linear algebra kernels. In: Workshop on Performance Optimization of High-level Languages and Libraries (POHLL08) (2009)
9. Marker, B.: Design by transformation: from domain knowledge to optimized program generation. Doctoral Dissertation,Department of Computer Science, The University of Texas at Austin (2014)
10. Marker, B., Smith, T., Batory, D., Van Zee, F., Van de Geijn, R.: Code generation to aid parallel code development. Technical report TR-14-08, The University of Texas at Austin, Department of Computer Science (2014)
11. Lam, M.: Software pipelining: an effective scheduling technique for VLIW machines. In: PLDI, pp. 318–328 (2008)
12. Whaley. C.R., Dongarra, J.: Automatically tuned linear algebra software. In: SIAM Conference on Parallel Processing for Scientific Computing (1999)

A Study on the Influence of Caching: Sequences of Dense Linear Algebra Kernels

Elmar Peise$^{(\boxtimes)}$ and Paolo Bientinesi

AICES, RWTH Aachen, Aachen, Germany
{peise,pauldj}@aices.rwth-aachen.de

Abstract. It is universally known that caching is critical to attain high-performance implementations: In many situations, data locality (in space and time) plays a bigger role than optimizing the (number of) arithmetic floating point operations. In this paper, we show evidence that at least for linear algebra algorithms, caching is also a crucial factor for accurate performance modeling and performance prediction.

1 Introduction

In dense linear algebra (DLA), very basic yet highly tuned kernels — such as the Basic Linear Algebra Subprograms (BLAS) — are used as building blocks for high level algorithms — such as those included in the Linear Algebra PACKage (LAPACK). The objective of our research is to develop performance models for those building blocks, aiming at predicting the performance of high level algorithms, without executing them. In a recent article [1], we introduced a methodology for modeling and predicting performance, and showed its effectiveness in ranking different algorithmic variants performing the same target operation. However, to accurately tune algorithmic parameters such as the block-size, predictions of significantly higher precision are required. Intuitively, one would attempt to resolve this issue through more accurate performance models. Unfortunately, beyond a certain level, higher accuracy in the models of the building blocks does not translate into more precise predictions. In this paper we illustrate that such a mismatch is due to the influence of CPU caching on the performance of the compute kernels.

Several other works on the influence of caching on DLA performance exist; some notable examples are given in the following. Whaley empirically tunes the block-size for LAPACK routines and emphasizes its impact on performance [2]. Lam et al. study caching in the context of blocking within DLA kernels [3]. Iakymchuk et al. model the number of cache misses analytically based on a very detailed analysis of kernel implementations [4].

The rest of this paper is structured as follows. We introduce the considered problem and setup in Sect. 2 and establish bounds for the kernel execution times in Sect. 3. Then, we develop a cache prediction model in Sect. 4 and apply it to a broader range of scenarios in Sect. 5.

© Springer International Publishing Switzerland 2015
M. Daydé et al. (Eds.): VECPAR 2014, LNCS 8969, pp. 245–258, 2015.
DOI: 10.1007/978-3-319-17353-5_21

2 The Problem

In order to better understand the influence of caching on the performance of compute kernels, we focus on a specific, yet exemplary algorithm and setup: On one core of a quadcore INTEL HARPERTOWN E5450, we analyze the performance of LAPACK's QR decomposition (dgeqrf) linked to OPENBLAS v. 0.2.8 [5], on a square matrix of size[1] $n = 1{,}568$. With a size of 18 MB, this matrix exceeds this CPU's largest cache (L2), consisting of 6 MBs per 2 cores.

The routine dgeqrf implements a blocked algorithm and traverses the input matrix diagonally from the top left to the bottom right corner in steps of a prescribed block-size b. We fix this block-size — this routine's only optimization parameter — at $b = 32$. Within each step of the blocked traversal, dgeqrf executes the following sequence of kernels on operands of decreasing size: dgeqr2(unblocked QR), dlarft(form triangular factor T for the compact representation of Q), b dcopys (together transpose a matrix panel), dtrmm$_{\text{RLNU}}$ (triangular matrix-matrix product)[2], dgemm$_{\text{TN}}$ (matrix-matrix product), dtrmm$_{\text{RUNN}}$, dgemm$_{\text{NT}}$, and dtrmm$_{\text{RLTU}}$.

To measure the execution time of these kernels within dgeqrf (henceforth called *in-algorithm timings*), we manually instrument this routine, and collect timestamps[3] between kernel invocations. The in-algorithm timings computed from these timestamps are presented in Fig. 1: Along the x-axis, we enumerate the 1,873 kernel invocations; along the y-axis we present timings of each invocation grouped by the type of kernels. The figure shows that the execution time is dominated by the two dgemms ($\times$ and $\circ$); notably, although the size of their operands is the same, the corresponding timings differ significantly. Our ultimate goal is to develop performance models that accurately predict such differences and all other features of the in-algorithm timings.

To focus on the cache related performance features, we here attempt to reconstruct the in-algorithm timings with a very elementary timing setup: repeated execution of the kernels independent from each other. In these executions, we use the same flags and matrix sizes as within dgeqrf and a well separated memory location for each operand. The relative error in execution time of the median of 100 such independent repetitions compared to the in-algorithm timings is shown in Fig. 2. While the relative error for dcopy () is rather large, the total contribution of its 1,536 invocations to the total runtime is below 1 %. Not considering these dcopys[4], the absolute errors of the instrumented timings relative to the in-algorithm timings averaged across kernel invocations (in the following simply referred to as error) is 4.48 %.

---

[1] With $n = 1{,}568 = 2^5 \cdot 7^2$, we choose a matrix size that is not a power of 2 to avoid performance artifacts due to the specific problem size.

[2] The subscripts R through U are the values of the flag arguments side, uplo, trans, and diag; they distinguish the form of the operation performed by the kernel.

[3] Read from the CPU's time stamp counter through the assembly instruction rdtsc.

[4] The system fluctuations cause variations of the dgeqrf timings of 0.057 % on average. With the exception of the tiny dcopys, these fluctuations are not significant.

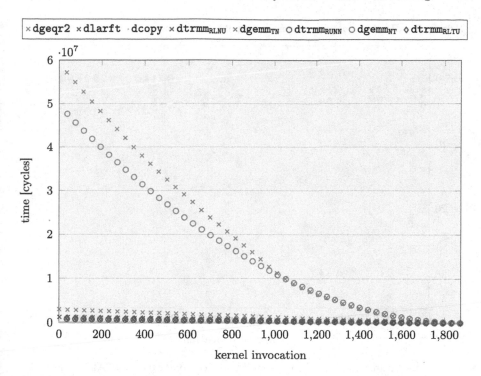

Fig. 1. In-algorithm timings. Along the x-axis, we enumerate the 1,873 kernel invocations within `dgeqrf` (Color figure online).

For most routines and especially for $\text{dtrmm}_{\text{RLTU}}$ ($\diamond$) and `dgeqr2` ($\times$), the repeated execution underestimates the in-algorithm timings for the first 1,000 kernel invocations. More surprisingly however, dgemm_{NT} is even overestimated — it is faster within `dgeqrf`.

3   Cache-Aware Timings

The change in behavior noticeable around the 1,000th kernel invocation (see Fig. 2) is directly linked to the size of the cache. While traversing the matrix, `dgeqrf` only operates on its bottom right quadrant, which becomes smaller at each iteration. Beyond the 1,000th invocation, the quadrant is small enough to fit in the L2 cache. As a result, the subsequent runtime measurements of repeated executions show only minimal differences with respect to the in-algorithm timings. This confirms the cache as the cause of the discrepancies.

To better understand the scope of this influence we now manipulate the cache locality of the kernel's operands in our independent executions. To do so, we assume a simplified cache replacement policy: a fully associative Least Recently Used (LRU) algorithm. We consider the two extreme scenarios in which the

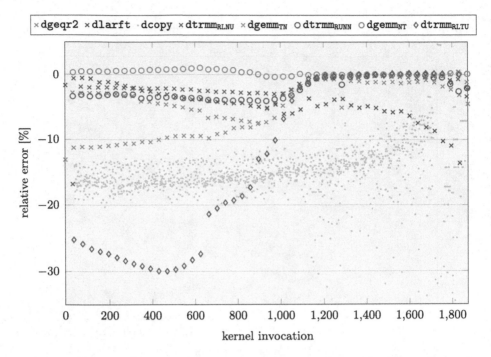

Fig. 2. In-algorithm timings and comparison with repeated execution. Along the x-axis, we enumerate the 1,873 kernel invocations within `dgeqrf`. (Colour figure online)

operands immediately required by the kernels are either entirely within the L2 cache or not at all. These in- and out-of-cache scenarios serve, respectively, as lower and upper bounds on the in-algorithm timings.

For kernels with operands whose size is smaller than 6 MBs, repeated execution suffices to guarantee that the operands are in cache prior to execution. By contrast, when the aggregate size of all kernel operands exceeds 6 MB (as for `dgemm`$_{NT}$ (○)), different kernel implementations (i.e. different libraries) may initially access different regions of the operands. An ideal in-cache setup would place exactly the immediately accessed regions in cache. However, since we do not assume knowledge about kernel implementation, we restrict our in-cache setup to fulfill the reasonable assumption that input-only operands are accessed before input/output and output-oly operands. In order to accordingly prepare the cache, we touch[5] all input operands just before the kernel invocation. This timing setup yields the runtime estimates shown in Fig. 3. Here, the estimates are in all cases equal to or underestimating the in-algorithm timings. The error is 4.51 %.

---

[5] By "touching", we mean a simple read+write access to the data, e.g. $x := x + \varepsilon$.

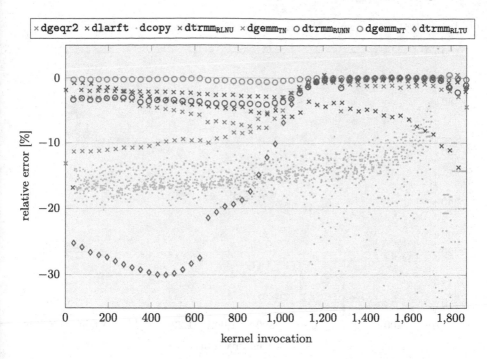

Fig. 3. In-cache compared to in-algorithm timings (Color figure online).

Under the assumption of a fully associative LRU cache, to ensure that the operands are not in the cache, it suffices to touch a section of the main memory larger than the cache size. This approach yields the runtime estimates presented in Fig. 4. Now, almost all estimates are equal to or overestimating the in-algorithm timings. The error is 29.1 %.

Not only do the established in-cache and out-of-cache timings indeed serve as lower and upper bounds on the in-algorithm timings, for most kernel invocations one of these two bounds is actually attained (see Figs. 3 and 4). Based on this observation, the next section introduces a cache model to use these in-core and out-of-core timings to estimate the in-algorithm timings.

4   Modeling the Cache

In order to predict the state of the cache throughout the execution of dgeqrf, we consider which parts of its operands are accessed by its kernel invocations. dgeqrf itself receives three operands: the input matrix $A \in \mathbb{R}^{n \times n}$, an output vector $\tau \in \mathbb{R}^n$, and auxiliary work space $W \in \mathbb{R}^{n \times b}$. Algorithm 1 shows where within these three memory regions the operands of the kernels invoked in one step of dgeqrf's blocked algorithm lie. Since we do not consider details of the kernel

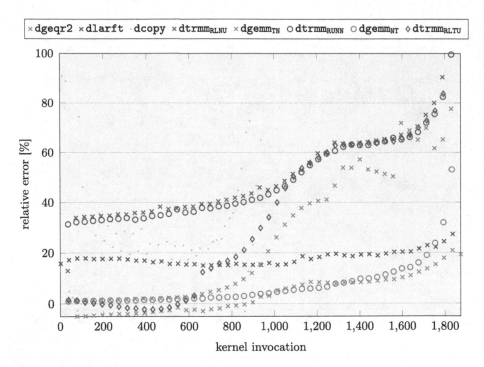

Fig. 4. Out-of-cache compared to in-algorithm timings. The error for dcopy (·) is around 1,000 %. (Color figure online)

implementations, we do not make any assumptions on the patterns in which the kernels access their operands.

For the assumed fully associative LRU cache replacement policy, identifying if a memory region is available in cache reduces to the task of counting how many other data elements were accessed since its last use. To determine this count (henceforth referred to as *access distance*), we scan the sequence of kernel invocations and keep a history of the memory regions they access[6]. We consider the cache line as the smallest accessible memory unit: An access to a single data element means an access to the entire surrounding cache line. For each operand of a kernel invocation, we go backward through the access history until (and including) we find its last access; thereby summing the sizes of the accessed memory regions yields the operand's access distance. (If the access history does not reveal a previous access, the access distance is set to ∞.)

---

[6] The length of the list can be safely restricted to the number of kernel calls per iteration of the blocked algorithm.

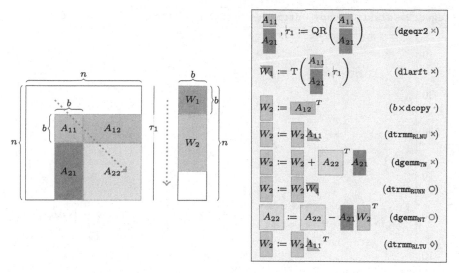

Algorithm 1. QR Decomposition `dgeqrf`. The shapes on the left illustrate `dgeqrf`'s traversal of its data arguments A, τ, and W.

By comparing the obtained access distances to the cache size, we determine whether the corresponding operand is expected in the cache or not. Given these expectations, we separately sum the sizes of the in-cache and out-of-cache kernel operands. These sums are then used to weight the runtime of the corresponding timings to yield initial estimates of the instrumentation timings, shown in Fig. 5. Comparing to Figs. 3 and 4, our mechanism chooses (or weights) the in-cache and out-of-cache timings correctly for most kernels. However, the error is still 4.65 %, because for dtrmm$_{\text{RUNN}}$ (○) out-of-cache is erroneously favored over in-cache.

The reason for this flaw is that (see Algorithm 1) dtrmm$_{\text{RUNN}}$ (○) is preceded by the large dgemm$_{\text{TN}}$ (×): This dgemm's operands, which are together larger than the cache, are accumulated into dtrmm$_{\text{RUNN}}$'s right-hand-side operand's access distance. However, since dtrmm$_{\text{RUNN}}$'s right-hand-side happens to be the output operand of the very matrix-times-vector-shaped dgemm$_{\text{TN}}$, it appears to be left in cache. We use this insight to extend our cache model with a crucial assumption: After a kernel, whose (input-)output operand is significantly smaller than its input-only operands, we expect the (input-)output operand to be in cache. This assumption is implemented by splitting the memory accesses of such a kernel into two parts: The first access contains the large input-only operand(s), while the second only involves the small (input-)output operand. Therefore, the backward traversal of the access history will encounter the latter separately and, in case it is the sought operand, terminates before processing the cache-exceeding accesses. The timing estimates from this modifications (called *splitting estimates*) are shown in Fig. 6. Here, *all* kernels are chosen correctly from the in-cache and out-of-cache timings. As a result, the error is reduced to 2.27 %.

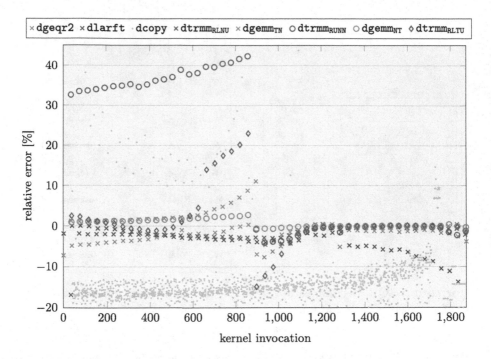

Fig. 5. Basic estimates compared to in-algorithm timings (Color figure online)

The only remaining deficiency of our estimates is in the form of severe spikes around the transition from out-of-cache to in-cache, around the 900th kernel invocation. To avoid such spikes, we apply smoothing of the association of operands with in-cache and out-of-cache. To determine whether an operator was in-cache (+1) or out-of-cache (−1), we previously used a step function. In terms of the relative access distance

$$r = \frac{(\text{cache size}) - (\text{access distance})}{\text{cache size}} \ ,$$

this function was sgn(r). We now replace it with

$$f(r) = \begin{cases} \tanh(\alpha r), \text{ for } r \geq 0 \\ \tanh(\beta r), \text{ for } r < 0 \end{cases} ,$$

where α and β are smoothing coefficients. As shown in Fig. 7, $f(r)$ converges toward sgn(r) for both large and small values of r and exhibits a smooth transition from −1 to +1 though the origin. When applied to our estimates with empirical values of $\alpha = 4$ and $\beta = 2$, we obtain the smoothed estimates shown in Fig. 8. With all estimates very close to the instrumentation timings, the error further decreases to 1.84 %.

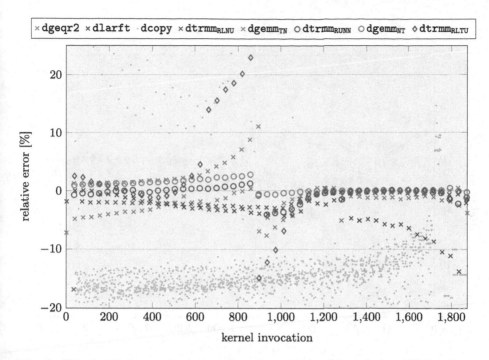

Fig. 6. Kernel-splitting estimates compared to in-algorithm timings (Color figure online)

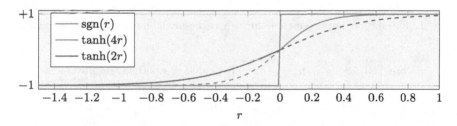

Fig. 7. Smoothing functions.

5 Results

In the previous sections we focused on one specific setup (see Sect. 2). To demonstrate that our observations and models are more broadly applicable, we now vary this setup. Before looking at alternative algorithms, we present the obtained accuracy improvements of our smoothed estimates over the repeated execution timings for a range of scenarios involving `dgeqrf` in Table 1.

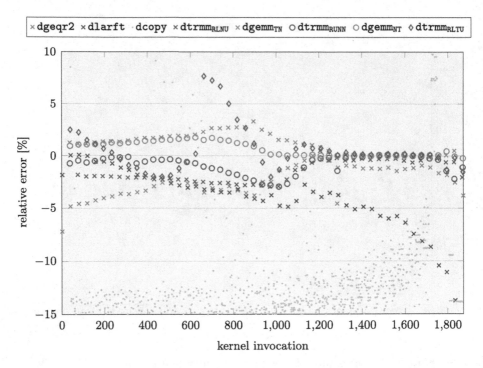

Fig. 8. Smoothing estimates compared to in-algorithm timings (Color figure online)

Table 1. Estimation improvements through cache-modeling for various scenarios of `dgeqrf`

#Cores	BLAS	n	b	Repeated execution	Smoothed estimates	Improvement
1	OPENBLAS	1,568	32	4.48 %	1.84 %	×2.44
1	OPENBLAS	1,568	**64**	3.15 %	1.64 %	×1.92
1	OPENBLAS	1,568	**128**	2.68 %	2.13 %	×1.26
1	OPENBLAS	**2,080**	32	5.11 %	1.84 %	×2.78
1	OPENBLAS	**2,400**	32	5.23 %	1.75 %	×2.99
1	**ATLAS**	1,568	32	3.55 %	1.98 %	×1.79
1	**ATLAS**	**2,400**	32	4.22 %	2.51 %	×1.68
1	**MKL**	1,568	32	8.58 %	4.40 %	×1.95
1	**MKL**	**2,400**	32	9.58 %	6.22 %	×1.54
1	**reference**	1,568	32	2.31 %	1.54 %	×1.50
2	OPENBLAS	1,568	32	9.58 %	4.63 %	×2.07
4	OPENBLAS	1,568	32	22.71 %	19.75 %	×1.15

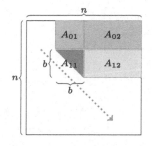

$$A_{11} := A_{11} - A_{01}{}^T A_{01} \quad (\text{dsyrk}_{UT} \,\times)$$
$$A_{11} := \text{Chol}(A_{11}) \quad (\text{dpotf2}_U \,\times)$$
$$A_{12} := \boxed{A_{12}} - A_{01}{}^T \boxed{A_{02}} \quad (\text{dgemm}_{TN} \,\times)$$
$$\boxed{A_{12}} := \boxed{A_{11}}{}^{-T} \boxed{A_{12}} \quad (\text{dtrsm}_{LUTN} \,\times)$$

Algorithm 2. Cholesky decomposition `dpotrf`$_U$ (upper triangular storage)

Although the error of our estimates remains above 1.5 %, it is in many cases an improvement of about a factor of 2 over the repeated execution timings. For both increasing block-size b and matrix size n, with a varying error for repeated executions timings, our estimates reliably yield an error of around 2 %. Changing the BLAS implementation, one can appreciate that with ATLAS the results are very much the same as with OPENBLAS; with MKL instead, the error in both the repeated execution timings and in our estimates increases significantly; however, the latter is still an improvement over the former. Even for the reference BLAS implementation, which is not designed for high performance, our estimates improve the already low error further by a factor of 1.5. When doubling the number of cores to 2, the error of both estimates increases significantly; yet, our smoothing estimates provide a factor of 2 improvement over repeated execution. When we use all 4 cores of our CPU however, the error increases drastically; this is because every two cores share a separate L2 cache, while our model is designed for a single large cache. To account for multiple top-level caches, would require detailed knowledge of the BLAS-implementation and thus substation changes in our models.

5.1   Cholesky Decomposition: `dpotrf`$_U$

Next, we consider LAPACK's Cholesky decomposition for symmetric positive definite (SPD) matrices: `dpotrf`$_U$. This routine works entirely in-place (i.e. no additional work space is required) on one of the symmetric triangular halves of the input matrix; we present results for the upper triangular case. The algorithm uses by `dpotrf`$_U$ in this case is shown in Algorithm 2: As in the QR decomposition `dgeqrf`, the input matrix is traversed along its diagonal in steps of the block-size b. As the algorithm unfolds, both the size and the shape of A_{02} (the largest operand) change noticeably, as this starts as row panel matrix, then grows to a square matrix and finally shrinks to a column panel. In each step, the following kernels are invoked: `dsyrk`$_{UT}$ (symmetric rank k update), `dpotrf`$_U$ (unblocked Cholesky), `dgemm`$_{TN}$, and `dtrsm`$_{LUTN}$ (triangular solve with multiple right hand sides), where `dgemm`$_{TN}$ is the most compute intensive.

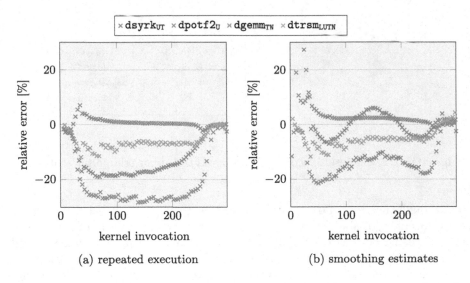

Fig. 9. dpotrf$_U$: Repeated execution and smoothing estimates compared to in-algorithm timings (Color figure online)

In our experiments, we execute dpotrf$_U$ with single-threaded OpenBLAS, a matrix size of $n = 2400$,[7] and block-size of $b = 32$. In Fig. 9, we present the relative performance difference with respect to instrumentation measurements for both repeated execution timings and our final estimates. Our estimates show improvements for the invocation of dsyrk$_{UT}$ (($\times$)) and dpotf2$_U$ (($\times$)) in the middle stage of the traversal involving large matrices. In the very beginning of the traversal, our estimates are generally too pessimistic, because some matrices are (partially) brought into cache by prefetching, which is not accounted for in our estimates yet. On average the relative error is reduced from 11.11 % to 7.87 % by a factor of 1.41.

5.2 Inversion of a Triangular Matrix: dtrtri$_{LN}$

We now take a closer look at the routine dtrtri$_{LN}$, which inverts a lower triangular matrix in place. The blocked algorithm employed by this LAPACK routine is presented in Algorithm 3 and uses the kernels dtrmm$_{LLNN}$, dtrsm$_{RLNN}$, and dtrti2$_{LN}$(unblocked inversion of a triangular matrix) on the operands of decreasing size revealed in each step of the blocked traversal. We invert a matrix of size $n = 2,400$ with block-size $b = 32$ using single-threaded OPENBLAS. Performance measurements from repeated execution and our estimates are compared to instrumentation-based timings in Fig. 10. The improvements in the estimation error are most significant in dtrmm$_{LLNN}$ (($\times$)) (which performs the most

---

[7] For $n = 2400$, the upper triangular portion of the matrix is about twice as large as the cache size.

computations) and $\mathtt{dtrti2_{LN}}$ (($\times$)) and are reduced from an average of 6.70 % to 3.37 % for a total improvement of 1.99$\times$.

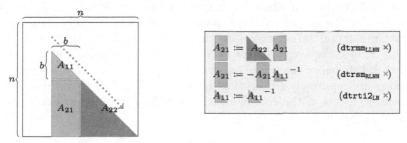

$$A_{21} := A_{22}\ A_{21} \qquad (\mathtt{dtrmm_{LLNN}}\ \times)$$
$$A_{21} := -A_{21} A_{11}^{-1} \qquad (\mathtt{dtrsm_{RLNN}}\ \times)$$
$$A_{11} := A_{11}^{-1} \qquad (\mathtt{dtrti2_{LN}}\ \times)$$

Algorithm 3. Inversion of a lower triangular matrix $\mathtt{dtrtri_{LN}}$

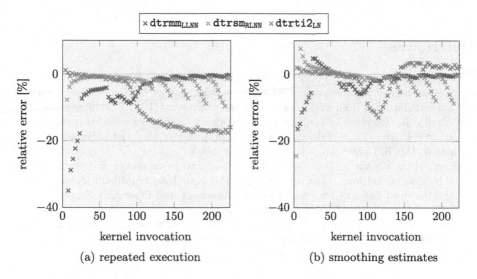

(a) repeated execution   (b) smoothing estimates

Fig. 10. $\mathtt{dtrtri_{LN}}$: Repeated execution and smoothing estimates compared to in-algorithm timings (Color figure online)

In our experiments, we invert a matrix of size $n = 2{,}400$ with block-size $b = 32$ using single-threaded OPENBLAS. Performance measurements form repeated execution and our estimates are compared to instrumentation-based timings in Fig. 10. The improvements in the estimation error are most significant in $\mathtt{dtrmm_{LLNN}}$ (($\times$)) and $\mathtt{dtrti2_{LN}}$ (($\times$)) and are reduced from an average of 6.70 % to 3.37 % by a factor of 1.99.

6   Conclusion

In this paper, we studied the influence of caching on the execution time of sequences of dense linear algebra kernels within blocked algorithms. We established in-cache and out-of-cache timings as lower and upper bounds on the kernel execution times within the algorithm. We then developed a cache tracking model that, based on a sequence of kernel invocations, identifies which memory regions are available in cache and which are not. With the help of this model, we were able to combine the in-cache and out-of-cache timings into highly accurate estimates for the actual kernel execution times. This methodology was shown to noticeably reduce the average error for our estimates. The insights and results presented in this paper constitute an important step towards our ultimate goal of selecting and optimally configuring dense linear algebra algorithms through performance models of the computational kernels, without ever executing the algorithms themselves.

Acknowledgments. Financial support from the Deutsche Forschungsgemeinschaft (DFG) through grant GSC 111 and the Deutsche Telekom Stiftung is gratefully acknowledged.

References

1. Peise, E., Bientinesi, P.: Performance modeling for dense linear algebra. In: Proceedings of the 3rd International Workshop on Performance Modeling, Benchmarking and Simulation of High Performance Computer Systems (PMBS12), November 2012
2. Whaley, R.: Empirically tuning lapack's blocking factor for increased performance. In: 2008 International Multiconference on Computer Science and Information Technology, IMCSIT 2008, pp. 303–310, October 2008
3. Lam, M.D., Rothberg, E.E., Wolf, M.E.: The cache performance and optimizations of blocked algorithms. In: Proceedings of the Fourth International Conference on Architectural Support for Programming Languages and Operating Systems, ASPLOS IV, pp. 63–74. ACM, New York (1991)
4. Iakymchuk, R., Bientinesi, P.: Modeling performance through memory-stalls. ACM SIGMETRICS Perform. Eval. Rev. **40**(2), 86–91 (2012)
5. OpenBLAS: http://www.openblas.net/

Toward Restarting Strategies Tuning for a Krylov Eigenvalue Solver

France Boillod-Cerneux[1](✉), Serge G. Petiton[1,2], Christophe Calvin[3], and Leroy A. Drummond[4]

[1] CNRS LIFL, Cité Scientifique, Bâtiment M3, 59655 Villeneuve d'Ascq, France
boillod.france@gmail.com
[2] Maison de la Simulation, Bâtiment 565 Digitéo, CEA Saclay,
91191 Gif-sur-Yvette, France
[3] CEA/DEN/DANS/DM2S, CEA Saclay, 91191 Gif-sur-Yvette, France
[4] Lawrence Berkeley National Laboratory, 1 Cyclotron Road,
Berkeley, CA 94720-8150, USA

Abstract. Krylov eigensolvers are used in many scientific fields, such as nuclear physics, page ranking, oil and gas exploration, etc. In this paper, we focus on the ERAM Krylov eigensolver whose convergence is strongly correlated to the Krylov subspace size and the restarting vector v_0, a unit norm vector. We focus on computing the restarting vector v_0 to accelerate the ERAM convergence. First, we study different restarting strategies and compare their efficiency. Then, we mix these restarting strategies and show the considerable ERAM convergence improvement. Mixing the restarting strategies optimizes the "numerical efficiency" versus "execution time" ratio as we do not introduce neither additionnal computation nor communications.

1 Introduction

In the large shade of nuclear physics applications and simulations, solving an eigenvalue problem is a common occurrence. One of the best example is the neutron transport equation which is the heart of physical processes in nuclear reactor simulations: Solving this equation requires to compute the dominant eigenpair of a large non-Hermitian matrix. In some other cases, we must compute a subset of eigenpairs as will be illustrated with the Fission matrix [3]: The more eigenpairs we compute, the more we improve the Monte-Carlo process convergence. The Krylov methods are good candidates for such problems [1,2], as they can compute either the dominant or a subset of eigenpairs.

In this paper, we will focus on the Explicitly Restarted Arnoldi Method (ERAM) [4,5]. The ERAM has some lacks to ensure the system convergence. Based on the Arnoldi process, it iteratively builds a Krylov subspace $\mathbb{K}_{m,v_0} = span\{v_0, Av_0, ..., A^{m-1}v_0\}$ and its associated matrices $H \in \mathbb{C}^{(m+1)\times m}$ and $V \in \mathbb{C}^{n\times(m+1)}$ that verify $AV = VH$, where H is the unitary projection of A onto $\mathbb{K}_{m,v_0}$ [4]. The H eigenvalues may be good approximations of A eigenvalues. We will denote by $s \in [1, m]_\mathbb{N}$ the number of desired eigenpairs and by

© Springer International Publishing Switzerland 2015
M. Daydé et al. (Eds.): VECPAR 2014, LNCS 8969, pp. 259–268, 2015.
DOI: 10.1007/978-3-319-17353-5_22

$\theta_j \in \mathbb{C}, j \in [1, s]_{\mathbb{N}}$ the jth real largest modulus eigenvalue. The H eigenvectors are projected onto $\mathbb{K}_{m,v_0}$ basis, leading to approximated eigenvectors corresponding to the θ_j eigenvalue. We will denote the approximated eigenvector associated to θ_j by $u_j \in \mathbb{C}^n$. We then compute the associated residual of approximated eigenpairs $res_j = \frac{\|Au_j - \theta_j u_j\|_2}{|\theta_j|}$: If this residual is small enough, for each desired eigenpairs, the ERAM has reached the convergence. Otherwise, we must restart the process until convergence, by using a new restarting vector v_0. This aims to force the new Krylov subspace $\mathbb{K}_{m,v_0}$ convergence to the desired eigensubspace. Throughout all the following paper, we will retain these notations. It is assumed that only a subset of m computed eigenpairs will be a good approximations of the non-Hermitian matrix $A \in \mathbb{C}^{n \times n}$ eigenpairs [4,5] and many scientific research has been conducted to fix the s and m values in consequence. The ERAM is parallel and efficient on actual supercomputers but its convergence is not ensured, depending on many parameters which are tricky to fix correctly before the ERAM execution.

One of the most influent parameter of the ERAM convergence is the size m of $\mathbb{K}_{m,v_0}$. It is assumed that the larger m is, the better the ERAM convergence is [4,5]. However, increasing m implies two issues that are part of the extreme-scale computing barriers. Firstly, we increase the data size, especially the $V \in \mathbb{C}^{n \times (m+1)}$ (dense matrix) size. Secondly, it implies more operations to execute the Arnoldi process leading to more blocking and global communications. A large m may rapidly lead to a bottleneck due to the Arnoldi process global/blocking communications. Many research has been done for the Krylov methods applied to the linear system resolution to fix the m value such as it optimizes the parallel time execution versus the numerical convergence ratio of the method [6,7]. In the context of up-coming exascale computing, optimizing this ratio is fundamental and implies to re-design algorithms, even mathematical methods themselves [8].

In this paper, we aim to improve the ERAM convergence by changing only the restarting vector v_0, as it influences the ERAM convergence. Regarding the ERAM, there is no general method to compute v_0 and especially no method to ensure the $\mathbb{K}_{m,v_0}$ convergence to the desired eigensubspace. Usually, v_0 is a linear combination of the Ritz vectors u_j that are uniformly weighted. In this paper, we choose to focus on finding a pertinent u_j combination to compute a new v_0 as it requires neither parallel communications nor complex operations. In the first part we propose to use different restarting strategies and study their influence on the ERAM convergence. In the second part, we will mix the restarting strategies and show the convergence improvement of the ERAM using mixed restarting strategies versus the ERAM using a single restarting strategy. All this work is the basis of an upcoming smart-tuning to dynamically mix the restarting strategies and obtain better numerical performances without being affected by additionnal operations neither additionnal communications.

2   The ERAM Restarting Strategies

The ERAM starts its $i + 1^{th}$ restart with a restarting vector $v_0^{(i+1)}$ using the last computed Ritz vectors $u_j^{(i)}, j \in [1, s]_{\mathbb{N}}$. Throughout all the following paper, we

Table 1. The ERAM restarting strategies: $j \in [1, s]_\mathbb{N}$ and $n >> m \geq s$ and $res_j^{(i)} = \frac{||Au_j^{(i)} - \theta_j^{(i)} u_j^{(i)}||_2}{|\theta_j^{(i)}|}$.

Restarting strategy	Abreviation	$\alpha_j^{(i)}$ Value
Default	α_{Def}	1
Residual	α_{Res}	$|1 - |res_j^{(i)}||$
Linear	α_{Li}	$(s - j + 1)$
Linear residual	α_{LiRes}	$(s - j + 1) \times |1 - |res_j^{(i)}||$
Lambda	α_{La}	$|\theta_j^{(i)}|$
Lambda residual	α_{LaRes}	$|\theta_j^{(i)}| \times |1 - |res_j^{(i)}||$

will annotate all previous definitions by their computing restart (i). We introduce $\alpha_j^{(i)}$, the restarting coefficient associated to j^{th} eigenpair for the ith restart. Then, the restarting vector $v_0^{(i+1)}$ is computed as follows:

$$v_0^{(i+1)} = \sum_{j=1}^{s} \alpha_j^{(i)} u_j^{(i)}, s \in [1, m]_\mathbb{N} \tag{1}$$

We summarize the restarting strategies used in this paper in the Table 1. Throughout all this paper, we will refer to the restarting strategy using their restarting strategies abreviations. In the scientific litterature, α_{Def} is the most commonly used restarting strategy [5].

3   The ERAM Restarting Strategies Influence

We summarize the target matrices properties in the Table 2. For each figures presented bellow, we use $s = 4$ and the Arnoldi process uses a CGSR orthogonalization process [9]. For all results presented in this section, the subspace size m is fixed during all the ERAM execution.

For each target matrix, we execute the ERAM with exactly the same parameters (m, s, the orthogonalization process, the threshold, the number of MPI

Table 2. The target matrices.

Matrix name	Size	nnz	Field
Bayer04	20545	85,537	chemical process simulation problem
Bcspwr09	1723	6,511	power network problem
Ex11	16614	1,096,948	computational fluid dynamics problem
Fission	10000	100,000,000	nuclear physics reactor simulation
Rim_dense	22560	508,953,600	computational fluid dynamics problem

Table 3. Target matrices, ERAM executions using a single restarting strategy, results. The fission and Rim_dense matrices ERAMs have been executed on the PRACE curie (TGCC saclay france) supercomputer thin nodes.

Matrix name/ERAM param	Target machine	α_{Def}	α_{Res}	α_{La}	α_{LaRes}	α_{Li}	α_{LiRes}
Bayer04, $m = 15, tol = 10^{-14}$	1 Intel i5-2430M	93	48	118	27	72	106
Bcspwr09, $m = 25, tol = 10^{-14}$	1 Intel i5-2430M	#	454	395	#	125	#
Bcspwr09, $m = 30, tol = 10^{-14}$	1 Intel i5-2430M	30	34	35	38	25	91
Ex11, $m = 20, tol = 10^{-14}$	1 Intel i5-2430M	148	45	55	10	229	#
Fission, $m = 10, tol = 10^{-13}$	400 Intel Sandy Bridge	46	#	80	#	300	#
Rim_dense, $m = 15, tol = 10^{-13}$	480 Intel Sandy Bridge	152	65	135	138	69	309

tasks and the same hardware , cf the Table 3) excepted the restarting strategy. Such executions will emphasize the ERAM convergence with respect to the restarting strategy used. In this context, we use the number of ERAM restarts until convergence as the reference metric to compare our results. The restarting strategies do not impact the execution time per restart but only the number of restarts to reach the convergence (therefore the global execution time of the ERAM solver). As an illustration, the execution time per restart using α_{Def} is the same as α_{La} and all other restarting strategies.

We first present the results for each ERAM using only one restarting strategy during its execution (Table 3). The # symbols means the ERAM did not reach the convergence after 500 restarts.

We will detail the restarting strategy efficiency of the Bayer04 matrix, as the conclusions can be generalized to all other target matrices. The best restarting strategy is α_{LaRes} leading to convergence with only 27 restarts. Then α_{Res} converges at restart 48, α_{Li} at restart 70 and α_{Def} at restart 93. α_{LaRes} saves 66 restarts compared to α_{Def} which is the most commonly used restarting strategy. For this configuration, all ERAMs converge with the parameters fixed before the runtime execution, however the convergence behavior between the best and the worst restarting strategies is really different.

The number of restarts to reach the convergence varies for each configuration and each target matrix: There are no tools to prevent before the runtime execution which restarting strategy will provide the best convergence scheme. We can not emphasize a "best restarting strategy" that ensure the best convergence scheme for all target matrices, as restarting strategies efficiency depends on the matrix itself and the ERAM execution.

4   The ERAM with Mixed Restarting Strategies

In this paper, we focus on the ERAM using α_{Def} during its complete execution. The same work has been done for all other restarting strategies, leading to the same conclusions. We choose to illustrate the mixed restarting strategies based on ERAM using α_{Def} as it is the most commonly used restarting strategy. All the following results considering the ERAM with mixed restarting strategies are executed on the same number of MPI tasks as presented in the Table 3.

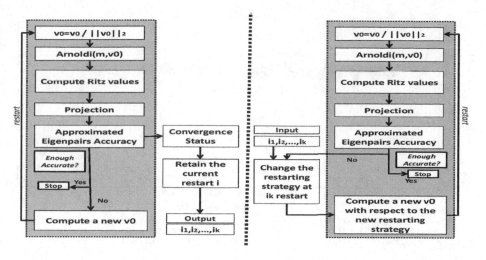

Fig. 1. We execute an ERAM using α_{Def} during the complete execution. We evaluate its convergence status and retain only the $i^{(th)}$ restarts (that we denote by i_k) when a stagnation or a divergence status has been detected. We re-execute the ERAM using α_{Def} and change α_{Def} by another restarting strategy at i_k restart.

We executed an algorithm that detects the ERAM convergence based on $res_j^{(i)}$ analysis (not detailed in this paper). The convergence algorithm detects the stagnation or divergence states for the ERAM using α_{Def}. We retain the restarts listed by the convergence algorithm where such divergence or stagnation status have been detected. We re-executed the ERAM using α_{Def} and switch α_{Def} by another restarting strategy at the corresponding restarts, leading to an ERAM using mixed restarting strategies. We experimented different combinations of mixed restarting strategies and operated the switch restarting strategies at different restarts. The Fig. 1 presents the process used to evaluate the ERAM with mixed restarting strategy performances versus an ERAM using only α_{Def}.

The following figure shows the ERAM with mixed restarting strategies applied to the Bayer04 matrix. The results regarding the other target matrices will be summarized in tables.

We present on Fig. 2 the ERAMs with mixed restarting strategies for the Bayer04 matrix. As a reference, we added the ERAM using α_{Def} during its complete execution (blue line). We experimented several switches: at restart 4, 5, 16, 18, 26 and finally 38 according to our convergence algorithm. A divergence or stagnation was detected at these restarts therefore we changed the restarting strategy in order to avoid such convergence behavior. We re-executed some of the ERAM with mixed restarting strategies and combined at maximum three restarting strategies. Throughout all the following tables showing restarting strategies mixed, we indicated the restart where we changed the restarting strategy in parenthesis. As an illustration, the "Res(4),Def(25)" legend on Fig. 2 means we started with α_{Def} restarting strategy, then switch it by α_{Res} at restart 4 and switch it again by α_{Def} at restart 25.

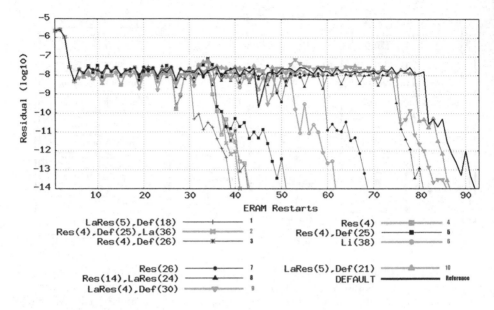

Fig. 2. Bayer04 Matrix, $m = 15, tol = 10^{-14}$. ERAMs (using initially α_{Def}) with mixed restarting strategies efficiency compared to the ERAM using α_{Def} only (black line, Reference). Results have been executed on a single Intel i5-2430M (Color figure online).

The best configuration is obtained by changing α_{Def} by α_{LaRes} at restart 5 and then by α_{Def} at restart 18 (LaRes(5), Def(18) on Fig. 2): we obtain the convergence in 39 restarts instead of 93 for the original convergence. All changes presented on Fig. 2 show a gain from 54 to 10 restarts. Using α_{La} combined with other restarting strategies saves 50 restarts (green line, Res(4), Def(25), La(36) on Fig. 2.), while α_{La} has the worst convergence on Table 3. One restarting strategy alone may not be very efficient while mixing it with the others may considerably improve the ERAM convergence.

In what follows, we will present the results for other matrices in tables. All the ERAM with mixed restarting strategies began their process using α_{Def}. For all matrices listed on Table 2, we will compare the ERAM using only α_{Def} as restarting strategy and the ERAM using mixed restarting strategies. One may note that we executed the same work with the other restarting strategies, id est ERAM using initially other restarting strategies than α_{Def}. This leads to the same conclusion that we present in this paper. As α_{Def} is the classic restarting strategy, we choose to focus on this one in this paper.

On Table 4 we present the mixed restarting strategies convergence for bcspwr09 matrix. As an illustration, $\alpha_{Li}(18,10,2)$ refers to three distinct ERAM switching their restarting strategy to α_{Li} at respectively restart 18, 10 and 2. We indicate their restart convergence: the ERAM using α_{Li} at restart 18 converges at restart 245, the ERAM using α_{Li} at restart 10 converges at restart 132 etc. Throughout all this paper, we will retain this notation. The best scheme is obtained by switching

Table 4. $Bcspwr09$ matrix, $m = 25, tol = 10^{-15}$, ERAMs (using initially α_{Def}) with mixed restarting strategy results. Results have been executed on a single intel i5-2430M.

bcspwr09	$\alpha_{Li}(18,10,2)$	$\alpha_{Res}(10)$	$\alpha_{La}(10)$
Restart CV	245,132,225	285	300

α_{Def} by restarting strategy α_{Li} at restart 10: the ERAM with these mixed restarting strategies converges in only 132 restarts, while the initial ERAM using α_{Def} did not reached the convergence (see Table 3). Choosing the right restart is important: one may note that switching α_{Def} by restarting strategy α_{Li} at restart 18 has poor performance compared to the same switch at restart 10. Another observation is that mixing α_{Def} with α_{Res} or α_{La} has good performance compared to results presented in Table 3: α_{Res} and α_{La} restarting strategies have poor performances while mixing them may accelerate considerably the ERAM convergence and turn them to efficient restarting strategies.

Table 5. $Bcspwr09$ matrix, $m = 30, tol = 10^{-15}$, ERAMs (using initially α_{Def}) with mixed restarting strategy results. Results have been executed on a single intel i5-2430M.

bcspwr09	$\alpha_{LaRes}(8,7,5)$	$\alpha_{LiRes}(8)$	$\alpha_{Li}(8,6,5)$	$\alpha_{La}(8,6,5)$	$\alpha_{Res}(8,6,5)$
Restart CV	30,29,34	31	30,26,29	29,29,34	30,34,28

We ran the same tests with a bigger subspace size. The Table 3 shows that α_{LiRes} has still difficulties to converge. We denote a difference of 10 restarts only between the best restarting strategy α_{Li} and α_{La}. α_{Li} provides a gain of 70 restarts compared to α_{LiRes} restarting startegy. On Table 5, we changed the α_{Def} restarting strategy to evaluate if we can still improve the convergence, even in case of a strong ERAM convergence. We have a maximum gain of 4 restarts with mixed restarting strategies compared to the initial configuration. By mixing restarting strategies, whether we have still a small gain, whether we converge in a comparable number of restarts.

On Table 3, the ERAM using α_{LaRes} converges in only 10 restarts, while α_{Def} converges at restart 148. We have a gain of 138 restarts, which is clearly not negligible in terms of execution time. By mixing the restarting strategies (Table 6), we obtain in the best configuration a convergence in 29 restarts. This offers a gain of 119 restarts compared to the original α_{Def} ERAM convergence.

Table 6. $Ex11$ matrix, $m = 20, tol = 10^{-15}$, ERAMs (using initially α_{Def}) with mixed restarting strategy results. Results have been executed on a single intel i5-2430M.

ex11	$\alpha_{Li}(8,2)$	$\alpha_{LaRes}(13,8,7,4,2)$	$\alpha_{La}(13,5)$	$\alpha_{Res}(8)$	$\alpha_{Li}(1),\alpha_{LaRes}(4)$
Restart CV	139,97	84,40,49,39,29	154,104	122	142

Table 7. *Fission* matrix, $m = 10, tol = 10^{-13}$, ERAMs with mixed restarting strategy results. Results have been executed on 400 Intel sandy bridge from the PRACE curie supercomputer.

Fission	$\alpha_{Res}(8,7,2)$	$\alpha_{Li}(7,2)$
Restart CV	46,46,56	32,32

Switching α_{Def} by α_{La} (see Table 6) provides comparable results as the original restarting strategy α_{Def}. The best restarting strategy we could find is α_{Li} initially and then α_{LaRes} at restart 3. This configuration converges in 6 restarts only, which is even better than α_{LaRes} on Table 3. This means that we can even accelerate the best ERAM convergence by mixing the restarting strategies all-together. For this matrix, mixing α_{LaRes} and α_{Def} restarting strategies provide very satisfiable results in terms of number of restarts to reach the convergence.

On Table 3, the ERAM using α_{Def} as a single restarting strategy converges in 46 restarts, which is the best configuration for each ERAM using a single restarting strategy. In Table 7, mixing α_{Def} with α_{Li} converges in 32 restarts only, which is a considerable gain. On Table 3, one may observe that the ERAM using α_{Li} restarting strategy alone has a poor convergence. Mixing α_{Li} with α_{Def} had turned it to a very efficient restarting strategy. Finally, mixing α_{Def} with α_{Res} at restart 2 considerably improve the ERAM using single restarting strategy α_{Res} presented in Table 3: the original configuration did not reached convergence while this new one could.

Table 8. *Rim_dense* matrix, $m = 15, tol = 10^{-13}$, ERAMs with mixed restarting strategy results. Results have been executed on 480 Intel sandy bridge from the PRACE Curie supercomputer.

Rim_dense	$\alpha_{LaRes}(25,4)$	$\alpha_{Li}(26,25)$	$\alpha_{LiRes}(25)$	$\alpha_{La}(25)$	$\alpha_{Res}(26,4)$
Restart CV	94,59	92,72	115	129	137,82

On Table 3, the ERAM using α_{Def} converges at the 152 restart, which is more than two times worst than α_{Res} (which converges at 65 restart). The best mixed restarting strategy mixed is obtained by switch α_{Def} by α_{LaRes} at restart 4 ($\alpha_{LaRes}(4)$ on Table 8). Firstly we improve considerably the initial ERAM configuration, secondly, this configuration is even better than the best ERAM using a single restarting strategy (Table 3, ERAM using α_{Res}).

5   Conclusion on the ERAM Restarting Strategies

We presented in Sect. 3 diverse ERAM configurations using several restarting strategies. Their efficiency clearly depends on the matrix and the ERAM convergence itself, as these restarting strategies are using residuals or computed eigenvalues. There are no tools to fix the ERAM parameters to ensure the convergence or have the optimal ERAM convergence. There is a need to change the ERAM parameters during the runtime execution. This first necessitates to study and affect to the ERAM convergence a status. Does the ERAM converges, diverges or stagnates? Such status must be detected earliest as possible so we can improve the ERAM convergence and save execution time. Based on the ERAM convergence study, we know at which restart we must change the ERAM restarting strategy. On the presented figures, we changed the restarting strategy when the ERAM using α_{Def} tends to stagnate or diverge. After an analysis of each ERAM restarting strategy, we mixed the restarting strategies and improved the efficiency of the original ERAM, used as a reference to compare our results. Mixing the ERAM restarting strategies has a great improvement on the ERAM convergence as we can find better convergence with mixed restarting strategies compared to the best ERAM using a single restarting strategy.

In this paper, we presented the changes for the ERAM using α_{Def} initially, but the same work has been done for every restarting strategies id est we ran the ERAMs using respectively α_{Res}, α_{La}, α_{LaRes}, α_{Li} and α_{LiRes} to begin their process, showing promising results, as each restarting strategy can be ameliorated. We ran the same tests on other matrices, leading to the same conclusion, meaning that we can considerably improve the ERAM convergence with respect to its initial parameters. The next step is to automate the restarting strategies mix. Our results show promising accelerations for the ERAM convergence with mixed restarting strategies, but we will automate this process to explore more restarting strategies mix based on the exisiting results. We are now focused on the restarting strategy choice: which restarting strategy should we choose when a stagnation or divergence is detected and how to optmize the restarting strategy mix? The presented work constitues the basis of the ongoing algorithm to tune the restarting strategies and ensure the ERAM convergence as fast as possible in terms of number of restarts and parallel execution time. The most important point for this amelioration is the abscence of additionnal operations nor communications. Computing the restarting vector $v_0^{(i+1)}$ is a parallel operation that requires vectors additions and especially scal operations to weight them with the restarting coefficients. The restarting coefficients are using already computed data, we only focus on reusing computed data to improve the ERAM convergence at no cost in terms of memory, number of operations and most of all, communications. This offers a very good ratio of the ERAM numerical convergence versus parallel time execution, as the first is ameliorated while the second is unchanged.

There is however a necessity to mix this restarting strategy tuning with the existing subspace-size tuning for the Krylov solver applied to linear system resolution [6,7], especially in the cases when ERAM do not reach convergence whatever the restarting strategy is. We aim to improve the ERAM convergence using as a priority the restarting strategy tuning and if it is inefficient, we increase the subspace size. With such restarting strategies and subspace size tuning, we aim to ensure the ERAM convergence as fast as possible with a low parallel communication scheme whatever the ERAM initial parameters are.

References

1. Davidson, G.G., et al.: Massively parallel, three-dimensional transport solutions for the k-eigenvalue problem. NSE **75**, 283–291 (2013)
2. Evans, T.M.: Full core reactor analysis: running Denovo on Jaguar. In: PHYSOR 2012 (2012)
3. Dufex, J., Gudowski, W.: The Semi-source Fission Matrix Method for Accelerating the Monte-carlo Eigenvalue Calculations. AlbaNova University Center, Stockholm (2007)
4. Châtelin, F.: Valeurs Propres De Matrices. Masson, Paris (1988)
5. Saad, Y.: Numerical solution of large nonsymmetric eigenvalue problems. Comp. Phys. comm. **53**, 71–90 (1989)
6. Baker, A., Jessup, E.R., Kolev, T.V.: A simple strategy for varying the restart parameter in GMRES(m). J. Comp. Appl. Math. **230**(2), 751–761 (2009)
7. Katagiri, T., Aquilanti, P.-Y., Petiton, S.: A smart tuning strategy for restart frequency of GMRES(m) with hierarchical cache sizes. In: Daydé, M., Marques, O., Nakajima, K. (eds.) VECPAR. LNCS, vol. 7851, pp. 314–328. Springer, Heidelberg (2013)
8. Dongara, J., et al.: Applied Mathematics Research for Exascale Computing. U.S. Department of Energy, March 2014
9. Dubois, J., Calvin, C., Petiton, S.: Performance and numerical accuracy evaluation of heterogeneous multicore systems for Krylov orthogonal basis computation. In: Palma, J.M.L.M., Daydé, M., Marques, O., Lopes, J.C. (eds.) VECPAR 2010. LNCS, vol. 6449, pp. 45–57. Springer, Heidelberg (2011)

Performance Analysis of the Householder-Type Parallel Tall-Skinny QR Factorizations Toward Automatic Algorithm Selection

Takeshi Fukaya[1,3]([⊠]), Toshiyuki Imamura[1,3], and Yusaku Yamamoto[2,3]

[1] RIKEN Advanced Institute for Computational Science, Kobe, Japan
`takeshi.fukaya@riken.jp`
[2] The University of Electro-Communications, Tokyo, Japan
[3] JST CREST, Tokyo, Japan

Abstract. We consider computing tall-skinny QR factorizations on a large-scale parallel machine. We present a realistic performance model and analyze the difference of the parallel execution time between Householder QR and TSQR. Our analysis indicates the possibility that TSQR becomes slower than Householder QR as the number of columns of the target matrix increases. We aim for estimating the difference and selecting the faster algorithm by using models, which falls into auto-tuning. Numerical experiments on the K computer support our analysis and show our success in determining the faster algorithm.

1 Introduction

The QR factorization of a tall and skinny matrix, which has many more rows than columns, appears in many numerical computations. In block subspace projection methods for sparse linear systems and eigenvalue problems, an orthogonal basis of the subspace is often calculated via the tall-skinny QR factorization [1–3]. As these methods are used on large-scale parallel machines, an efficient algorithm for computing tall-skinny QR factorizations on such machines is strongly required.

Because of the increasing costs of transferring data among distributed processors over a network, referred to as *communication*, on today's parallel machines, so-called *communication-avoiding* algorithms have been extensively studied [4]. The lower bound for communication costs of a QR factorization has been derived by Demmel et al. [5,6]. They have also presented an algorithm named TSQR for tall-skinny QR factorizations and shown that their algorithm attains the lower bound but the conventional Householder QR algorithm [7], widely used due to its excellent numerical stability, does not. In addition, several cases where TSQR outperforms Householder QR have been actually reported [8,9].

Their [5,6] main interest lies in showing the optimality of TSQR from the theoretical point of view, and not in estimating the performance of TSQR or Householder QR accurately. They therefore use a rather simple and abstract model that does not necessarily reflect the characteristics of actual machines; e.g. they assume that the effective performance of a floating-point operation

M. Daydé et al. (Eds.): VECPAR 2014, LNCS 8969, pp. 269–283, 2015.
DOI: 10.1007/978-3-319-17353-5_23

is independent of the computing kernel. However, such a simple model would not be sufficient to estimate the performance difference between TSQR and Householder QR for a give situation: matrix size and platform. In fact, in our evaluation of TSQR on the K computer, we have observed several cases where TSQR is slower than Householder QR, which can hardly be expected from their performance model.

In this paper, we present a more realistic performance model and analyze the difference of the parallel execution time between Householder QR and TSQR based on the model. Our model reflects the dependency of the effective floating-point performance on computational kernels. Moreover, by using models, we aim for estimating the difference and selecting the faster algorithm for given situations, which realizes the automatic algorithm selection. The results of the numerical experiments on the K computer were in general agreement with our analysis.

The rest of the paper is organized as follows: in Sect. 2, the problem setting of computing a tall-skinny QR factorization and two algorithms, namely Householder QR and TSQR, are briefly introduced. We then discuss the performance modeling and analyze the parallel execution time of both algorithms in Sect. 3. In this section, we also mention a way of estimating the difference of the execution time and selecting the faster algorithm in advance. We show the results of the numerical experiments on the K computer in Sect. 4. The conclusion remarks are give in Sect. 5.

2   Tall-Skinny QR Factorization

Let A be an $m \times n$ real matrix and $m \gg n$, meaning A is tall and skinny. The factorization $A = QR$, where Q is an $m \times n$ matrix with orthonormal columns and R is an $n \times n$ upper triangular matrix, is called the thin QR factorization of A [7]. In this paper, P distributed processors connected by a network are assumed to store A in a one-dimensional block row layout: $A = [A_1^\top \ A_2^\top \ \cdots \ A_P^\top]^\top$. Then, the QR factorization of A is considered to be calculated by a parallel algorithm. Note that m_i denotes the number of rows of A_i and that $m_i \geq n$ is assumed for the reason that A is tall and skinny.

In this paper, we discuss the two parallel algorithms of computing tall-skinny QR factorizations: the Householder QR algorithm [7], which has been used widely in practical computations, and the TSQR algorithm, which has been presented as a communication-avoiding algorithm by Demmel et al. [5,6].

2.1   The Householder QR Algorithm

In the Householder QR algorithm, the target matrix A is transformed into the upper triangular matrix R by a sequence of the Householder transformations $H_i := I - t_i \boldsymbol{y}_i \boldsymbol{y}_i^\top$ $(i = 1, \ldots, n)$, which implicitly represents Q. This algorithm consists of the iteration of the two steps: generation of the Householder transformation from the target column vector, and application of the Householder

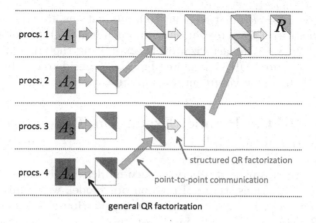

Fig. 1. Sketch of the parallel computation of the binary tree based TSQR algorithm when $P = 4$.

transformation: $H_i A = A - t_i \boldsymbol{y}_i (\boldsymbol{y}_i^\top A)$. For details of the Householder QR algorithm, see [7].

In parallel computing, where A is distributed as previously mentioned, at least two global collective communications, namely MPI_allreduce, are required per iteration: in calculating the 2-norm of a column vector and in a matrix-vector multiplication $\boldsymbol{y}_i^\top A$. Thus, $2n$ global communications are totally necessary in the whole algorithm.

2.2 The TSQR Algorithm

The TSQR algorithm [5,6] is an algorithm based on the idea that $A = QR$ can be given by

$$A = \begin{bmatrix} A_1 \\ A_2 \end{bmatrix} = \begin{bmatrix} Q_1 R_1 \\ Q_2 R_2 \end{bmatrix} = \begin{bmatrix} Q1 & O \\ O & Q_2 \end{bmatrix} \begin{bmatrix} R_1 \\ R_2 \end{bmatrix} = \left(\begin{bmatrix} Q1 & O \\ O & Q_2 \end{bmatrix} Q_{12} \right) R,$$

where Q_1 and Q_2 are $\frac{m}{2} \times n$ orthogonal, R_1 and R_2 are $n \times n$ upper triangular and Q_{12} is $2n \times n$ orthogonal. It is clear that this idea is recursively applicable to the QR factorizations of A_1 and A_2 while $m \geq 2n$. In TSQR, the QR factorizations of the partitioned matrices are first calculated, and the obtained R_i ($i = 1, \ldots, P$) are then reduced into R by the QR factorizations of the matrices built by coupling two upper triangular matrices, which we call *structured* QR factorization. Although there are several reduction trees to calculate R [5,6], we assume that P is a power of two (for the simplicity) and that a binary reduction tree is used.

In parallel computing (see Fig. 1), each processor first compute the QR factorization $A_i = Q_i R_i$, which we call *generall* QR factorization. After that, one repeats sending/receiving one's R to/from one's neighbor by a point-to-point communication and calculating a structured QR factorization. Through this computation, only $\log_2 P$ point-to-point communications are required. One can calculate a structured QR factorization with much less floating-point operations than a general $2n \times n$ QR factorization by exploiting the triangular structure appearing in structured QR. Therefore, TSQR is considered to be superior to Householder QR from the viewpoint of communication-avoiding.

3   Analysis of the Parallel Execution Time Based on a Realistic Model

In this section, we first present a performance model that is more realistic than the model used in previous studies. Based on our model, we mainly analyze the difference of the parallel execution time between the Householder QR algorithm and the TSQR algorithm. We also aim for estimating the difference and selecting the faster algorithm for given situation

3.1   Performance Modeling

In the previous studies by Demmel et al. [5,6], an execution time on a parallel machine is modeled as

$$T = \gamma \cdot (\#flops) + \alpha_{1to1} \cdot (\#msgs) + \beta_{1to1} \cdot (\#words), \tag{1}$$

where

- γ: the time per floating-point operation (i.e., the inverse of the floating-point performance)
- α_{1to1}: the setup cost of a point-to-point communication,
- β_{1to1}: the inverse of the network bandwidth of a point-to-point communication.

This model seems to be appropriate enough to theoretically discuss the communication optimality of algorithms, which is their main interest.

When comparing the execution time of different algorithms, which is our main interest in this study, the model in Eq. (1) is less realistic; it ignores the practical aspect that γ usually depends on the computing kernel. Actual (effective) γ is some larger than the theoretical one (i.e., the inverse of the theoretical peak floating-point performance) because of the cost for memory access. Since the memory access pattern is not necessarily same in each computing kernel, taking this aspect into account is often crucial for modeling the execution time.

TSQR has at least two kinds of computing kernel: general QR factorization and structured QR factorization. The memory access pattern (including its optimization) in these kernels seem to be significantly different. Both the size and structure of the target matrix in each computation are obviously not same. In addition, high-performance implementation for small and structured matrices is much less studied than that for general matrices. These facts indicate that γ for these two kernels should be distinguished.

The computing kernel in Householder QR is easily verified to be almost equivalent to the kernel of general QR factorization; the main difference is whether it involves MPI functions or not. Thus, if the cost for MPI fuctions is individually modeled, as in Eq. (1), it is reasonable to suppose that γ for Householder QR equals to γ for general QR factorization in TSQR.

As a result of the above discussion, we introduce two kinds of γ:

- γ_{ge}: the effective time per floating-point operation in general QR factorization,
- γ_{st}: the effective time per floating-point operation in structured QR factorization,

and modify Eq. (1) as

$$T = \gamma_{ge} \cdot (\#flops\_ge) + \gamma_{st} \cdot (\#flops\_st) + \alpha_{1tol} \cdot (\#msgs) + \beta_{1tol} \cdot (\#words). \quad (2)$$

By referring to [5,6], we list the value of $\#flops\_ge$, $\#flops\_st$, $\#msgs$ and $\#words$ in each algorithm in Table 1, where measurements are made along the critical path of the parallel execution.

Table 1. Performance models of the Householder QR algorithm and the TSQR algorithm: supposing the case of computing the QR factorization (Q factor is implicit) of an $m \times n$ matrix by using P processes. Measurements are made along the critical path of the parallel execution. Most lower-order terms are omitted.

Algorithm	$\#flops\_ge$	$\#flops\_st$	$\#msgs$	$\#words$
Householder	$2\dfrac{m}{P}n^2$	0	$2n \cdot \log_2 P$	$\dfrac{n^2}{2} \cdot \log_2 P$
TSQR	$2\dfrac{m}{P}n^2 - \dfrac{2}{3}n^3$	$\dfrac{2}{3}n^3 \cdot \log_2 P$	$\log_2 P$	$\dfrac{n^2}{2} \cdot \log_2 P$

3.2 Analysis

Using the modified model in Eq. (2), we analyze the difference of the parallel execution time between Householder QR and TSQR. From Table 1, the difference T_{diff} is given as

$$T_{diff} := T_{HouseholderQR} - T_{TSQR}$$

$$= \frac{2}{3}n^3(\gamma_{ge} - \gamma_{st} \cdot \log_2 P) + \alpha_{1tol} \cdot (2n - 1) \log_2 P. \quad (3)$$

We then consider the case that

$$\gamma_{st} \cdot \log_2 P - \gamma_{ge} \simeq \gamma_{st} \cdot \log_2 P \quad (4)$$

and that

$$2n - 1 \simeq 2n. \quad (5)$$

The approximation in Eq. (4), which nearly means $\gamma_{st} \gg \gamma_{ge}$, is acceptable; the high-performance implementation of structured QR seems to be more difficult than that of general QR for the following reasons:

- hiding the latency of memory access is difficult because the total arithmetic cost is not large,
- exploiting parallelism (i.e., using SIMD architecture and OpenMP) is difficult because of the matrix size and structure,
- its implementation has not been studied enough because it is regarded as an operation peculiar to TSQR.

Note that if one does not exploit the triangular structure of the matrix appearing in structured QR, the number of floating-point operations becomes about quintuple.

In the case that Eqs. (4) and (5) are satisfied, Eq. (3) can be approximated as

$$T_{\text{diff}} \simeq (\alpha_{1\text{to}1} \cdot 2n - \gamma_{\text{st}} \cdot \frac{2}{3} n^3) \cdot \log_2 P. \tag{6}$$

Equation (6) indicates that

- the sign of T_{diff} (i.e., TSQR is faster or not) does not depend on m or P,
- T_{diff} is proportional to $\log_2 P$,
- T_{diff} becomes negative (i.e., TSQR is slower) as n grows.

To understand the essence of Eq. (6), it is worth interpreting that TSQR is an algorithm based on the global reduction on a triangular matrix (see Fig. 1) [10]; the reduction operator is a structured QR factorization. As shown in Table 1, the amount of data transferred in TSQR is equivalent to that in Householder QR. TSQR decreases the number of global reduction ($2n \rightarrow 1$) but makes the reduction operation more complicated (addition $\rightarrow$ structured QR). This essential difference between Householder QR and TSQR is clearly described in Eq. (6).

3.3   Estimation Toward Automatic Algorithm Selection

Equation (6) indicates an important fact that T_{diff} becomes negative, meaning TSQR becomes slower than Householder QR, as n grows. Precisely,

$$T_{\text{diff}} < 0 \quad \Leftrightarrow \quad n > \sqrt{\frac{3\alpha_{1\text{to}1}}{\gamma_{\text{st}}}}. \tag{7}$$

This is understandable because the arithmetic cost of structured QR factorization in TSQR grows in $O(n^3)$ whereas the setup cost of the communication in Househlder QR increases in $O(n)$.

We now consider estimating T_{diff} and selecting the faster algorithm between Householder QR and TSQR in advance when a target matrix is given. In the following discussion, the platform and implementation of each algorithm are supposed to be fixed.

Estimation of T_{diff} based on Eq. (6) (or estimating only the threshold from Eq. (7)) is facile because the parameters in the equation are easily measured; $\alpha_{1\text{to}1}$ can be measured by executing a simple benchmark program (e.g. ping-pong program), which needs only two processes. γ_{st} can be obtained through running the kernel of structured QR factorization on one node. The cost for both measurements is obviously small.

Although the model written in Eq. (2) and resulting Eq. (6) are more realistic than the model in Eq. (1), they still does not reflect the characteristics of actual computations sufficiently, which occasionally makes the estimation inaccurate. First, the effective performance of floating-point operations generally depends on the size of target matrix. Thus, it is practical to replace the constant parameter γ_{st} with $\gamma_{st}(n)$ that is a function of n. Alternatively using the execution time of a structured QR factorization $T_{st}(n)(= \gamma_{st}(n) \cdot \frac{2}{3}n^3)$ itself is also a feasible idea.

Second, the MPI_(all)reduce function is sometimes implemented not with a binary reduction tree based point-to-point communications but in a specialized way. In such case, modeling the setup cost of the function as $\alpha_{1to1} \cdot \log_2 P$ is not always appropriate. A straightforward solution to this issue is measuring the setup cost of the function among P processes, which we denote by $\alpha_{all}(P)$. The cost for measuring $\alpha_{all}(P)$ is of course not small, especially when P is large. However, this idea is less of impractical because the results of MPI benchmarks are highly reusable.

Based on the above discussion, we present another model

$$T_{\mathrm{diff}} \simeq \alpha_{all}(P) \cdot 2n - T_{st}(n) \cdot \log_2 P. \tag{8}$$

Although the cost using this model is much larger than that for the model in Eq. (6), the former is expected to give more reliable estimations than the latter.

By using either model, T_{diff} can be estimated, and the faster algorithm will be selected in advance. These procedures are easily automatized, in other words, an automatic algorithm selection is realizable. To construct this mechanism, the cost for measuring the parameters (and function values) in each model is initially required. The mechanism works before the target computation (i.e., tall-skinny QR factorization) and does nothing while the computation, which means this approach falls into so-called off-line auto-tuning.

4   Numerical Experiments

We verify our analysis described in the previous section through numerical experiments on the K computer. We examine the behavior of the difference of the parallel execution time between TSQR and Householder QR. We also estimate the difference based on our models and attempt to select the faster algorithm for a given condition: m, n and P.

4.1   Experimental Conditions

Our experiments were conducted on the K computer[1]. It consists of 88129 computational nodes; each node has one SPARC64 VIIIfx processor (2.0 GHz,

---

[1] Operated at the RIKEN Advanced Institute for Computational Science.

8 cores) and 16 GB memory. The nodes are connected by the 6D mesh/torus network (5 GB/sec, per link, bidirectional), namely Tofu interconnect.

We implemented a parallel Householder QR program (denoted *House* in the followings) with BLAS routines including DGEMM. We also coded two kinds of parallel TSQR programs (*TSQR 1* and *TSQR 2*); the difference between TSQR 1 and TSQR 2 resides in the implementation of the kernel of structured QR factorization. The code of TSQR 1 is based on the BLAS routines, while that of TSQR 2 calls no BLAS routines and is written with simple loop blocking techniques. The kernel of general QR factorization in both TSQR programs employ BLAS routines. All programs are parallelized with the MPI functions.

As test matrices, we used real matrices whose elements are random numbers because execution times are independent with the matrix elements. We measured the execution time for computing the QR factorization of an $m \times n$ matrix by using P nodes, in which the Q factor is implicitly obtained (i.e., the explicit Q is not formed). We assigned one MPI process per node and used thread parallelized BLAS routines (eight threads per process). Note that we used the MPI and BLAS libraries provided from Fujitsu on the K computer.

4.2   Verification of the Analysis on T_{diff}

We first present the results on the basic computation kernels and communication routines. Figure 2(a) shows the effective time per floating-point operation in each computation kernel in TSQR. The graph clearly indicates the necessity of distinguishing γ in performance modeling; the effective floating-point performance in both implementations of structured QR are far from that of general QR. The difficulty of high-performance implementation of structured QR can be seen from the graph, which justifies our assumption in Eq. (4). Figure 2(b) shows the setup cost of the MPI_allreduce function on the K computer. Although the actual setup cost is a little different from its approximation with point-to-point

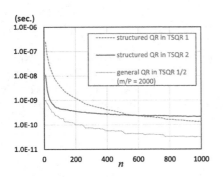
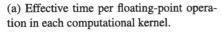
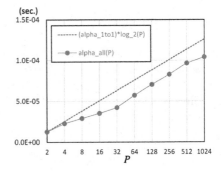

(a) Effective time per floating-point operation in each computational kernel.

(b) Setup cost of the MPI_allreduce routine and point-to-point communication.

Fig. 2. Basic performance data in the numerical experiments on the K computer.

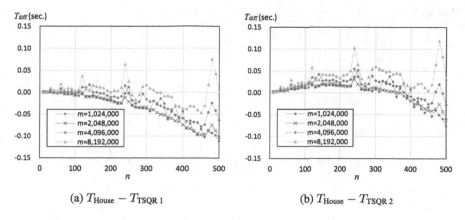

(a) $T_{\text{House}} - T_{\text{TSQR}\,1}$ (b) $T_{\text{House}} - T_{\text{TSQR}\,2}$

Fig. 3. Dependency of T_{diff} on m: $P = 256$ and $n = 10, 20, \ldots, 490, 500$.

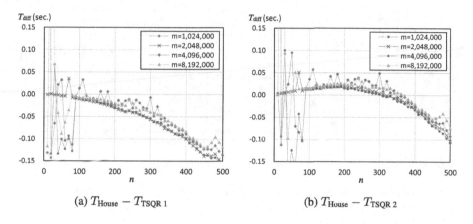

(a) $T_{\text{House}} - T_{\text{TSQR}\,1}$ (b) $T_{\text{House}} - T_{\text{TSQR}\,2}$

Fig. 4. Dependency of T_{diff} on m: $P = 1024$ and $n = 10, 20, \ldots, 490, 500$.

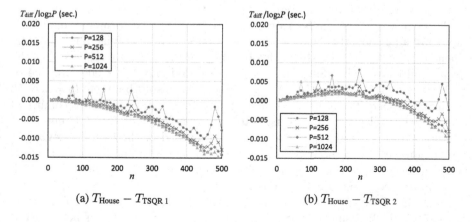

(a) $T_{\text{House}} - T_{\text{TSQR}\,1}$ (b) $T_{\text{House}} - T_{\text{TSQR}\,2}$

Fig. 5. Dependency of T_{diff} on P: $m = 2,048,000$ and $n = 10, 20, \ldots, 490, 500$.

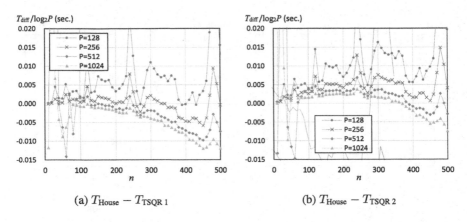

(a) $T_{\text{House}} - T_{\text{TSQR 1}}$

(b) $T_{\text{House}} - T_{\text{TSQR 2}}$

Fig. 6. Dependency of T_{diff} on P: $m = 8,192,000$ and $n = 10, 20, \ldots, 490, 500$.

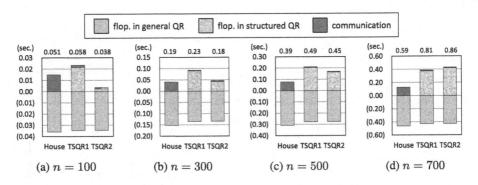

(a) $n = 100$

(b) $n = 300$

(c) $n = 500$

(d) $n = 700$

Fig. 7. Breakdown of parallel execution time ($P = 256, m = 2,048,000$): the figure on each bar is the total execution time.

communication, it is almost proportional to $\log_2 P$. Both figures totally support that our assumptions are reasonable.

We next verify the analysis on T_{diff} based on our performance modeling. Figures 3 and 4 show the dependency of T_{diff} on m. These graphs illustrate little dependency of T_{diff} on m, which corroborates our analysis. We suspect that the exceptional cases in Fig. 3, namely $m = 4,096,000$ and $8,192,000$, are due to the largeness of $\frac{m}{P}$. Figures 5 and 6 give the relationship between T_{diff} and P. We analyzed that T_{diff} is proportional to $\log_2 P$, and it is basically confirmed by the graphs. However, Fig. 6 indicates another factor of the difference; the behavior shown in the graphs seems to be composed of the factor analyzed in the previous section and another factor that becomes dominant as p decreasing. Every figure illustrates that T_{diff} actually becomes negative as n increases, which is of importance in the practical use of TSQR.

For more detailed evaluation, we present the breakdown of typical cases in Fig. 7 ($P = 256$) and Fig. 8 ($P = 1024$). In all cases, it can be observed that T_{diff} mainly comes from the difference between the communication time

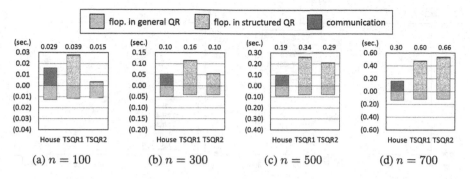

Fig. 8. Breakdown of parallel execution time ($P = 1024, m = 2,048,000$): the figure on each bar is the total execution time.

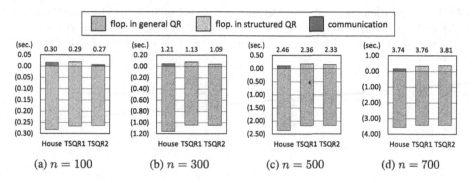

Fig. 9. Breakdown of parallel execution time ($P = 128, m = 8,192,000$): the figure on each bar is the total execution time.

in Householder QR and the arithmetic cost of structured QR factorization in TSQR. This observation strongly supports our analysis on T_{diff}. However, we need to mention the exceptional cases; their breakdown are shown in Fig. 9. In these cases, the above factor is much small because P is not large. On the other hand, the gap in the cost of general QR factorization is not negligible; the arithmetic cost for this part grows in $O(\frac{m}{P})$. This result suggests that more careful modeling and analysis are required for cases that $\frac{m}{P}$ in extremely large.

4.3   Evaluation of the Estimation of T_{diff} and Algorithm Selection

We examine the estimation of T_{diff} by using models. We compared three ways of estimating T_{diff}:

- estimation 1: using Eq. (6),
- estimation 2: using Eq. (8),
- estimation 3: using Eq. (6) but replacing γ_{st} with γ_{ge}.

In estimation 1, we used $\gamma_{\text{st}} = 1.43 \times 10^{-10}$ for TSQR 1 and $\gamma_{\text{st}} = 2.22 \times 10^{-10}$ for TSQR 2. In estimation 3, we used $\gamma_{\text{ge}} = 3.30 \times 10^{-11}$ for TSQR 1 and 2.

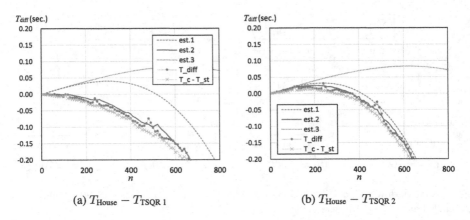

Fig. 10. Estimations of T_{diff} ($P = 256$ and $m = 2,048,000$): est. 1 (by Eq. (6)), est. 2 (by Eq. (8)) and est. 3 (by Eq. (6) but replacing γ_{st} with γ_{ge}). T_{c} means the communication time in Householder QR and T_{st} means the arithmetic time of structured QR in TSQR.

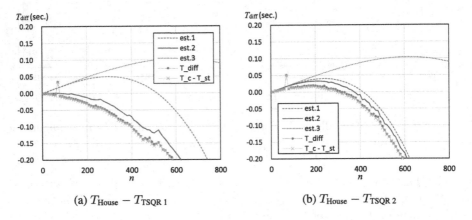

Fig. 11. Estimations of T_{diff} ($P = 1024$ and $m = 2,048,000$): est. 1 (by Eq. (6)), est. 2 (by Eq. (8)) and est. 3 (by Eq. (6) but replacing γ_{st} with γ_{ge}). T_{c} means the communication time in Householder QR and T_{st} means the arithmetic time of structured QR in TSQR.

We present typical results of the estimation in Fig. 10 ($P = 256$) and Fig. 11 ($P = 1024$). In these graphs, we compare each estimation with actual T_{diff} and the difference between the communication time in Householder QR (T_{c}) and the arithmetic cost of structured QR factorization in TSQR (T_{st}). From these graphs, we have the following observations:

- Estimation 3 is much far from the actual T_{diff}, which is reasonable considering the gap γ_{st} and γ_{ge}. This estimation often mislead that TSQR is almost always faster than Householder QR until n is unpractically large.
- Estimation 2 gives very accurate results as we expected in the previous section.

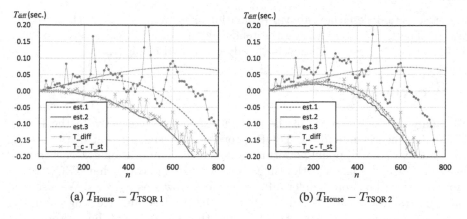

(a) $T_{\text{House}} - T_{\text{TSQR 1}}$

(b) $T_{\text{House}} - T_{\text{TSQR 2}}$

Fig. 12. Estimations of T_{diff} ($P = 128$ and $m = 8,192,000$): est. 1 (by Eq. (6)), est. 2 (by Eq. (8)) and est. 3 (by Eq. (6) but replacing γ_{st} with γ_{ge}). T_{c} means the communication time in Householder QR and T_{st} means the arithmetic time of structured QR in TSQR.

Table 2. The number of correct selection of the faster algorithm (House v.s. TSQR 1 or 2): in each combination of P and m, 100 cases ($n = 10, 20, \ldots, 990, 1000$) were tested. The left figure in each cell is the results of est. 1 (by Eq. (6)) and the right one is that of est. 2 (by Eq (8)).

(a) House v.s. TSQR 1   (b) House v.s. TSQR 2

$m\backslash P$	128	256	512	1024	$m\backslash P$	128	256	512	1024
1,024,000	54 94	51 92	56 93	51 94	1,024,000	93 98	97 96	86 90	83 87
2,048,000	59 93	51 94	50 91	51 94	2,048,000	93 94	95 100	92 96	90 94
4,096,000	82 72	62 93	55 92	55 90	4,096,000	92 87	95 96	89 93	89 93
8,192,000	76 40	80 75	58 91	51 94	8,192,000	82 77	92 86	91 93	94 98

– Estimation 1 is more accurate for TSQR 2 than TSQR 1. Figure 2(a) clearly illustrates the reason for this difference; regarding γ_{st} as constant is deemed to be acceptable for TSQR 2 but not for TSQR 1.

We also show the results for the case that $\frac{m}{P}$ is much large in Fig. 12. In this case, estimation 1 (for TSQR 1) and estimation 2 (for TSQR 1 and 2) give the good estimation of $T_{\text{c}} - T_{\text{st}}$, which are as accurate as in Figs. 10 and 11. However, T_{diff} is far from $T_{\text{c}} - T_{\text{st}}$. Thus, even estimation 2 is much less accurate from the viewpoint of estimating T_{diff}.

We finally present the results of the algorithm selection based on estimation 1 and 2: for TSQR 1 (Table 2(a)) and for TSQR 2 (Table 2(b)). For each combination of P and m, we tested 100 cases ($n = 10, 20, \ldots, 990, 1000$) and list the number of the collect selections. The results in both tables agree with the observation described above; estimation 2 is more accurate than estimation 1. The tables also indicate the tendency that the accuracy degenerates as $\frac{m}{P}$ becomes large, whose reason is already confirmed in Fig. 12.

5   Conclusion

We analyzed the difference of the parallel execution time between the Householder QR algorithm and the TSQR algorithm. We suggest that distinguishing the effective floating-point performance for each computational kernel is vital in modeling and analyzing the time. Our analysis indicates that TSQR is not always faster than Householder QR; it becomes slower as n grows. We also considered estimating the difference based on the models and selecting the faster algorithm for given matrices, which falls into the category of off-line auto-tuning. The numerical experiments on the K computer totally supported our analysis and the realizability of the automatic algorithm selection between Householder QR and TSQR.

This work implies that structured QR factorization plays a key role in TSQR, which is reasonable because its arithmetic cost is the drawback in avoiding communication. As shown in this paper, considering its effective performance is vital when analyzing the performance of TSQR. Practically, improving the performance of structured QR factorization directly makes TSQR more efficient.

To strengthen our results described in the paper, numerical experiments on other systems than the K computer are required. Particularly, we need to conduct the experiments on multi-core clusters where the cost of intra-node communication differs from that of inter-node communication. In addition, comparison with other algorithms than Householder QR is also vital; there are several state-of-the-art communication avoiding algorithms for computing the QR factorization [11–13]. In the case that n in not small, these algorithms are possibly more efficient than both Householder QR and TSQR. Besides, it has been reported [14] that applying the TSQR algorithm for computing local QR factorizations sometimes improves the performance. It requires an additional analysis on the difference in general QR factorizations, which we did not deal with in this study.

Acknowledgments. The authors would like to thank the anonymous referees for their valuable comments. The first author appreciates the fruitful discussion with Dr. Mark Hoemmen at iWAPT2014. This research was supported by JST, CREST and used computational resources of the K computer provided by the RIKEN AICS through the HPCI System Research project (Project ID:hp120170).

References

1. Bai, Z., Demmel, J., Dongarra, J., Ruhe, A., van der Vorst, H.: Templates for the Solution of Algebraic Eigenvalue Problems: A Practical Guide. SIAM, Philadelphi (2000)

2. Gutknecht, M.H.: Block Krylov space methods for linear systems with multiple right-hand sides: An introduction (2006)
3. Sakurai, T., Sugiura, H.: A projection method for generalized eigenvalue problems using numerical integration. J. Comput. Appl. Math. **159**, 119–128 (2003)
4. Ballard, G., Demmel, J., Holtz, O., Schwartz, O.: Minimizing communication in numerical linear algebra. SIAM J. Matrix Anal. Appl. **32**, 866–901 (2011)
5. Demmel, J., Grigori, L., Hoemmen, M., Langou, J.: Communication-avoiding parallel and sequential QR factorizations. CoRR abs/0806.2159 (2008)
6. Demmel, J., Grigori, L., Hoemmen, M., Langou, J.: Communication-optimal parallel and sequential QR and LU factorizations. SIAM J. Sci. Comp **34**, 206–239 (2012)
7. Golub, G.H., Van Loan, C.F.: Matrix Computations, 4th edn. The Johns Hopkins University Press, Baltimore (2012)
8. Agullo, E., Coti, C., Dongarra, J., Herault, T., Langou, J.: Qr factorization of tall and skinny matrices in a grid computing environment. In: 24th IEEE International Parallel and Distributed Processing Symposium, pp. 1–11. IEEE (2010)
9. Constantine, G., Gleich, D.: Tall and skinny qr factorizations in mapreduce architectures. In: 2nd international workshop on MapReduce and its applications. pp. 43–50 (2011)
10. Langou, J.: Computing the r of the qr factorization of tall and skinny matrices using MPI_Reduce. arXiv:1002.4250 (2010)
11. Song, F., Ltaief, H., Hadri, B., Dongarra, J.: Scalable tile communication-avoiding QR factorization on multicore cluster systems. In: Proceedings of the 2010 ACM/IEEE International Conference for High Performance Computing, Networking, Storage and Analysis (SC 2010). pp. 1–11 (2010)
12. Dongarra, J., Faverge, M., HéRault, T., Jacquelin, M., Langou, J., Robert, Y.: Hierarchical QR factorization algorithms for multi-core clusters. Parallel Comput. **39**, 212–232 (2013)
13. Ballard, G., Demmel, J., Grigori, L., Jacquelin, M., Nguyen, H.D., Solomonik, E.: Reconstructing Householder vectors from tall-skinny QR. Technical Report UCB/EECS-2013-175, EECS Department, University of California, Berkeley (2013)
14. Hoemmen, M.: A communication-avoiding, hybrid-parallel, rank-revealing orthogonalization method. In: 23th IEEE International Parallel and Distributed Processing Symposium, pp. 966–977. IEEE (2011)

Automatic Parameter Tuning
of Three-Dimensional Tiled FDTD Kernel

Takeshi Minami[1], Motoharu Hibino[1], Tasuku Hiraishi[2],
Takeshi Iwashita[3](✉), and Hiroshi Nakashima[2]

[1] Graduate School of Informatics, Kyoto University, Kyoto, Japan
[2] Academic Center for Computing and Media Studies, Kyoto University,
Kyoto, Japan
[3] Information Initiative Center, Hokkaido University, Sapporo, Japan
iwashita@iic.hokudai.ac.jp

Abstract. This paper introduces an automatic tuning method for the tiling parameters required in an implementation of the three-dimensional FDTD method based on time-space tiling. In this tuning process, an appropriate range for the tile size is first determined by trial experiments using cubic tiles. The tile shape is then optimized by using the Monte Carlo method. The tiled FDTD kernel was multi-threaded and its performance with the tuned parameters was evaluated on multi-core processors. When compared with a naively implemented kernel, the performance of the tuned FDTD kernel was improved by more than a factor of two.

1 Introduction

The three-dimensional (3D) finite-difference time-domain (FDTD) method is widely used in high-frequency electromagnetic field analysis for the design of electrical devices [1,2]. The method is based on iterative stencil computations composed of nested outer temporal and inner spatial loops. In iterative stencil computations, the number of floating point operations is often relatively small when compared with the total amount of data transferred between the CPUs and the main memory. Consequently, the computation time of the 3D FDTD kernel is often determined by the (effective) memory bandwidth rather than by the processing core performance. In particular, on computational nodes based on multi-core processors, the speed-up obtained is seldom proportional to the number of cores used, despite the fact that the computational kernel of the 3D FDTD method can be parallelized in a straightforward manner [3]. This is due to the well-known "memory-wall problem". In iterative stencil computations, "time-space" tiling is one possible solution to this problem [4,5].

In time-space tiling, the domain to be analyzed is divided into a number of smaller regions, i.e. *tiles*, each of which is so small that it is accommodated by the cache, and the electromagnetic field variables in each tile are updated for a certain number of time steps successively. Because most calculations on the tile are performed in the cache, this technique increases the cache hit ratio

© Springer International Publishing Switzerland 2015
M. Daydé et al. (Eds.): VECPAR 2014, LNCS 8969, pp. 284–297, 2015.
DOI: 10.1007/978-3-319-17353-5_24

and thus produces better performance. The application of this technique to a two-dimensional (2D) FDTD kernel was reported in [6,7]. However, because the 3D FDTD kernel has a more complex stencil shape and loop structure than the 2D case, there are few reports on time-space tiling for the 3D FDTD method. Under these circumstances, we previously introduced a time-space tiling method for the 3D FDTD kernel without redundant calculations (3D tiled FDTD), and confirmed its effectiveness when using a multi-core processor system [8].

In this paper, to achieve further improvements in performance, we describe a parameter tuning method for the 3D tiled FDTD kernel. The parameters to be tuned are the shapes of the tiles and the number of time steps of the successive updates of a tile. We tune these parameters based on the experimental results from relatively small-sized jobs in which the kernel performs for a small number of time steps. Numerical tests on multi-core processors of the latest generation exhibit that the tuned 3D FDTD kernel attains a more than two-fold performance improvement when compared to the naive implementation of the kernel.

2   Temporal Tiling of 3D FDTD Method

2.1   Basic Equations and Standard Implementation of 3D FDTD Method

From Maxwell equations and constitutive equations, the basic equations describing electromagnetic field phenomena are given by

$$\nabla \times \boldsymbol{E} = -\mu \frac{\partial \boldsymbol{H}}{\partial t} \tag{1}$$

and

$$\nabla \times \boldsymbol{H} = \epsilon \frac{\partial \boldsymbol{E}}{\partial t} + \sigma \boldsymbol{E}, \tag{2}$$

where $\boldsymbol{E}$, $\boldsymbol{H}$, ϵ, μ, and σ denote are the electric field, magnetic field, permittivity, magnetic permeability, and conductivity, respectively. In FDTD method, the space and time partial derivatives are approximated by the finite difference scheme. In both space and time domains, the centered finite difference method based on a staggered grid is used. Consequently, in the time-dependent calculation, the electric and magnetic field variables are alternatively updated as follows:

$$\boldsymbol{E}^n = \frac{1 - \{\sigma \Delta t/2\epsilon\}}{1 + \{\sigma \Delta t/2\epsilon\}} \boldsymbol{E}^{n-1} + \frac{\Delta t/\epsilon}{1 + \{\sigma \Delta t/2\epsilon\}} (\nabla \times \boldsymbol{H}^{n-1/2}), \tag{3}$$

and

$$\boldsymbol{H}^{n+1/2} = \boldsymbol{H}^{n-1/2} - \frac{\Delta t}{\mu} (\nabla \times \boldsymbol{E}^n), \tag{4}$$

where the superscripts denote the time step and Δt is the time interval. In the computation of the space derivatives, a staggered grid, called the Yee cell, is used

```
for(t=0;t<total_t;t++){
  for(i=1;i<=nx;i++){
    for(j=1;j<=ny;j++){
      for(k=1;k<=nz;k++){
        m=id[i][j][k];
        Ex[i][j][k]=Cex[m]*Ex[i][j][k]
                    +Cexry[m]*(Hz[i][j][k]-Hz[i][j-1][k])
                    +Cexrz[m]*(Hy[i][j][k]-Hy[i][j][k-1]);
        Ey[i][j][k]=Cey[m]*Ey[i][j][k]
                    +Ceyrz[m]*(Hx[i][j][k]-Hx[i][j][k-1])
                    +Ceyrx[m]*(Hz[i][j][k]-Hz[i-1][j][k]);
        Ez[i][j][k]=Cez[m]*Ez[i][j][k]
                    +Cezrx[m]*(Hy[i][j][k]-Hy[i-1][j][k])
                    +Cezry[m]*(Hx[i][j][k]-Hx[i][j-1][k]);
}}}
  for(i=1;i<=nx;i++){
    for(j=1;j<=ny;j++){
      for(k=1;k<=nz;k++){
        m=id[i][j][k];
        Hx[i][j][k]=Hx[i][j][k]
                    +Chxry[m]*(Ez[i][j+1][k]-Ez[i][j][k])
                    +Chxrz[m]*(Ey[i][j][k+1]-Ey[i][j][k]);
        Hy[i][j][k]=Hy[i][j][k]
                    +Chyrz[m]*(Ex[i][j][k+1]-Ex[i][j][k])
                    +Chyrx[m]*(Ez[i+1][j][k]-Ez[i][j][k]);
        Hz[i][j][k]=Hz[i][j][k]
                    +Chzrx[m]*(Ey[i+1][j][k]-Ey[i][j][k])
                    +Chzry[m]*(Ex[i][j+1][k]-Ex[i][j][k]);
}}}
}
```

Fig. 1. Outline of 3D FDTD kernel based on naive implementation

to discretize the electric and magnetic fields. Figure 1 shows an outline of the program code of the 3D FDTD method naively implemented. In the program, the spatial coefficients in (3) and (4) for each grid point are given through the array **id**, which prescribes the type of medium (i.e. material) involved at each grid point. An analytical model is usually formed by only a few materials even in practical simulations. In this case, the implementation using indirect addressing is usually advantageous in regard to total simulation time.

2.2   Time-Space Tiling for 3D FDTD Method

In this subsection, we introduce time-space tiling for the 3D FDTD method. In this paper, we discuss the single level tiling, i.e. the tiling for the last level cache. Here, we consider a grid of $n_x \times n_y \times n_z$ for the analysis, where the subscripts represent the direction of each axis in the 3D coordinate system. In the tiling method, the grid is divided into tiles, each of which consists of $t_x \times t_y \times t_z$

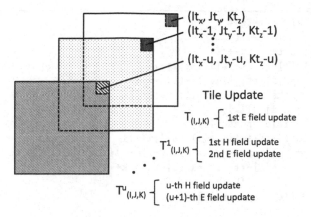

(a) Movement of tile for time steps

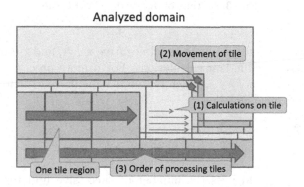

(b) Order of calculations of field variables

Fig. 2. Time-space tiling for FDTD kernel

grid points. For the purposes of the discussions that follow, we introduce 3D coordinates for the grid points and the tiles, which are written as (i, j, k), $(i = 1, \ldots, n_x, j = 1, \ldots, n_y, z = 1, \ldots, n_z)$ and $(I, J, K) = (\lceil i/t_x \rceil, \lceil j/t_y \rceil, \lceil k/t_z \rceil)$, respectively. Then, the tile located at (I, J, K), which is denoted here by $T_{I,J,K}$, is given by

$$T_{I,J,K} = \{(i,j,k) | (I-1)t_x + 1 \leq i \leq I \cdot t_x, \ (J-1)t_y + 1 \leq j \leq J \cdot t_y,$$
$$(K-1)t_z + 1 \leq k \leq K \cdot t_z\}. \tag{5}$$

The time-space tiling involves updating the electric and magnetic field variables for multiple time steps for each tile. The method is expected to reduce the number of main memory accesses and increase both the cache hit ratio and the kernel performance. In the analysis, we use the time-skew algorithm which has been shown to be applicable to the 3D FDTD method in [8]. In the technique,

```
for(t=0;t<total_t;t+=ts){
  for(l=1;l<=ntile;l++){
    for(tt=0;tt=ts-1;tt++){
      for(i=is[l]-tt;i<=ie[l]-tt;i++){
        for(j=js[l]-tt;j<=je[l]-tt;j++){
          for(k=ks[l]-tt;k<=ke[l]-tt;k++){
            m=id[i][j][k];
            E-field update }}}
      for(i=is[l]-tt-1;i<=ie[l]-tt-1;i++){
        for(j=js[l]-tt-1;j<=je[l]-tt-1;j++){
          for(k=ks[l]-tt-1;k<=ke[l]-tt-1;k++){
            m=id[i][j][k];
            H-field update }}}
}}}
```

Fig. 3. Outline of 3D tiled FDTD kernel

the tile is moved by one grid point in all negative x-, y-, and z-directions at each time step to avoid redundant calculations. For a more precise explanation, we introduce the notation $T_{I,J,K}^{u}$, which represents the moved tile for the u-th step update and is defined as follows:

$$T_{I,J,K}^{u} = \{(i, j, k)|(i + u, j + u, k + u) \in T_{I,J,K},$$
$$1 \le i \le n_x,\ 1 \le j \le n_y,\ 1 \le k \le n_z\}, \tag{6}$$

where $1 \le u \le t_s$ and t_s is the number of successive updates of the tile. In this technique, the electric field in $T_{I,J,K}^{u-1}$ and the magnetic field in $T_{I,J,K}^{u}$ are updated in the u-th time step for $T_{I,J,K}$, as shown in Fig. 2 (a). It should be noted that the tiles are processed in lexicographical order. Figure 2 (b) shows the order in which the calculation is performed for each grid point.

Figure 3 shows the simplified sample program of the tiled 3D FDTD method, where the process for the domain boundaries is excluded for simplicity. In the present work, the calculations on each tile are parallelized. Because most calculations on the tile are executed on the cache, sufficient speedup is expected when using a multi-core processor. The spatial loop for the field variable on each tile is parallelized using OpenMP directives. In our program, the outer-most spatial loop with respect to i, which corresponds to the x-direction, is parallelized.

3   Parameter Tuning for 3D Tiled FDTD Kernel

3.1   Tuning Parameters and Preliminary Experiments

In the 3D tiled FDTD kernel, the tuned parameters are given by the number of grid points of each tile in the x-, y-, and z-directions, which are t_x, t_y, and t_z, respectively, and the number of time steps of the successive updates of a tile, t_s.

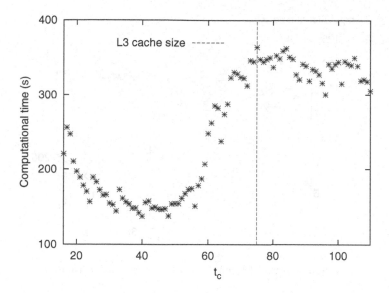

Fig. 4. Computation times for various cubic tiles

Before determining the tuning strategy, we first consider the general characteristics of the parameters in terms of their effects on the kernel performance. Since the performance improvement owing to the tiling technique is derived from the increased cache hit ratio, the tile size must be smaller than the last level cache size. In addition, as the effect of increasing t_s gradually declines, the effects of setting too many time steps in a tile need not be examined. Therefore, we set the maximum number for t_s at 30.

Next, we conducted preliminary numerical experiments for a better understanding of the kernel characteristics, where the experimental condition is the same as described in Sect. 4.1. Figure 4 shows the computation time for the tiled FDTD program with cubic tiles, where t_c is the number of grid points in a single tile in one direction. This figure indicates that tiles that are either too small or too large are ineffective. This is because the use of tiles that are too small causes increased management costs for the tiles. On the other hand, since the cache is not usually occupied only by the data for the tile, the appropriate tile size range spreads below the size perfectly matching the (last-level) cache size.

The preliminary experimental results have also revealed some characteristics of the tiled FDTD program. In our initial research, although we tried using a local optimization technique such as the steepest descent method to tune the parameters, we could not obtain acceptable results. Figure 5 shows the computation times when t_y and t_z are varied with $t_x = 40$ and $t_s = 10$. In the figure, many local minimum points can be observed, which implies that the local optimization techniques have not worked well. Consequently, we chose to use a Monte Carlo (MC) method in the tuning step, while the search space for the tile shape was limited to reduce the tuning cost.

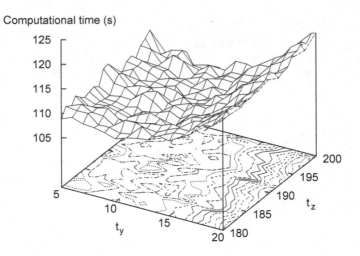

Fig. 5. Effects of t_y and t_z on the computation time

3.2   Design of Automatic Tuning Method

In using the MC simulation for the kernel tuning, we first consider the limitation on the tile size in the simulation. In our method, we find the appropriate tile size range, which depends on various factors such as cache architecture and problem size, from trial experiments using cubic tiles on the computer that is to be used for the FDTD calculations. Although an analytical model can replace the trial experiments, the computational cost for the cubic tile optimization is much smaller than that for the following MC simulation. After determining the optimal cubic tile, we only search the tile from 95 % to 135 % as large as the optimized cubic tile in the MC simulation. This limitation is derived from our experiences of tuning tests. When the tile shape is optimized, the tile tends to be larger than the optimized cubic tile. This is because the optimized tile uses the (last-level) cache more efficiently, and thus the tile size ratio with respect to the cache size increases when compared with the optimized cubic tile. Consequently, we mainly search a tile larger than the optimized cubic one in the MC simulation.

Although the size of the optimal tile is not guaranteed to be in the range (between 95–135 % of the optimized cubic tile), the limitation is considered to be practically useful for the following reason. Generally, the performance of a memory intensive kernel can vary on a parallel computer. In 3D-tiled FDTD analyses, roughly 5 % fluctuations in the performance are often observed even when the same parameter set is used. Therefore, it is practically impossible to define the best (optimal) parameter set. Considering this, the objective of the tuning process is given by finding a sufficiently good parameter set. The preliminary experiments indicated that a multiple of good parameter sets satisfying the above tile-size limitation could be found.

Next, we consider the constraint on the tile shape in the MC simulation. In our program, the elements of the arrays for the field variables are contiguous on the main memory in the z-direction. Therefore, a tile having relatively long edges along the z-direction has an advantage for contiguous memory access, which is expected to increase the cache hit ratio. Another requirement is given by the x-direction spatial loop to be multi-threaded. That is, t_x is required to be larger than the number of threads, and if it is sufficiently large, the load imbalance among the threads is kept small. Accordingly, we can narrow down the search space for the tile shape to tiles that satisfy $t_x > t_y$ and $t_z > t_y$.

The numerical results also imply that the number of time steps t_s can be tuned separately from the tile shape parameters. It was noted that the effect of t_s on the calculation time is smaller than that of the tile shape. The numerical results also indicate that the optimized value of t_s for a specific tile also works well for tiles that are similar to the specified tile. Considering the aspects discussed above, we finally propose the following tuning strategy:

STEP 1. (Cubic tile optimization) Fix t_s at 10. Optimize the tile size, t_c, subject to $t_c = t_x = t_y = t_z$. Let t_c' denote the optimized value obtained for t_c.

STEP 2. Optimize $t_s \in \{1, 2, \ldots, 30\}$ using the tile size $(t_c')^3$. Here, let t_s' denote the optimized value of t_s.

STEP 3. (Tile shape optimization) Fix t_s at t_s'. Perform a Monte Carlo search in the tile-shape space satisfying the following criteria:
  (a) $0.95 \times (t_c')^3 \le t_x \times t_y \times t_z \le 1.35 \times (t_c')^3$
  (b) $t_y < t_x, t_y < t_z$

STEP 4. Optimize t_s by using the tile that was obtained in STEP 3.

STEP 5. (Optional) Perform a local search around the parameters obtained in Step 4.

Although we believe that STEP 5 increases the robustness of the tuning process, the performance improvement obtained from STEP 5 is often too small to compensate the computational cost required for the step. Therefore, in the following numerical experiments, we did not conduct STEP 5 and used the results of STEP 4 for the final tuned parameters.

4   Numerical Results

4.1   Test Conditions

To examine the performance of the tuned 3D FDTD kernel, we conducted numerical tests using a single processor (socket) of a node of Appro/Cray GreenBlade 8000 in Academic Center for Computing and Media Studies, Kyoto University. The processor is an 8-core Intel Xeon E5-2670 processor. Because the computational node has four DDR3 (1600 MHz) channels for each processor, the peak memory bandwidth per processor is 51.2 GB/s. Moreover, the effective memory and L3 (last level) cache bandwidth evaluated by the STREAM Triad benchmark are 35.8 and 174 GB/s, respectively.

In the numerical tests, all array data are allocated to the memory modules connected to the processor (socket). The Intel C compiler version 12.1.6 was

Table 1. Performance evaluation of 3D tuned FDTD kernel

(a) Sequential computation using one thread

	# trial jobs	t_x, t_y, t_z, t_s	Elapsed time (s)
Naive implementation	-	-	1586
Tuned kernel by Step 2 (cubic tile)	103	48, 48, 48, 23	875
Tuned kernel	633	35, 8, 414, 29	686

(b) Parallel computation using eight threads

	# trial jobs	t_x, t_y, t_z, t_s	Elapsed time (s)
Naive implementation	-	-	237
Tuned kernel by Step 2 (cubic tile)	96	40, 40, 40, 17	135
Tuned kernel	626	24, 10, 320, 15	104

used with the option of "-restrict -O3 -xHost -static_intel -openmp". In practical simulations, the problem size is often large enough to use most of the main memory spaces of the computer. Thus, the domain to be analyzed is given by a cubic grid space of 800^3, which is much larger than the last level cache size. The total number of time steps is 90, which is much less than that for the practical simulation, but large enough to evaluate the kernel performance. The number of trial tasks in STEP 3 of the tuning process is 500.

In the proposed tuning method, a number of trial jobs should be managed. For this purpose, we used Xcrypt [9], which is a Perl-based script language for job-level parallel processing on batch-queuing parallel computers. The Xcrypt (Perl) program automatically generates the set of parameters for each tuning step, to be swept by many jobs in parallel, from the results of the previous step.

4.2   Numerical Results

Table 1 provides a performance comparison between the tuned tiled 3D FDTD program and the program based on the naive implementation. In both sequential and parallel computation, the tuned version of the program is approximately 2.3 times as fast as the naive implementation. The number of trial jobs is about 600 and the tuned parameter is expected to be effective for all analyses with the same problem size and the same number of threads. Consequently, the proposed tuning method is expected to be useful for practical simulations in which many time steps are solved.

The tuned FDTD program is about 1.3 times as fast as the tiled FDTD program with the optimized cubic tile. The final tuned tile has relatively long edges in the z-direction, which corresponds to the contiguous memory access. The final tuned tile is larger than the optimized cubic tile obtained in STEP 2. In parallel computation, the sizes of the optimized cubic and tuned tiles are 16.3 % and 19.5 % of the cache size, respectively. This result indicates that the tile shape optimization promotes efficient use of the cache.

Table 2. Evaluation results of the tuned kernel on various computational conditions

Machine	# Threads	Problem size	Naive implementation (s)	Tuned version (s)
XE6	16	600^3	124	73.3
XE6	16	800^3	292	140
XE6	16	1000^3	632	268
Xeon E5	8	600^3	97.5	44.5
Xeon E5	8	1000^3	503	213

Figure 6 shows the relative (parallel) speedup when compared with the sequential FDTD kernel based on the naive implementation. Figure 6 confirms that the tuned kernel attains good parallel performance achieving 8-thread speedup of 6.6-fold and 15.2-fold over the sequential executions of its own and of the naive kernel. In the 8-thread execution, the data transfer (load and store) rate of the tuned program measured 106 GB/s.

Next, we examine the quality of our tuning method and the obtained result. In the analysis, there are about 46 billions candidates (search space) for the parameter set. In our tuning algorithm, the search space of MC simulation is greatly narrowed down to about 6.6 million times as small as the whole search space. Although we have conducted 60K MC iterations with the only limitation being that the tile size is smaller than the last level cache size, the obtained result (elapsed time=186 s) is still inferior to our tuning result. Moreover, in performing our tuning method several times with an increased number of trial jobs in

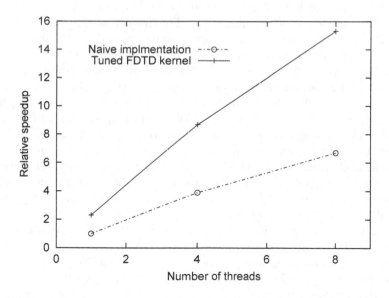

Fig. 6. Comparison of the tuned kernel with naive sequential implementation

STEP 3, the best elapsed time was 102 s. Although this result is the best result among many trial jobs that we have conducted in our research period, it is only 2 % better than the result shown in Table 1. When considering the fluctuations of the kernel performance on a parallel computer, the proposed tuning method is thought to be useful for getting a sufficiently good parameter set.

Finally, for further confirmation of the effectiveness of the proposed tuning method, we conducted additional numerical tests using different problem sizes and other computational nodes. One of the computational nodes of the Cray XE6 system was used for these tests. The XE6 node consists of two 16-core AMD Opteron processors, and one of these processors was used in the analysis. The Cray C compiler version 5.26 was used with the option of "-h omp -O3 -h pic -dynamic". Table 2 shows a comparison between the results from the tuned 3D FDTD program and those from the naive implementation. It was confirmed that the tuned kernel outperformed the naive kernel in all cases. The tuned kernel was mostly more than twice as fast as the naive implementation. The numerical results indicate that the use of time-space tiling with the tuned parameters is effective in obtaining better performance on multi-core processors.

5   Related Works

Tuning of the computational kernel that is used in various application programs is an important research target for high performance computing. Here, we briefly introduce recent research activities on tuning for computational kernels, especially for the (iterative) stencil computation, in which 3D FDTD method is categorized.

Software tools such as ATLAS [10] and OSKI [11] that automatically tune a computational kernel are well known. The former is the auto-tuning tool for the BLAS library, and the latter tunes sparse matrix vector multiplication kernels. In the area of high-performance stencil computation, K. Datta et al. reported on their developed automatic optimizer for various multicore processors in [12]. Their optimizer for the stencil computation includes NUMA-aware allocation, multi-level blocking, and loop unrolling. These optimization (tuning) processes are divided into four parts (1) Problem decomposition, (2) Data allocation, (3) Bandwidth optimizations, and (4) In-core optimizations. The optimizer attains significant performance improvement on various multicore processors and a GPU. Moreover, in [13], K. Datta et al. reported auto-tuning of more complex 27 stencil computations. In [14], J. Shirako presented analytical bounds for appropriate tile size which are effectively used to prune the search space in empirical tile optimization for general tiling techniques.

To develop a high-performance stencil code for multiple computational nodes, which should be based on multi-process programming model, Maruyama et al. presented a high-level programming framework based on DSL, Physis [15]. The framework specially focused on the execution of the generated code on a GPU cluster. Because the communication between GPU nodes involves additional data copies between GPU and CPU memory, the communication greatly affects the stencil kernel performance. The Physis framework is equipped with an optimizer for communications, which effectively overlaps computations and communications.

Next, we describe research works on temporal tiling (blocking) for the iterative stencil computation. Although there are many reports dealing with the topic, we introduce here only recent activities. G. Wellein et al. presented a pipelined wavefront parallelization approach for the stencil computation. Their technique well combines temporal blocking and multithreading on a multicore processor having a shared cache [16]. Moreover, M. Wittmann et al. discussed the pipelined temporal blocking method more generally and reported results obtained on a distributed memory parallel environment in [17]. In [18], D. Orozco and G. Gao reported an efficient implementation of FDTD method on the many-core chip architecture of a IBM Cyclops-64. They discussed temporal tiling and its optimization of a 1D FDTD method application. In [7,19], U. Bondhugula et al. presents an automatic source-to-source transformation framework including parallelization and locality optimization, PLuto. PLuto can automatically generate multithreaded and optimized iterative stencil code using temporal tiling and other optimization techniques. In [6], R. Strzodka proposed the cache oblivious algorithm, called CORALS, for iterative stencil computations. In their method, the entire time-space grid is recursively divided into parallelograms. In each parallelogram, computations are performed by the time-skewing scheme. In 2D and 3D stencil computations, the program code based on CORALS outperforms the optimized code automatically generated by PLuto.

Temporal blocking is also examined in the context of GPU computing. In [20], A. Nguyen presented 3.5-D (2.5D-spatial and temporal) blocking method for stencil computations and a lattice Boltzmann simulation run on CPUs or GPUs. Moreover, in [21], G. Jin examined the effectiveness of temporal blocking on minimizing the cost for data transfer between CPU and GPU in a stencil computation with a large analysis domain, the size of which exceeded the GPU memory capacity.

The difference between our work and the above related works is in examining the temporal tiling and its optimization in the context of the 3D FDTD method. Most of the techniques introduced above can be applied in principle to the 3D FDTD method. However, the effectiveness of temporal blocking in 3D FDTD kernel has not been fully examined, although results for 2D FDTD have been reported for example in [6,7]. Because the 3D FDTD method has more complex stencil shape and requires more memory footprints than the 2D version, its performance prediction is not straightforward. Consequently, temporal tiling in the 3D FDTD method is worth examining.

6   Conclusions

In this paper, we summarized a 3D tiled FDTD method and proposed an auto-tuning technique for the tiling parameters to enhance the performance of the method. In the tuning step, we first find the appropriate tile size range by performing trial jobs with cubic tiles. Next, the tile shape is optimized by using the

Monte Carlo method. To reduce the total number of trial jobs, we enforce limitations with respect to the tile size and shape in this tuning step. The resulting tuned FDTD kernel is about twice (up to 2.36 times) as fast as the naive kernel in both serial and parallel computations on two types of multi-core processors of the latest generation.

In future works, we target improving the kernel performance and the tuning process. Although we exploited inner-tile parallelism in the present study, there is a possibility to improve the performance by introducing inter-tile parallelism. Moreover, in our program code, SIMD instructions are not used effectively because of indirect addressing. With future processors equipped with a wider SIMD width than in current generation processors, strategies for effective use of SIMD instructions in a practical 3D FDTD program need developing. Furthermore, we will examine the hierarchical tiling for other levels cache or TLB. To improve the tuning process, we will compare our tuning method with other tuning algorithms and examine it in the context of other iterative stencil computations.

References

1. Lu, J., Thiel, D., Saario, S.: FDTD analysis of dielectric-embedded electronically switched multiple-beam (DE-ESMB) antenna array. IEEE Trans. Magn. **38**, 701–704 (2002)
2. Ala, G., Di Piazza, M.C., Tine, G., Viola, F., Vitale, G.: Numerical simulation of radiated EMI in 42 V electrical automotive architectures. IEEE Trans. Magn. **42**, 879–882 (2006)
3. Chew, K.C., Fusco, V.F.: A parallel implementation of the finite difference time-domain algorithm. Int. J. Numer. Model. **8**, 293–299 (1995)
4. Wolf, M.: More iteration space tiling. In: Proceedings of the Supercomputing 1989, pp. 655–664 (1989)
5. Wonnacott, D.: Using time skewing to eliminate idle time due to memory bandwidth and network limitations. In: Proceedings of the IPDPS 2000 (2000)
6. Strzodka, R., et al.: Cache oblivious parallelograms in iterative stencil computations. In: Proceedings of the ICS 2010, pp. 49–59 (2010)
7. Bondhugula, U., Hartono, A., Ramanujam, J., Sadayaooan, P.: A practical automatic polyhedral parallelizer and locality optimizer. In: Proceedings of the 2008 ACM SIGPLAN Programming Language Design and Implementation (PLDI), pp. 101–113 (2008)
8. Minami, T., et al.: Temporal and spatial tiling method without redundant calculations for three-dimensional FDTD method. IPSJ Tran. Adv. Comput. Syst. (In Japanese) (to appear)
9. Hiraishi, T., et al.: Xcrypt: a perl extension for job level parallel programming. In: Proceedings of the WHIST 2012 (2012)
10. Whaley, R.C., Petitet, A., Dongarra, J.: Automated empirical optimization of software and the ATLAS project. Parallel Comput. **27**, 3–35 (2001)
11. Vuduc, R., Demmel, J., Yelick, K.: OSKI: a library of automatically tuned sparse matrix kernels. In: Proceedings of the SciDAC 2005, Journal of Physics: Conference Series, vol. 16, pp. 521–530 (2005)

12. Datta, K., et al.: Stencil computation optimization and auto-tuning on state-of-the-art muticore architectures. In: Proceedings of the SC 2008 (2008)
13. Datta, K., et al.: Auto-tuning the 27-point stencil for multicore. In: Proceedings of the iWAPT 2009 (2009)
14. Shirako, J., Sharma, K., Fauzia, N., Pouchet, L.-N., Ramanujam, J., Sadayappan, P., Sarkar, V.: Analytical bounds for optimal tile size selection. In: O'Boyle, M. (ed.) CC 2012. LNCS, vol. 7210, pp. 101–121. Springer, Heidelberg (2012)
15. Maruyama, N., Nomura, T., Sato, K., Matsuoka, S.: Physis: an implicitly parallel programming model for stencil computations on large-scale GPU-accelerated supercomputers. In: Proceedings of the SC 2011 (2008)
16. Wellein, G., et al.: Efficient temporal blocking for stencil computations by multicore-aware wavefront parallelization. In: Proceedings of the COMPSAC 2009, pp. 579–586 (2009)
17. Wittmann, M., Hager, G., Wellein, G.: Multicore-aware parallel temporal blocking of stencil codes for shared and distributed memory. In: Proceedings of the 2010 IEEE International Symposium on Parallel and Distributed Processing. WS and Phd Forum (IPDPSW) (2010)
18. Orozco, D., Gau, G.: Mapping the FDTD application to many-core chip architectures. In: Proceedings of the 2009 International Conference on Parallel Processing (ICPP), pp. 309–316 (2009)
19. PLUTO - An automatic parallelizer and locality optimizer for multicores. http://pluto-compiler.sourceforge.net
20. Nguyen, A., Satish, N., Chhugani, J., Changkyu, K., Dubey, P.: 3.5-D blocking optimization for stencil computations on modern CPUs and GPUs. In: Proceedings of the SC 2010 (2010)
21. Jin, G., Endo, T., Matsuoka, S.: A multi-level optimization method for stencil computation on the domain that is bigger than memory capacity of GPU. In: Proceedings of the 2013 27th IEEE International Symposium on Parallel and Distributed Processing. WS and Phd Forum (IPDPSW), pp. 1080–1087 (2010)

Automatic Parameter Tuning of Hierarchical Incremental Checkpointing

Alfian Amrizal[1,3], Shoichi Hirasawa[1,3],
Hiroyuki Takizawa[1,3(✉)], and Hiroaki Kobayashi[1,2,3]

[1] Graduate School of Information Science,
Tohoku University, Sendai, Miyagi 980-8578, Japan
{alfian,hirasawa}@sc.isc.tohoku.ac.jp
[2] Cyberscience Center, Tohoku University, Sendai, Japan
[3] Japan Science and Technology Agency, CREST, Kawaguchi, Japan
{tacky,koba}@cc.tohoku.ac.jp

Abstract. As future HPC systems become larger, the failure rates and the cost of checkpointing to the global file system are expected to increase. Hierarchical incremental CPR is a promising approach to solve this problem. It utilizes a hierarchical storage system of local and global storages and performs incremental checkpointing by writing only updated memory pages between two consecutive checkpoints. In this paper, we response to an open question; how to optimize the checkpoint interval when the checkpoint overheads are changing with time as in hierarchical incremental CPR. We propose a runtime checkpoint interval autotuning technique to optimize the efficiency of hierarchical incremental CPR. Evaluation results show that the efficiency can be significantly increased if the storage hierarchy can be exploited with appropriate checkpoint intervals.

1 Introduction

The computational power of high-performance computing (HPC) systems is exponentially growing every year and hence enables finer-grained scientific simulations. However, the exponentially-increasing number of components of the HPC systems causes an increase in the overall failure rate. Future HPC systems are predicted to experience a failure every tens of minutes [1]. Thus, fault-tolerance has become more important than ever for future HPC systems.

Checkpointing and rollback recovery (CPR) is the most widely-used fault-tolerance mechanism for HPC systems. CPR writes the state of a running process to a checkpoint file, and recovers the previous state later by loading the checkpoint file when necessary. These checkpoint files are generally stored in a stable storage, typically a global file system. In this work, we define the *efficiency* of a CPR mechanism as the ratio of *original execution time* to *total execution time*. Here, *original execution time* is the execution time of an application itself if the application encounters no failures and takes no checkpoints, while *total execution time* is the total execution time of the application and CPR when a CPR mechanism is implemented and the occurrence of failures is considered.

© Springer International Publishing Switzerland 2015
M. Daydé et al. (Eds.): VECPAR 2014, LNCS 8969, pp. 298–309, 2015.
DOI: 10.1007/978-3-319-17353-5_25

Since the computational capabilities are increasing faster than the bandwidth to the global file system, the checkpointing overhead to the global file system still can dominate the total execution time even if the checkpointing is performed infrequently. The incremental checkpointing [2] can be one of the promising technologies to decrease the huge overhead caused by checkpointing. The incremental checkpointing reduces the data size to be written into a checkpoint file at every checkpoint by writing only updated data or updated memory pages between two consecutive checkpoints. The changed data/memory pages will be marked as dirty and only the dirty memory pages are transferred during the checkpointing.

Another promising approach to efficient CPR is the hierarchical checkpointing [3,15] that exploits a hierarchical storage system of local and global storages. Each storage has different degrees of resiliency and checkpointing cost. The hierarchical checkpointing relies on node-local storages for restarting from more common transient failures and the global file system for less frequent hardware failures. In hierarchical checkpointing, failures are clasified into two; local failures and global failures. Local failures include any transient failures (a.k.a., software failures) and can be recovered by local checkpoints stored in local storages. On the other hand, global failures include all hardware failures and must be recovered by global checkpoint stored in the global file system. By frequently taking inexpensive node-local checkpoints and less frequently taking expensive system-wide global checkpoints, applications can achieve both high reliability and efficiency.

Both the incremental checkpointing and the hierarchical CPR mechanism can potentially reduce the timing overhead of CPR if some parameters are adjusted properly. One of the important parameters that can significantly affect performance is the checkpoint interval [3]. The optimal checkpoint interval depends not only on the system configuration but also on the application to be checkpointed. This is because some information which is required to determine the optimal checkpoint interval, such as the growth speed of the number of dirty memory pages, is application-specific and dynamic. Therefore, a runtime autotuning technique is required because this information is unknown in advance of the execution.

In this paper, we consider a hierarchical incremental CPR mechanism that consists of two different checkpoint implementations, *local incremental checkpointing* and *global incremental checkpointing*. We focus on how to optimize the checkpoint intervals of each checkpoint implementation by constructing a performance model to find the optimal solution through an in-depth analysis of the performance model. Then, a runtime autotuning technique that can fit the hierarchical incremental CPR mechanism is introduced.

The rest of this paper is organized as follows. Section 2 describes the hierarchical incremental CPR mechanism and a performance model that is to be used for its optimization. Section 3 discusses an optimization technique and a runtime autotuning technique to find the optimal checkpoint interval for the hierarchical incremental CPR mechanism. Section 4 shows the evaluation results. The conclusion of this paper is stated in Sect. 5.

2   A Hierarchical Incremental Approach for High Performance Checkpointing

In this section, we describe an incremental approach as a CPR mechanism to reduce the amount of data transferred to a checkpoint file on a system with a hierarchical storage of local storages and global storages. Then, we formulate our research as an optimization problem based on a hierarchical CPR model.

2.1   A Hierarchical Incremental CPR Mechanism

This paper focuses on a hierarchical incremental CPR mechanism. The CPR mechanism, which uses local and global storages. The local storage is used for *local incremental checkpointing* and the global storage for *global incremental checkpointing*.

When the initial checkpoint request comes, a *full checkpoint*, i.e., the whole memory data of an application, is first taken and dumped to both local and global storages. After this initial checkpoint, the type of checkpointing conducted, i.e., local incremental checkpointing or global incremental checkpointing, is determined by their intervals.

When the local incremental checkpoint is performed, only the dirty memory pages are transferred to the local storage. The full checkpoint file that has previously been saved at the local storage during the initial checkpoint is then updated using the transferred dirty memory pages. Similarly, global incremental checkpointing is performed by updating the full checkpoint file in the global storage with the global incremental data.

2.2   Performance Model

We first consider a general hierarchical CPR model of local and global checkpoints that is used for our proposed autotuning technique. This model is related to the two-level recovery schemes studied by Vaidya [4] and Dong et al. [5].

Suppose the total execution time of an application running with hierarchical CPR as T_E and divide it into the following four parts:

$$T_E = T_S + T_C + T_L + T_G. \tag{1}$$

Here T_S is the original execution time, T_C is the total time spent for checkpointing, T_L is the time lost due to local failures and T_G is the time lost due to global failures.

We assume that, in general, an application user can estimate with a certain high precision his/her own application original execution time, T_S, based on experiential analysis [6]. Note that T_S does not include any other failure-related costs such as checkpoint overhead and time lost due to failures.

Figure 1 presents the overall scheme of the hierarchical CPR model with two failure cases (local failure and global failure), where n_L refers to the number

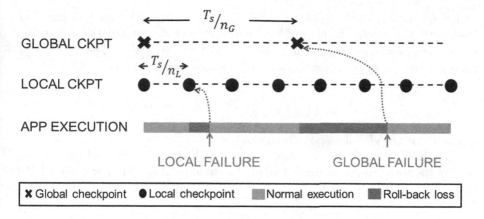

Fig. 1. Scheme of the hierarchical checkpointing model

of local checkpoints and n_G refers to the number of global checkpoints during the period of T_S. For simplicity of explanation, this figure does not present checkpoint overheads and restart overheads.

In this model, the total time spent for checkpointing, T_C, can be expressed as follows,

$$T_C = C_L n_L + C_G n_G, \tag{2}$$

where C_L is the local checkpoint overhead and C_G is the global checkpoint overhead.

The expected number of local failures (denoted by F_L) and the expected number of global failures (denoted by F_G) during the execution of an HPC application can be predicted from the historical statistic data of the mean time between local failures ($MTBF_L$) and the mean time between global failures ($MTBF_G$) as the following equations.

$$F_L = \frac{T_S}{MTBF_L}, \tag{3}$$

$$F_G = \frac{T_S}{MTBF_G}. \tag{4}$$

On average, failures will occur halfway through any checkpoint interval [7]. Hence, the time lost due to local failures, T_L, can be expressed as follows,

$$T_L = \left(\frac{T_S}{2n_L} + R_L \right) \times F_L, \tag{5}$$

where R_L is the local restart overhead.

The time lost due to global failures is different from the time lost due to local failures. The time lost due to global failures has to also include all local checkpoint overheads. For example, as shown in Fig. 1, when a global failure happens, the application must return further back to its latest global checkpoint and hence two local checkpoint overheads should also be counted in the time lost. Here, the average number of local checkpoints between a global failure and the latest global checkpoint can be represented as $\frac{T_S/(2n_G)}{T_S/n_L} = \frac{n_L}{2n_G}$. Hence, the time lost due to global failures, T_G, can be expressed as follows,

$$T_G = \left(\frac{T_S}{2n_G} + \frac{n_L}{2n_G}C_L + R_G \right) \times F_G, \tag{6}$$

where R_G is the global restart overhead.

Finally, by substituting terms of Eq. (1) with Eqs. (2), (5), and (6), we can predict the total execution time of an application running with hierarchical CPR by

$$T_E = T_S + C_L n_L + C_G n_G + \left(\frac{T_S}{2n_L} + R_L \right) \times F_L + \left(\frac{T_S}{2n_G} + \frac{n_L}{2n_G}C_L + R_G \right) \times F_G. \tag{7}$$

Our objective is to minimize T_E for each given application. In this work, the minimization problem is divided into two; one is to minimize the checkpoint overhead, and and the other is to minimize the time lost due to failures. The local and global incremental checkpoints are used to minimize the checkpoint overheads, C_L and C_G, by transferring only dirty memory pages to a checkpoint file. Note that the restart overheads, R_L and R_G, cannot be minimized since a full checkpoint is required for restarting from any failure. Then, to minimize the time lost due to failures, we need to determine the optimal values of n_L and n_G. We will discuss this in the following section.

3   Runtime Autotuning for Hierarchical Incremental CPR

3.1   Checkpoint Interval Optimization for Hierarchical CPR

We first need to find the optimal values of n_L and n_G to minimize the total execution time given by Eq. (7). To this end, we check whether T_E has a unique minimum point by checking its second derivatives. The second derivatives of T_E are as follows,

$$\frac{\partial^2 T_E}{\partial n_L{}^2} = \frac{T_S}{n_L{}^3} \times F_L > 0, \frac{\partial^2 T_E}{\partial n_G{}^2} = \frac{T_S + C_L n_L}{n_G{}^3} \times F_G > 0. \tag{8}$$

Hence, there exist a unique minimum point where $\frac{\partial T_E}{\partial n_L} = 0$ and $\frac{\partial T_E}{\partial n_G} = 0$. We can derive two first derivative equations as follows,

$$\frac{\partial T_E}{\partial n_L} = C_L - \frac{T_S}{2n_L{}^2} \times F_L + \frac{C_L}{2n_G} \times F_G, \tag{9}$$

$$\frac{\partial T_E}{\partial n_G} = C_G - \frac{T_S + C_L n_L}{2n_G{}^2} \times F_G. \tag{10}$$

Therefore, we can get the optimal interval of local checkpoints and global checkpoints as long as the following two simultaneous equations are solved.

$$C_L - \frac{T_S}{2n_L{}^2} \times F_L + \frac{C_L}{2n_G} \times F_G = 0, \tag{11}$$

$$C_G - \frac{T_S + C_L n_L}{2n_G{}^2} \times F_G = 0. \tag{12}$$

Solving the above equations directly will lead us to a polynomial equation with degree five as shown in Eq. (13). The Abel-Ruffini theorem (also known as Abel's impossibility theorem) states that there is no general algebraic solution to polynomial equations with a single variable of degree five or higher [8].

$$2C_L{}^3 n_L{}^5 + \left(2T_S C_L{}^2 - C_G F_G C_L{}^2\right) n_L{}^4 - 2T_S F_L C_L{}^2 n_L{}^3 - 2T_S{}^2 F_L C_L n_L{}^2$$
$$+ T_S{}^2 F_L{}^2 C_L n_L + T_S{}^3 F_L{}^2 = 0. \tag{13}$$

Since there is no formula to directly solve Eq. (13), we use an iterative method [9], which is commonly used to solve a system of linear equations. First, we convert Eqs. (11) and (12) to Eqs. (14) and (15).

$$n_L = \sqrt{\frac{F_L \times T_S}{C_L \times \left(2 + \frac{F_G}{n_G}\right)}}. \tag{14}$$

$$n_G = \sqrt{\frac{F_G \times (T_S + C_L n_L)}{2C_G}}, \tag{15}$$

From the above equations, we can derive iterative functions, $n_L{}^{(k+1)}$ and $n_G{}^{(k+1)}$, as follows,

$$n_L{}^{(k+1)} = \sqrt{\frac{F_L \times T_S}{C_L \times \left(2 + \frac{F_G}{n_G{}^{(k)}}\right)}}, \tag{16}$$

$$n_G{}^{(k+1)} = \sqrt{\frac{F_G \times (T_S + C_L n_L{}^{(k)})}{2C_G}}. \tag{17}$$

The computation of $n_L{}^{(k+1)}$ and $n_G{}^{(k+1)}$ requires the values of $n_L{}^{(k)}$ and $n_G{}^{(k)}$. Hence, we need to guess an initial values of $n_L{}^{(0)}$ and $n_G{}^{(0)}$, and then iteratively compute Eqs. (16) and (17) until Eqs. (11) and (12) approximately hold with small errors. We use Young's optimal checkpoint interval approximation [10] as our initial guesses of $n_L{}^{(0)}$ and $n_G{}^{(0)}$. Young's equation stated that the optimal checkpoint interval is actually a function of systems MTBF (Mean Time Between Failures) and the time to write checkpoint file C, as shown in Eq. (18).

$$interval_{opt} = \sqrt{2 \times C \times MTBF}. \tag{18}$$

Hence, we set $n_L{}^{(0)}$ and $n_G{}^{(0)}$ as follows,

$$n_L{}^{(0)} = \frac{T_S}{\sqrt{2C_L MTBF_L}}, \tag{19}$$

$$n_G{}^{(0)} = \frac{T_S}{\sqrt{2C_G MTBF_G}}. \tag{20}$$

Suppose that we are running a large scale HPC application with the original execution of two weeks with local and global checkpoints. The overheads for local and global checkpoints are 200 and 3000 s, respectively. We assume that the restart overheads are equal to the checkpoint overheads and the numbers of local failures and global failures are 30 and 10, respectively. Under these conditions, the number of iterative steps used in the iterative method so that Eqs. (11) and (12) hold with an error less than 10^{-15} is only 9 steps. This indicates that such an iterative method is simple but fast enough for computing suboptimal checkpoint intervals at runtime.

3.2   Runtime Autotuning for Hierarchical Incremental CPR

In Sect. 3.1, we have discussed a method to compute the suboptimal checkpoint intervals by computing the optimal numbers of local checkpoints and global checkpoints. In Sect. 3.2, we mainly focus on how to apply this optimization technique to our target CPR mechanism, i.e., the hierarchical incremental CPR.

In the case of incremental checkpointing, C_L and C_G are actually functions of time since the number of dirty memory pages transferred to a checkpoint file is changing with time. Hence, in order to minimize the total execution time, the growth rate of the dirty memory pages must be monitored and the optimal number of local and global checkpoints must be automatically tuned at runtime based on the monitored information. In this work, we measure the checkpoint overhead during the incremental checkpointing processes to obtain the growth rate of the dirty memory pages. The pseudo-code of our autotuning technique is shown in Algorithm 1.

In this algorithm, first we input the original execution time T_S, the expected number of local failures F_L, and the expected number of global failures F_G. As discussed in Sect. 2.1, when the initial checkpoint request comes, a full checkpoint is dumped to both local and global storages. We record the time overheads for performing this full checkpoint, C_{L0} and C_{G0}, and use these obtained information to set the initial values of n_L and n_G.

After the initial checkpoint, the type of checkpointing conducted will be either local incremental checkpointing or global incremental checkpointing. When the i-th local incremental checkpointing is taken, we record its time overhead, C_{Li}, and calculate $C_L{}^*$. Here, $C_L{}^*$ is a weighted moving average (WMA) [11] of C_L. We use WMA to give a greater weight to more recent C_L so that it can react more quickly to a sudden change in the growth rate of the dirty memory pages. Similar computation is also conducted for global incremental checkpointing to calculate $C_G{}^*$.

Algorithm 1. Autotuning Algorithm for Hierarchical Incremental CPR

1: **Input:** original execution time T_S, expected number of local failures F_L, expected
   number of global failures F_G
2: **while** application is running
3:      **if** Initial checkpoint taken
4:          Record C_{L0} and C_{G0}
5:          Compute optimal n_L and n_G based on C_{L0} and C_{G0}
6:          Set initial n_L and n_G
7:      **if** i-th local incremental checkpoint taken
8:          Record C_{Li}
9:          $C_L{}^* = \frac{iC_{Li}+(i-1)C_{L(i-1)}+\dots+2C_{L2}+C_{L1}}{i+(i-1)+\dots+2+1}$
10:          $T_S{}^* \leftarrow$ remaining time of T_S
11:          Recompute optimal n_L based on $C_L{}^*$ and $T_S{}^*$
12:          Adjust n_L
13:      **if** i-th global incremental checkpoint taken
14:          Record C_{Gi}
15:          $C_G{}^* = \frac{iC_{Gi}+(i-1)C_{G(i-1)}+\dots+2C_{G2}+C_{G1}}{i+(i-1)+\dots+2+1}$
16:          $T_S{}^* \leftarrow$ remaining time of T_S
17:          Recompute optimal n_G based on $C_G{}^*$ and $T_S{}^*$
18:          Adjust n_G

Then, we recompute optimal n_L and n_G using the iterative method for the remaining time left of the original execution time, $T_S{}^*$, and readjust their values at runtime. This process is repeated until the application reachs its end.

4 Evaluation

In this paper, the efficiency defined in Sect. 1 is used as the performance metric to evaluate the impact of the proposed autotuning technique on the hierarchical incremental CPR mechanism for a particular application and system configuration. We present the obtained results by analyzing the growth rate of the dirty memory pages of an HPC application called CTH [12], which frequently runs at very large scale for weeks at a time, and hence is prone to failures.

Figure 2 shows the percentage of dirty memory pages transferred to a checkpoint file when running CTH for two days and saving its state every 15 m using hash-based incremental checkpoint [13]. The upper bound of the shaded region in Fig. 2 represents the average percentage of dirty memory pages transferred to a checkpoint file using a page-protection based incremental checkpoint [14]. In this work, we assume that hash-based incremental checkpoint is used rather than page-based incremental checkpoint due to its higher accuracy on detecting dirty memory pages. Also, we assume that no same memory page is updated twice before the whole memory pages are all updated.

The evaluation is conducted via a simulation running CTH for two days with 16 GB memory footprint by assuming that a RAM disk and a Lustre system are used as the local storage and the global storage, respectively. The write

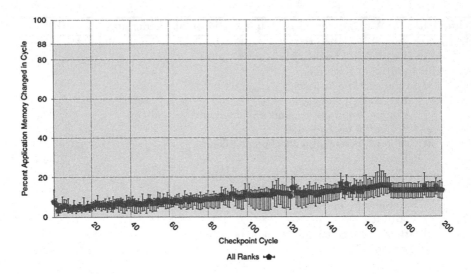

Fig. 2. Percent of dirty memory pages transferred to a checkpoint file during incremental checkpointing at CTH [13].

bandwidth data in [15] and the failure data in [16] are used as a reference to set several parameters in this evaluation. Accordingly, the bandwidth to local storage, the bandwidth to global storage, the mean time between local failure, and the mean time between global failure are set to 8 Gbps, 250 Mbps, 15 h and 3 days, respectively.

In order to show that the autotuning technique can increase the efficiency of hierarchical incremental CPR, we perform an efficiency comparison of the hierarchical incremental CPR with and without the proposed autotuning technique. When autotuning is not used, the values of n_L and n_G are set using the Young equation based on the checkpoint overheads recorded during the initial full checkpoint.

Figure 3 presents the efficiency comparison of the hierarchical incremental CPR mechanism with and without the autotuning technique. The results with the autotuning technique implemented are labeled auto-inc and those without the autotuning technique as inc. As future HPC systems become larger, the expected number of failures and the cost of accessing the global storage are expected to increase. To explore these effects, the number of failures and the global checkpointing overheads are increased by factors of two, ten, and fifty. The groups of bars along the x-axis correspond to the number of failures that are one, two, ten, and fifty times higher than the base value. Within each group, the global checkpointing overhead is increased to be one, two, ten, and fifty times higher than the base value.

In all the cases, the autotuning technique results in a higher efficiency as shown in Fig. 3. Moreover, the advantage increases with either increasing the number of failures or a higher global storage checkpoint overhead. The gain in

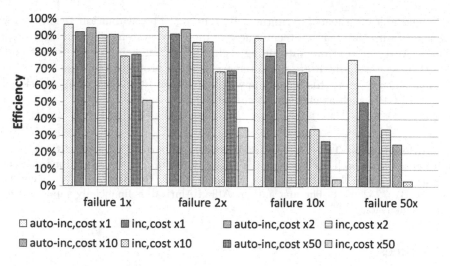

Fig. 3. Hierarchical incremental CPR's efficiency comparison with and without auto-tuning.

efficiency ranges from 4 % to 35 %. When, the number of failures is low (failure 1× and failure 2×), the autotuning technique successfully detected that the mean time between global failures is much longer than the original execution time of the application and decided not to perform any global checkpoint. Hence, it boosts the efficiency of the system to an extremely high value of more than 90 %. These results highlight the benefits of the proposed autotuning technique.

The results in Fig. 3 show that even with systems that are 50× less reliable, the efficiency achieved by the proposed approach exceeds 70 % as long as the global storage performance is unchanged. On the other hand, a higher number of failures cannot be tolerated if the overhead of global checkpointing increases. In particular, if a system becomes 50× less reliable and if the overhead of saving application state to the global storage rises by 50×, even with the proposed approach, the application will not be able to finish its computation (efficiency = 0).

Accordingly, our proposed autotuning technique is essential for future HPC systems. Even if the systems are ten times less reliable compared to current systems, our autotuning technique can guarantee the completion of their computation even though their efficiency degrades below 10 %. However, a more sophisticated technique is required for systems that are extremely prone to failures.

5   Conclusions

This paper proposed a runtime autotuning technique to reduce the timing over-head of hierarchical incremental CPR mechanism. The autotuning technique uses a fast iterative method to calculate optimal local and global intervals and adjusts them at runtime. The iterative method converges only in several steps resulting in an ignorable overhead for runtime autotuning. The evaluation results show that the runtime autotuning technique can increases system efficiency from 4 % to 35 %. Also, the runtime autotuning technique is beneficial for future extreme scale HPC systems which are predicted to be more prone to failures.

In the future work, this technique will be extended to account for additional features of CPR such as checkpoint compression.

Acknowledgments. This research is partially supported by JST CREST "An Evolutionary Approach to Construction of a Software Development Environment for Massively-Parallel Heterogeneous Systems" and Grant-in-Aid for Scientific Research(B) #25280041. The first author, Alfian Amrizal, is financially supported by Monbukagakusho.

References

1. Schroeder, B., Gibson, G.A.: Understanding failures in petascale computers. J. Phys. Conf. Ser. **78**, 012022 (2007)
2. Sancho, J.C., Pertini, F., Johnson, G., Fernandez, J., Frachtenberg, E.: On the feasibility of incremental checkpointing for scientific computing. In: Proceedings of IPDPS 2004, pp. 58–67 (2004)
3. Amrizal, A., Hirasawa, S., Komatsu, K., Takizawa, H., Kobayashi, H.: Improving the scalability of transparent checkpointing for GPU computing systems. In: Proceedings of the 2012 IEEE Region 10 Conference, pp. 989–994, 19–22 November 2012
4. Vaidya, N.H.: A case for two-level recovery schemes. IEEE Trans. Comput. **47**(6), 656666 (1998)
5. Dong, X., Muralimanohar, N., Jouppi, N., Kaufmann, R., Xie Y.: Leveraging 3D PCRAM technologies to reduce checkpoint overhead for future exascale systems. In: Proceedings of SC 2009 (2009)
6. Di, S., Bouguerra, M.S., Gomez, L.B., Cappello, F.: Optimization of multi-level checkpoint model for large scale HPC applications. In: Proceedings of IPDPS 2014, pp. 1181–1190 (2004)
7. Daly, J.T.: A higher order estimate of the optimum checkpoint interval for restart dumps. Future Gener. Comput. Sys. **22**(3), 303–312 (2006)
8. Dehn, E.: Algebraic Equations: An Introduction to the Theories of Lagrange and Galois. Columbia University Press, New York (1930)
9. Balakrishnan, N., Childs, A.: Outlier. In: Hazewinkel, M. (ed.) Encyclopedia of Mathematics. Springer (2001). ISBN 978-1-55608-010-4
10. Young, J.W.: A first order approximation to the optimum checkpoint interval. Commun. ACM **17**(9), 530–531 (1974)
11. Dash, S.: A comparative study of moving averages: simple, weighted, and exponential. http://www.tradestation.com/education/labs/analysis-concepts/a-comparative-study-of-moving-averages
12. Brun, R., Dumitrescu, L.Z.: CTH: a software family for multi-dimensional shock physics analysis. In: Hertel Jr., E.S., Bell, R.L., Elrick, M.G., Farnsworth, A.V., Kerley, G.I., McGlaun, J.W., Petney, S.V., Silling, S.A., Taylor, P.A., Yarrington, L. (eds.) Shock Waves @ Marseille, pp. 377–382. Springer, Heidelberg (1995)

13. Ferreira, K.B., Riesen, R., Brighwell, R., Bridges, P., Arnold, D.: ibhashckpt: hash-based incremental checkpointing using GPU's. In: Cotronis, Y., Danalis, A., Nikolopoulos, D.S., Dongarra, J. (eds.) Recent Advances in the Message Passing Interface. LNCS, vol. 6960, pp. 272–281. Springer, Heidelberg (2011)
14. Elnozahy, E.N., Alvisi, L., Wang, Y.M., Johnson, D.B.: A survey of rollback-recovery protocols in message-passing systems. ACM Comput. Surv. **34**(3), 375–408 (2002)
15. Moody, A., Bronevetsky, G., Mohror, K., Supinski, B.R.: Design, modeling, and evaluation of a scalable multi-level checkpointing system. In: Proceedings of SC 2010 (2010)
16. Sato, K., Maruyama, N., Mohror, K., Moody, A., Gamblin, T., de Supinski, B.R., Matsuoka, S.: Design and modeling of a non-blocking checkpointing system. In: Proceedings of SC 2012 (2012). http://portal.acm.org/citation.cfm?id=2389022

Author Index

Printed in the United States
By Bookmasters